LEGO® MINDSTORMS® NXT HACKER'S GUIDE

DAVE PROCHNOW

McGRAW-HILL

NEW YORK | CHICAGO | SAN FRANCISCO | LISBON | LONDON | MADRID | MEXICO CITY
MILAN | NEW DELHI | SAN JUAN | SEOUL | SINGAPORE | SYDNEY | TORONTO

The McGraw-Hill Companies

Library of Congress Cataloging-in-Publication Data

Prochnow, Dave.
Lego Mindstorms NXT hacker's guide / Dave Prochnow.
p. cm.
Includes index.
ISBN 0-07-148147-8 (alk. paper)
1. Robots—Design and construction—Juvenile literature. 2. Robots—Programming—Juvenile literature. 3. LEGO toys. I. Title.
TJ211.2P76 2007
629.8'92—dc22 2006033799

1 2 3 4 5 6 7 8 9 0 DOC/DOC 0 1 2 1 0 9 8 7 6

ISBN-13: 978-0-07-148147-2
ISBN-10: 0-07-148147-8

The sponsoring editor for this book was Judy Bass and the production supervisor was Pamela A. Pelton. It was set in ITC Officina Sans by Cindy LaBreacht. The art director for the cover was Anthony Landi.

Printed and bound by RR Donnelley.

This book is printed on acid-free paper.

ABOUT THE AUTHOR

Dave Prochnow is an award-winning writer and editor whose articles have appeared in *MAKE*, *Nuts and Volts*, *Popular Science*, and *SERVO Magazine*. In 2001, he won the Maggie Award for writing the best how-to article in a consumer magazine. Dave is also the author of *Take This Stuff and Hack It!*, *The Official Robosapien Hacker's Guide, PSP Hacks, Mods, and Expansions*, and *Experiments with EPROMs*, all available from McGraw-Hill. To learn more about his books and other projects, visit his Web site: www.pco2go.com.

CONTENTS

PART III. Katherine's Best Hacking Projects

PART IV. Katherine's Design Fun House

Appendixes

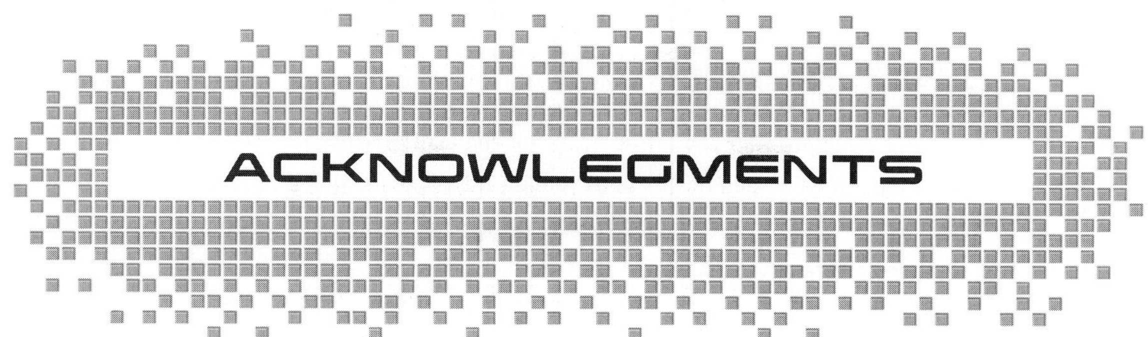

ACKNOWLEGMENTS

FIRST AND FOREMOST it is with a tremendous amount of pleasure that I acknowledge the incredible design skills of Kathy. Kathy's brilliant architectural design eye provided all of the gorgeous renderings, plans, and building instructions that make this book a LEGO building tour de force.

Supporting Kathy's magnificent design eye are the meticulous building skills of Katherine. Whenever I got stumped on how to rationalize a new building challenge, Katherine was quick with an inventive idea. Her thoughtful, uncompromising, sharing attitude completely embodies the spirit of the LEGO experience.

Finally, a book like this one would be impossible without the unselfish support from several corporate sponsors. Atmel, Flambeau, LaCie, Metzger Communications, Plano, Princeton University Press, The Rooster Group, and University of California Press made significant contributions to the development of this text.

Thank you one and all.

DISCLAIMER

Before you attempt to do any of the hacks for the LEGO® MINDSTORMS® NXT robot design kit described in this book, please read, understand, and accept the following warnings, precautions, and disclaimers regarding the disassembly of electronics. Thank you.

Precautions

Disassembling electronic devices will void your warranty. There is no authorization for the disassembly or modification of any equipment. There could be a risk of electrical shock or fire by disassembling electronics equipment.

Warnings

Monitors and LCD screens contain dangerous, high-voltage parts. Always remove the battery and disconnect any power cord(s) prior to disassembling any electronics equipment.

Disclaimers

The McGraw-Hill Companies and Dave Prochnow will neither assume nor be held liable for any damage caused to anyone or anything that is associated with the disassembly, modification, and hacking of any gadget, gizmo, or electronics equipment. The warranty for this equipment will be considered null and void if any associated warranty seal has been altered, defaced, or removed.

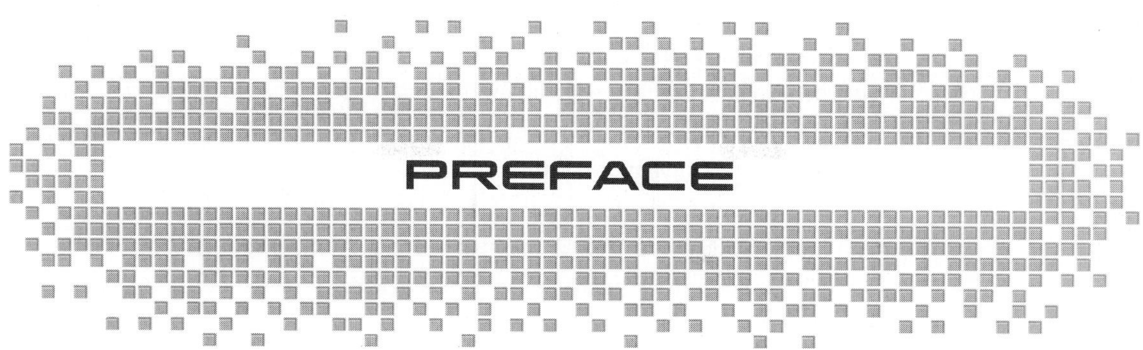

PREFACE

HI, MY NAME IS Dave Prochnow. And this is Katherine. Together we have assembled some of the best hacks, projects, plans, tips, techniques, and suggestions that you can find for the LEGO MINDSTORMS NXT robot design kit anywhere. Bar none.

No, you won't find an endless rehash of the well-known NXT robot designs in this book. Sorry. Nor will you be belittled with a mindless exploration of a silly Central American jungle site. My apologies. You can get all of that stuff elsewhere.

Rather, you will be introduced to some of the great hidden features that make the MINDSTORMS experience a mind-blowing one. Furthermore, you will become acquainted with some of the finest architecture, engineering, electronics, and robotics design work in the world. Inside these pages you will learn about the works of George Ferris, R. Buckminster Fuller, Rem Koolhaas, Picasso, and Alberto Santos-Dumont. Oh, and if these names are already familiar to you, that's great, but I still think that you just might see a different perspective on the work of one of your "old friends."

So strap your helmet on tightly, this ride is going to be fast and furious... and fun too.

There are a couple of structural points that I would like to mention before we begin our adventure, however. Book layout structure, that is.

Following a short trip down the Robotics Invention System memory lane, we'll take a quick look at the NXT system. These two introductory sections are just here to ensure that we are all on the same sheet of music regarding MINDSTORMS. Who knows, you might even learn something that you didn't already know about this inventive kit from the LEGO Group.

Now we get to the real meat of this book: hacks and projects. Oh boy, I can hardly wait. There are two massive sections for holding these two topics. And these sections are bursting at the seams with an incredible assortment of entertaining and enriching projects.

Rest assured that none of the hacks, projects, or designs in these two sections are linear in their presentation. In other words, you don't have to start with the first hack to understand the second hack. Nope, feel free to jump into any topic that tickles your fancy.

Think of this book as a grocery store. Inside this "store" you only have to shop and buy what your mind is hungry for. You can always come back later for more fun and food for thought.

The first section is Section III. Inside this section we have assembled a collection of Katherine's 10 best hacking projects. While none of these projects is extremely complex, some require a competency with soldering. If hardware hacking isn't your forte, don't fret. We've also included a couple of projects that don't require you to get your hands dirty. These are more software-oriented hacks that will exercise your mind more than your fingers.

If you'd rather not hack your NXT system, but you are still looking for some LEGO MINDSTORMS design challenges, then Section IV is for you. Katherine has really outdone herself with this section. She has assembled a veritable smorgasbord of plans, projects, and ideas for exploring the elements that constitute good design. This section is so remarkable that Katherine calls it her Design Fun House. But I'll bet that you'll call it, "the best darn section on LEGO building that you've ever seen." Actually, I like Katherine's title better. It's not as wordy.

No matter, call it what you want, but I'm hoping that you will find yourself both entertained and enriched by the hacks, projects, plans, tips, techniques, and suggestions that are packed inside this book. Oh, and if you're finding yourself missing a couple of the LEGO bricks, pieces, and elements that are mentioned in this book, please look inside the back cover for a terrific LEGO Factory kit that we've designed just for you.

By using the latest version of LEGO Digital Designer (LDD; Version 1.6), we were able to build a beautiful scale model of a shipping container (please refer to Section IV if you'd like more insightful information about the significance of shipping containers and architecture). But that's not all. Inside this handsome shipping container we've stashed a huge assortment of bricks, pegs, connectors, and elements that are sure to make your LEGO building less stressful. We even threw in a couple of minifigs just to keep things in perspective.

Finally, in this modern day of online information exchange, I encourage you to consult our special Web site for building instructions, brick parts lists, and iconoclastic LEGO models: www.pco2go.com/lego. Nose around this Web site a bit and you might even discover a special treat just for readers of this book. I don't think you will be disappointed.

Again, thank you for purchasing this book. And please be sure to wear your LEGO wristwatch with pride (see Section IV)—everywhere.

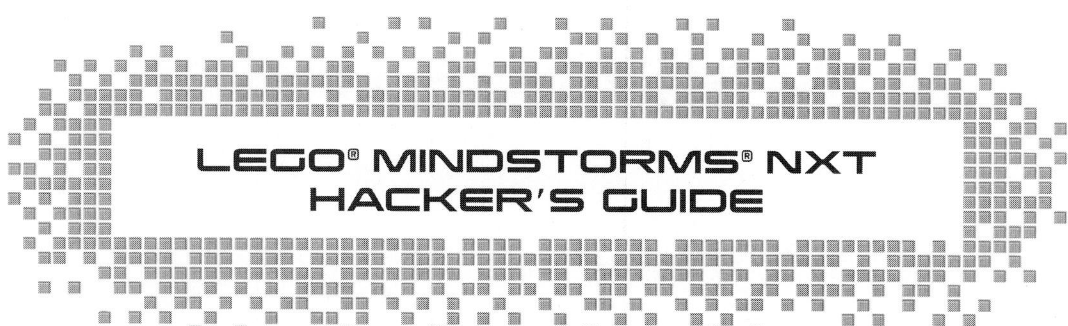
LEGO® MINDSTORMS® NXT
HACKER'S GUIDE

INTRODUCTION
The Birth of a Legend: Nirvana; the Brain and the Brick

AH, IN LOVE, lost, on the frozen plains.

Specifically, I was sitting in Love Library of the University of Nebraska campus on a gray and frigid January afternoon in 1984 staring at a book written by Seymour Papert. I was consumed with attempting to understand a strange package that I had received several days prior to the initiation of this little research outing.

Hidden deep in the stacks of Love Library I had found the first of many unique puzzle pieces that would ultimately lead me to a wild-eyed enlightenment of that odd package holding a single floppy disk. First, I had to wrap my brain around the interrelationship between epistemology and pedagogy.

This was the domain of Dr. Papert's research. A disciple of the studies of Swiss psychologist Jean Piaget, Dr. Papert had recognized a nexus between the nature of knowledge (i.e., epistemology) and the art of teaching (i.e., pedagogy). The conclusions from this research were clearly defined inside his book, *Mindstorms: Children, Computers, and Powerful Ideas*.

Born in Pretoria, South Africa in 1928, Dr. Seymour Papert is a mathematician by education, a computer scientist by training, and a noted educator by vocation. He is a recognized hinge pin in artificial intelligence at Massachusetts Institute of Technology (MIT).

It was during his early work at MIT that Papert, along with Marvin Minsky, founded the MIT Artificial Intelligence Laboratory. And from this laboratory, Papert gave birth to one of the oddest life forms.

No, this laboratory wasn't filled with bubbling beakers, coils of mysterious tubing, and bolts of electricity flashing across the room. Rather Papert's lab contained racks of computers and stacks of "green bar" fanfold paper printouts and the creature that Dr. Papert and his colleagues produced was a little turtle called Logo.

Logo only lived inside a computer—it was a programming language. Derived from another programming language called Lisp, Logo was different. This programming language was created for helping children to learn.

Following the creation of Logo in 1967 (officially released in 1968), Papert, MIT, and other researchers began to nurture this fledgling programming language. One early Logo innovation was the development of a small robot that was tethered to the same computer that was running Papert's new programming language.

This robot was called a turtle.

Controlled by commands written in Logo, the robotic turtle would execute these commands and drive around the floor. These commands in Logo consisted of 100 primitives or fundamental word elements (see sidebar; **A CONCISE CHEAT SHEET FOR SOME LOGO PRIMITIVES**). When these primitives were typed on the computer's monitor, the turtle would perform actions that were equivalent to these typed Logo primitives.

GET YAR MAMAMEDIA ON

During his tenure at MIT, Dr. Seymour Papert collaborated with Idit Harel on the development of computerized pedagogical methodologies. One of the products from this collaboration was a "tech-savvy" Web site for children called MaMaMedia.com. After going online in 1995, MaMaMedia.com became a leading site for delivering "exploration, expression, and exchange" of creative projects between children. Today, with Dr. Idit Harel functioning as the CEO of MaMaMedia.com, the Web site continues to expand its rich, creative content and now supports nearly 5 million registered users.

WHAT'S IN A NAME?

According to Seymour Papert's Web site, the name for the LEGO MINDSTORMS robot design kit series (i.e., both Robotics Invention System and NXT) was derived from the title of his book, *Mindstorms: Children, Computers, and Powerful Ideas.*

For example, if you wanted the turtle to drive around in a circle you could write the following Logo program:

```
REPEAT 360 [RT 1 FD 1]
```

In this example, "RT" is shorthand notation for the "RIGHT" in Logo primitive. Likewise, "FD" is shorthand for "FORWARD."

Computer scientists and educators alike were very excited by these developments in Logo. Logo had become a flexible, modular programming language that was capable of being extended into tangible manifestations of user interactivity.

In other words, users could write a program on a computer and then see a "real world" execution of that program with the robotic turtle. This educational duo would become a powerful pedagogical combination.

Logo also became that final puzzle piece for helping me unravel the mystery behind that floppy disk that I had received in early 1984.

It was a 5 1/4-inch floppy disk bearing a handwritten label that read: "Logo—Final Build—Confidential." I had received this disk from Commodore Business Machines (see Figure I-1). Some earlier work on their landmark Commodore VIC-20 computer had led to me receiving this unique disk.

I-1 Commodore Logo floppy disk.

My assignment was to use this pre-release version of Commodore Logo (actually Commodore Logo was a combined language version derived from Terrapin, Inc. and MIT) for co-authoring a book about programming Logo on the newly released Commodore 64 personal computer.

This secretive collaboration resulted in the publication of the groundbreaking book, *Commodore 64 Tutor for Home and School* (Scott, Foresman and Company, 1985). One of the most endearing elements of this book was the (then unheard of) combination of programming in Logo, PILOT, and BASIC languages on the Commodore 64 (see Figure I-2).

While this book has been out of print for many years, I would like to reprint the following passage from its *Introduction*:

> The approach to incorporating computer use in education has been misdirected and in some cases created unnecessary long-term problems. Take the case of Laura, for example. Laura has always enjoyed school. Each day provided a challenge that she greeted with enthusiasm and intensity. One Friday, she and her fourth-grade classmates were thrilled to learn that on Monday a new computer would occupy a corner in their classroom. Laura's teacher later recounted that the Monday the computer arrived was the first time the entire class had come to school early.
>
> The computer provided Laura with a firm background in programming—or so she thought! One weekend before Christmas, her parents took her shopping for a new home computer. They were extremely proud of their computer-literate daughter and decided to reward her academic excellence. "This one is just like the one in school," Laura bragged, whereupon she quickly began a futile attempt to demonstrate her programming prowess. Each failed try was met with increasing frustration and embarrassment. Even her parents began to wonder what her A's in computer class signified.
>
> Poor Laura. No one told her that many computers can speak several languages. While she learned Logo at school, the store model only understood BASIC. Laura's computer education should have been balanced with more emphasis on computer language traits. Hers is a classic case of well-intentioned administrators providing a misapplied dollar investment.

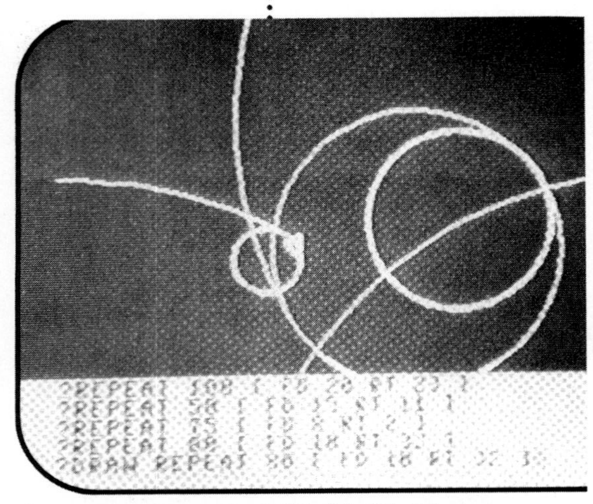

```
?REPEAT 100 [ FD 20 RT 23 ]
?REPEAT 50 [ FD 15 RT 11 ]
?REPEAT 75 [ FD 8 RT 2 ]
?REPEAT 80 [ FD 10 RT 32 ]
?DRAW REPEAT 80 [ FD 10 RT 32 ]
```

I-2 CRT flashback with Commodore Logo running on an Amdek color monitor.

How's that for a trip down (computer) memory lane?

LEGO OF MY LOGO

Now fast-forward one year after that passage was written and let's return to MIT. Two researchers in the MIT Media Lab, Mitchel Resnick and Steve Ocko, are working on a system to integrate Logo and a turtle equipped with sensors.

The result is LEGO Logo (commercially released as LEGO TC Logo) (see Figure I-3). Why was LEGO incorporated into this product's name? Because LEGO bricks were used for building this sensory-aware turtle (see Figure I-4).

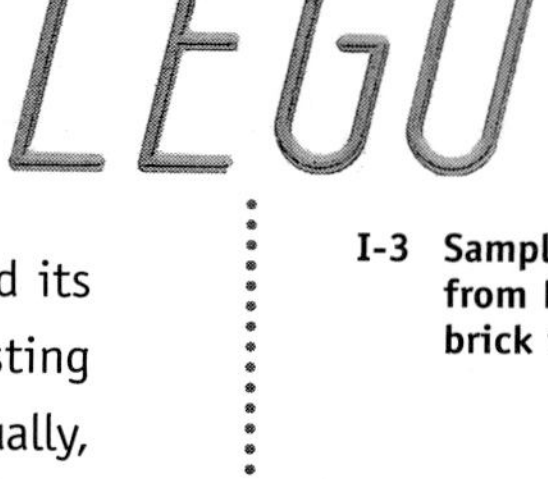

I-3 Sample art from LEGO brick tub.

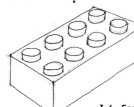

These MIT researchers became frustrated with the tethered turtle and its self-imposed limitation of being constantly attached to a computer. Enlisting the talents of Fred G. Martin, MIT developed the Programmable Brick. Actually, an embedded microcontroller, the Programmable Brick was "programmed" via a host computer in a programming language called Brick Logo. Once it was loaded with a program, the Programmable Brick could be disconnected from the host computer.

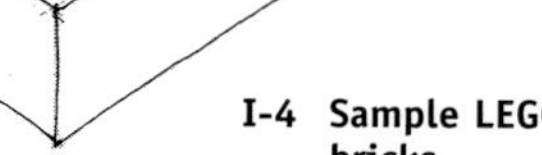

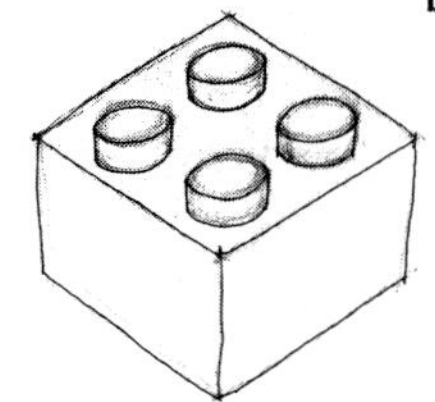

I-4 Sample LEGO bricks.

Now the Logo turtle could roam free—carrying a pre-programmed computer-like circuit on its back. The LEGO RCX was born (see Figure I-5). Originally defined by the LEGO Group as Robotics Command System, RCX has been more widely defined as Robotics Command Explorer.

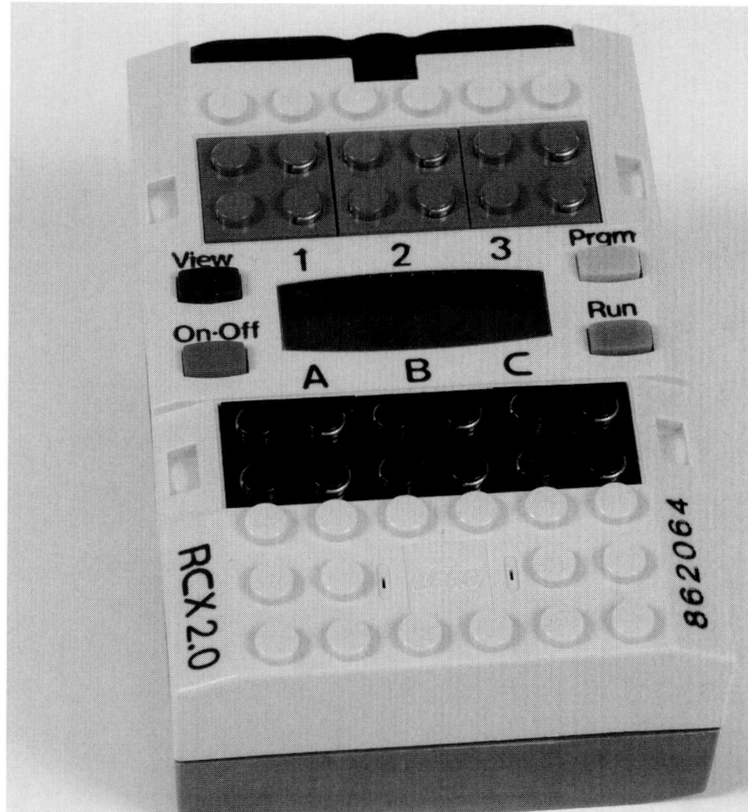

I-5 LEGO MINDSTORMS RCX brick.

SEYMOUR PAPERT TODAY

What can you do after you invent a computer programming language? Well, if you're Seymour Papert, you take your concept of learning to a whole new global marketplace—a marketplace that is educationally hungry, yet technologically starved.

Serving along with fellow MIT researcher Mitch Resnick as advisors to the One Laptop per Child (OLPC), they hope to provide all of the world's children with an opportunity to "explore, experiment, and express" themselves (see Figure I-14).

OLPC is an initiative that was introduced by Nicholas Negroponte at the World Economic Forum at Davos, Switzerland in January 2005 (see Figure I-15).

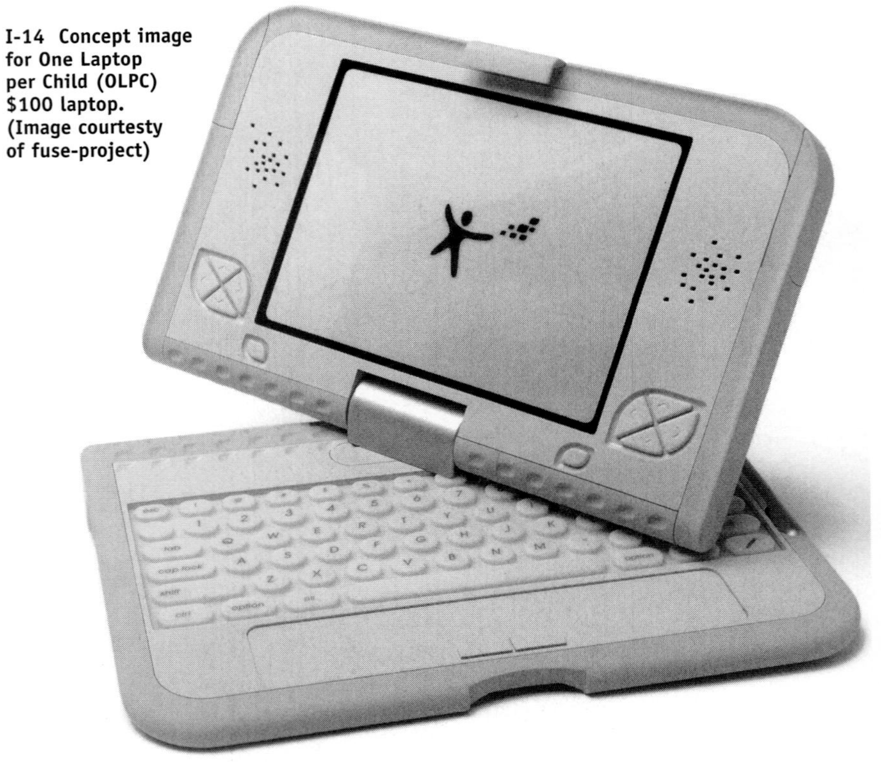

I-14 Concept image for One Laptop per Child (OLPC) $100 laptop. (Image courtesty of fuse-project)

I-15
Concept image for alternative OLPC $100 laptop. (Image courtesty of fuse-project)

Although maligned in the press by Microsoft Corp. Chairman and Chief Software Architect Bill Gates (as reported in a Reuters report), the OLPC is a worthy effort that commands corporate compassion and not corporate greed (see Figure I-16).

I-16 Another concept image for an OLPC $100 laptop. (Image courtesy of Design Continuum)

Although Martin was integral in the development of the Programmable Brick, he is better known for the design of his Handy Board and Cricket robot control systems. Remarkably, his Cricket (also called Handy Cricket) is a direct descendent from the original Programmable Brick.

The Handy Board is a 68HC11-based microcontroller board designed by MIT. In a remarkable gesture to the robotics community, MIT licensed the Handy Board design as "freeware," for educational, research, and industrial use. While you can readily download design information about building your own Handy Board on the Web (handyboard.com), there are a couple of manufacturers who sell assembled Handy Boards.

Since 1995, Gleason Research has been selling the MIT Handy Board to robot builders worldwide. The Handy Cricket Version 1.1 is a low-cost module based on the Microchip PIC® microprocessor featuring a built-in Logo interpreter. Equipped with two motor ports, two sensor ports, two bus ports, 4 kilobytes of static memory, and a piezo speaker, the Handy Cricket connects with the host PC via a serial port IR interface (see Figure I-6).

I-6 **Handy Cricket.**

HEY, LET'S NAME OUR COMPANY LEG GODT

According to the LEGO Group, their name, LEGO, is derived from a Danish term: "Leg Godt;" which means "play well." Or, for you Latin majors, the meaning is "I put together." I don't know about you, but LEGO, to me, means "hours of creative fun."

A unique IR transmitter/receiver circuit built into the Handy Cricket enables communication between two or more Handy Crickets. Just like the chirping of a cricket, the Handy Cricket is able to "chirp" IR signals at a 50-kilobytes data rate between another Handy Cricket.

Imagine this—you could have various Handy Cricket robots "talking" between each other. Oh, and just like a biological cricket, the Handy Cricket has a small footprint. The overall dimensions are just a bit under 2 1/2 inches per side.

Sure, all of this hardware stuff is exciting, but the part of the Handy Cricket that is really remarkable is the implementation of the Logo programming language, called "Cricket Logo," that's built into the Handy Cricket.

The Handy Cricket is programmed in a language that is a simplified version of the powerful yet easy-to-learn Logo language. Unlike most programming languages, Cricket Logo (as well as its ancestral Logo) is short, simple, sweet, and primitive. I'll grant you, Logo is a little archaic, but the payback can be terrific for robot builders. For example, commonsense commands like BRAKE and SETPOWER don't require a lot of human smarts to figure 'em out (see sidebar; **CRICKET LOGO PRIMITIVES**).

Individual Handy Crickets cost $59 or, you can purchase a complete Handy Cricket Starter Kit for $99 from Gleason Research.

Just like the Handy Cricket's ability to communicate en masse, Mitch Resnick created a large parallel version of Logo called StarLogo. By using StarLogo, thousands of Logo-driven turtles were able to carry on independent

processes, as well as interact with one another. This form of decentralized systems development was outlined in Resnick's 1997 book, *Turtles, Termites, and Traffic Jams: Explorations in Massively Parallel Microworlds*.

MEANWHILE, ACROSS THE POND

OK, the contributions of those overachieving eggheads from MIT to the development of global robotics is now well documented, but what about the people who invented the glueless assembly process? You know, that process that we *all* generically call LEGO®.

Just like the story of the refrigerator, facial tissue, and the photocopy machine, the story of LEGO is a fascinating discovery that has evolved into a ubiquitous trademark name becoming a generic symbol for creative design.

If you have any children or you frequent your local toy store, then you already know about the most visible symbol of the LEGO Company—the LEGO brick. A family-owned company founded in Denmark in 1932, the LEGO Company set a corporate goal of becoming the leading global brand for families by 2005.

While the jury is still out regarding the achievement of this corporate goal, the LEGO Company does have a viable brand recognition. Prior to the arrival of this deadline, however, the LEGO Company made some strategic moves in enhancing the image of their creative design concept.

On December 8, 2004, Charlotte Simonsen, Head of Corporate Communications, LEGO Group announced that, "With a view to streamlining the structure of the LEGO Group and achieving greater transparency, changes are announced in the ownership of the Group's companies. In future, LEGO Holding A/S will own both the Danish and the Swiss parts of the group.

IS THAT A CRYSTAL BALL YER HOLDIN', OR YOU JUST HAPPY TO SEE ME?

While many insiders at the LEGO Group fancy the LEGO MINDSTORMS product line to be the toy for the twenty-first century, the evolution of this product line from RIS to NXT has produced at least one very eerie coincidence. Whereas the RIS RCX was based on the Hitachi H8 microcontroller, the NXT is based on the 32-bit ARM7 microcontroller. During the sale of the RIS, the Hitachi slogan was "Inspire the Next." Next, NXT; coincidence? I think not.

I-7 The LEGO Center in Billund, Denmark. (Photograph courtesy of The LEGO Group, ©2004)

"The change means that the LEGO Group now coincides exactly with the LEGO Holding Group. It has therefore been decided that in future the group will be known as the LEGO Group instead of the LEGO Company," concluded Simonsen.

Thus, the now familiar corporate name of the LEGO Group was established (see Figure I-7). This transformation didn't go completely without a hitch.

Rumors had been circulating that the LEGO Group was planning to phase out the LEGO® MINDSTORMS® Robotics Invention System™ (RIS) (see Figure I-8). This planned obsolescence outraged many robot enthusiasts. So one month after its reorganization, the LEGO Group held a press conference.

In Billund, Denmark, on January 14, 2004, Kjeld Kirk Kristiansen, owner and CEO (and grandson of the LEGO Company's founder, Ole Kirk Christiansen) met with the press. He began his remarks with a rather confusing statement that unveiled a "change in direction" which the newly renamed company was going to follow. In short, Kristiansen stated that, "the future growth strategy will not be based on big, movie-related IPs such as Harry Potter."

Kristiansen then added, "This does not mean that the company will exclude that kind of stories and themes, but just that the growth should be based on the fundamental products, where sales do not to the same extent go up and down, depending on whether or not there is a new movie this year."

CRICKET LOGO PRIMITIVES FOR CONTROLLING MOTORS

PRIMITIVE	FUNCTION
A	selects motor A
B	selects motor B
AB	selects motors A and B
ON	turns selected motors on
OFF	turns motors off
BRAKE	brakes motors off
ONFOR	turns motor on for specified amount of time
THISWAY	sets motor direction; green motor LED lights
THATWAY	sets motor direction; red motor LED lights
RD	reverses motor direction
SETPOWER	sets motor speed
WAIT	waits for specified amount of time

FOR READING SENSORS

PRIMITIVE	FUNCTION
SWITCHA	returns state of switch A
SWITCHB	returns state of switch B
SENSORA	returns value of sensor A
SENSORB	returns value of sensor B

A CONCISE CHEAT SHEET FOR SOME LOGO PRIMITIVES

PRIMITIVE	SHORTHAND
BACK number	BK
BUTFIRST word	BF
BUTLAST word	BL
CLEARSCREEN	CS
CONTINUE	CO
EDIT	ED
ERASE procedure	ER
FORWARD number	FD
HIDETURTLE	HT
IFFALSE action	IFF
LEFT angle	LT
OUTPUT result	OP
PRINTOUT procedure	PO
READCHARACTER	RC
REQUEST	RQ
RIGHT angle	RT
SENTENCE word word	SE
SETHEADING angle	SETH
SHOWTURTLE	ST

I-8 **Original RIS packaging.**

While that type of strategy was logical and financially sound, Kristiansen softened the blow of his statements by concluding that, "Hearsay has it that a product range like LEGO MINDSTORMS is no longer in focus. This is not true. On the contrary, MINDSTORMS, CLIKITS and Bionicle are all good examples of products the company wants to stake on."

Whew. It looked like, at least on paper, that robots were in the LEGO Group's future. In fact, LEGO® MINDSTORMS® NXT was already in development. Thank goodness.

But if you're looking to become "the leading global brand for families by 2005," then you have to do more than just restructure your corporate flowchart. Armed with a bold vision, the LEGO Group sought to establish a foothold in three areas of commerce: retail, entertainment, and the Web.

REAL BRICK AND REAL MORTAR

The world's first LEGO Brand Store was officially opened by CEO Kjeld Kirk Kristiansen on October 1, 2002 on Kölner Hohe Strasse in Cologne, Germany. Lining the shelves of this store's 350-square-meter floor space was a selection from the entire LEGO Group product line including LEGO MINDSTORMS RIS, LEGO Wear clothing and accessories, and LEGO brand electronic video games (see Figure I-9).

I-9 **LEGO Brand Store in New Jersey. (Photograph courtesy of The LEGO Group, ©2004)**

Following this ceremonial grand opening, other LEGO Brand Stores were opened around the world. In fact, the second store in Milton Keynes, England opened one month after the Cologne, Germany store's ribbon-cutting ceremony.

Designed to give families a more team-oriented shopping experience, each 200-square-meter LEGO Brand Store features buckets of bricks, the latest products, and interactive sections where parents and kids can build the latest LEGO kits.

I'M GOIN' TO LEGOLAND

Who would think that you could successfully compete against The Walt Disney Company theme park juggernaut? Well, that's exactly what the LEGO Group attempted to do with LEGOLAND (see Figure I-10).

Operating unique theme parks in Billund (Denmark, 1968), Windsor (England, 1996), Carlsbad (California, 1999), and Deutschland (Günzburg, Germany, 2002), the LEGO Group has made a significant dent in the attendance figures at competitive parks throughout the world.

According to its annual financial reports, the combined four LEGOLAND Parks had an attendance of 5,651,316 in 2002 and a slight decline to 5,535,961 in 2003. Yet revenue from these figures were higher during each year of operation.

It isn't just about the money, however. The LEGO Group considers the overall theme park "experience" to be a prime determinant for rating each LEGOLAND Park. As such, when LEGOLAND California was selected as "one of the best theme parks in the world" by Forbes.com for the second year in a row (2002 and 2003), the staff in Billund was ecstatic.

I-10 The entrance to LEGOLAND Deutschland, Germany. (Photograph courtesy of The LEGO Group, ©2004)

More importantly, however, this type of "good press" helps to drive foot traffic to the doors of all LEGOLAND Parks. In fact, LEGOLAND Deutschland recorded one of its biggest 30-day attendance records between April 12, 2003 and May 19, 2003, when over 200,000 people came through the turnstile and pounded the bricks.

BUILD IT AND THEY WILL BUY AND BUY AND BUY

In the summer of 2005, the LEGO Group brought brick kit building into the digital age. Armed with a brand new Web site called the LEGO Factory and a free 3-D modeling program called LEGO Digital Designer (available for both PC and Mac OS X), the LEGO Group enabled anyone to design their own LEGO brand brick kit.

Once the user has completed a design, which, in itself, can take several hours to several days to complete, the user's brick creation can then be uploaded to a special LEGO Web page where it is prepared for purchase. Yes, you can then buy your own design which will be bundled into a distinctly unique LEGO brand kit and shipped to you. This entire bundling and shipping process can take one to two weeks.

If you're really in need of an ego boost, you can even share your new 3-D creation with others in a special LEGO Factory Gallery. Inside the Gallery other Web users can also purchase your kit (see Figure I-11).

I-11 It's your box. A example of a LEGO Factory kit.

THE TOY OF WHICH CENTURY?

At the conclusion of the first month of the new millennium, the Council of the British Association of Toy Retailers (BATR) recognized the LEGO Company as the winner of the *BATR Toy of the Century* award (see Figure I-17).

According to BATR, the LEGO Company brick had edged out both the teddy bear and Barbie® for winning this award. Hmm, that sounds a little fishy, doesn't it? The teddy bear *and* Barbie beaten by a brick?

Well, according to BATR the "voting" for this award was conducted via a special Web site (see Figure I-18).

I-17 The LEGO brand is an internationally recognized leader in modeling design toys.

On this site, visitors could "vote" for their pick for the"Toy of the Century." Anyone in favor of online voting for presidential elections? Who knows, maybe the teddy bear would win that competition.`

I-18 Just a handful of LEGO bricks can become a robot in a matter of moments. Then you can enjoy hours of play.

Bad designers beware, however. There is one small caveat in designing brick kits with LEGO Factory. All of the bricks come in predefined palettes or collections of bricks. During the design phase with LEGO Digital Designer, you select the palette or palettes that you need for building your brick kit. Whether you use a single brick from a palette or all of the bricks within a palette, you will still pay for the entire palette during the bundling of your kit.

Granted, a savvy kit designer should be able to use this limitation to his or her advantage. But using just one brick from a palette of 100 bricks can be a costly lesson in design frugality.

In less than six months of operation, the LEGO Factory had received over 30,000 kit uploads. And the cost of these user-created kits is comparable with a LEGO-designed kit. For example, I made a robot kit (Pogodae) from 267 pieces and the resultant cost for purchase of my kit was under $26 (exclusive of shipping and handling).

ONE OF THE SELECT NXT FEW

On March 6, 2006, the LEGO Group officially released background information about the selection process for its MINDSTORMS® Developer Program (MDP). First of all, if you weren't selected, don't feel slighted.

According to the LEGO Group, the MDP was an exclusive cadre of 100 enthusiasts who were charged with "helping guide the product development process for LEGO® MINDSTORMS® NXT, the next generation of LEGO robotics."

A total of 9,610 robotics enthusiasts, ranging from 18 to 75 years of age and representing 79 countries responded to the one-month online application process that was sponsored by LEGO. Following is a thumbnail view of the 100 selected candidates.

- Range in age from 18 to 75 (50 percent under age 35)
- More than 20 percent work in the software/QA/DBA sector
- Nearly 20 percent are teachers or educators
- 13 are architects or engineers
- 40 percent are from the United States

Unbelievably, I was one of those chosen few selected to the MDP. And, no, I'm not the 75-year-old MDP member, either. Inside the pages of this book you will benefit from this exclusive nomination privilege that I was honored to receive from the LEGO Group (see Figure I-12).

I-12 Original MDP LEGO MINDSTORMS NXT robot design kit. Footprints courtesy of DHL.

AND NOW FOR SOMETHING COMPLETELY NXT

Finally, one of the more humorous fallouts from the MIT development of the Programmable Brick was the creation of a Web site that listed "Twenty Things to Do with a Programmable Brick." Ironically, many of the items on this 1994 list would be great projects to explore with your LEGO® MINDSTORMS® NXT robot design kit...today.

So let's get hacking (see Figure I-13).

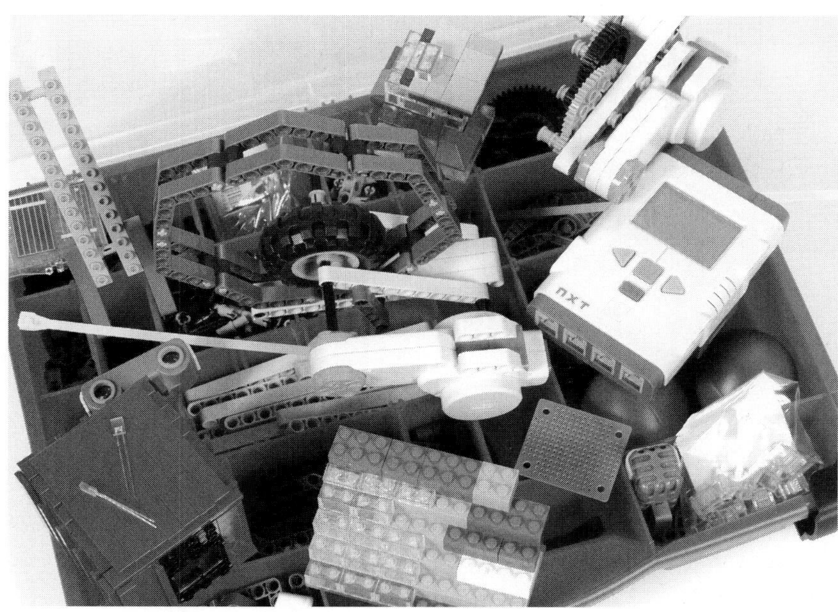

I-13 All of these wonderful projects are described inside this book.

HOW TO BUILD YOUR OWN MODEL WITH LEGO DIGITAL DESIGNER

Welcome to LEGO Digital Designer.

1 Pick an existing model or start a new one.

2 The opening base plate grid contains tools, brick palettes, and a brick selector.

3 Bricks are grouped into categories.

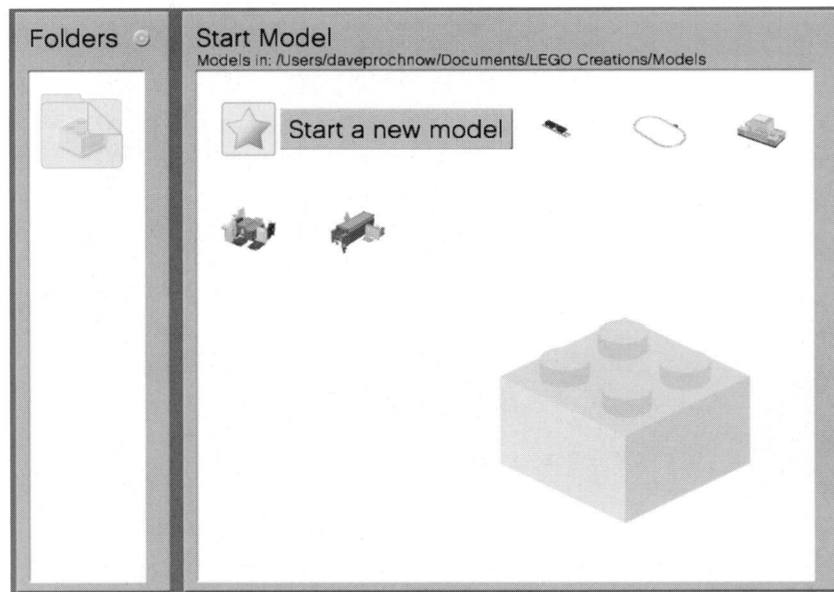

1

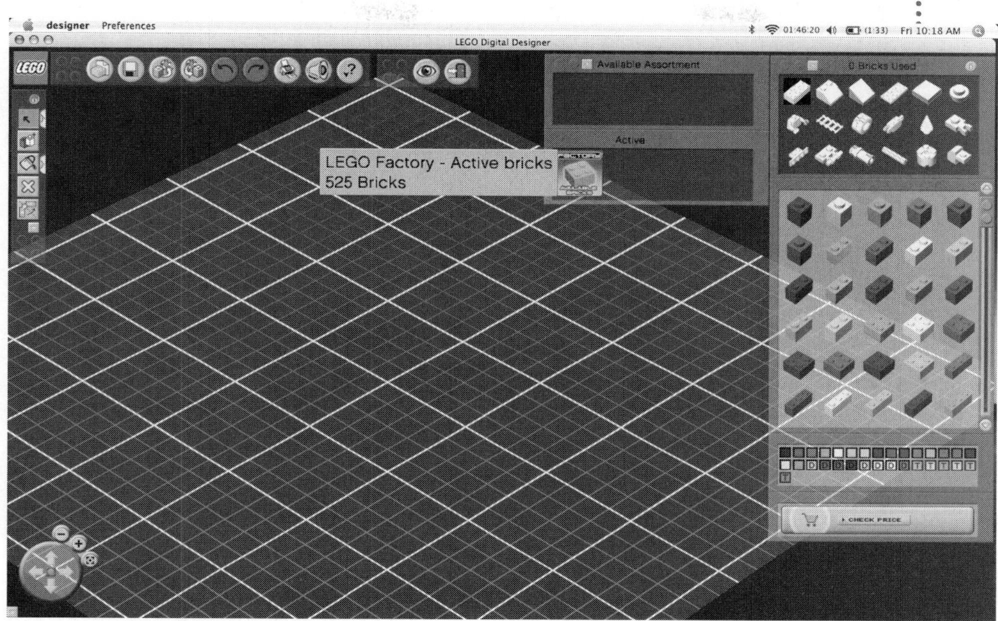
designer Preferences
01:46:20 (1:33) Fri 10:18 AM
LEGO Digital Designer
Available Assortment
0 Bricks Used
Active
LEGO Factory - Active bricks
525 Bricks
CHECK PRICE

2

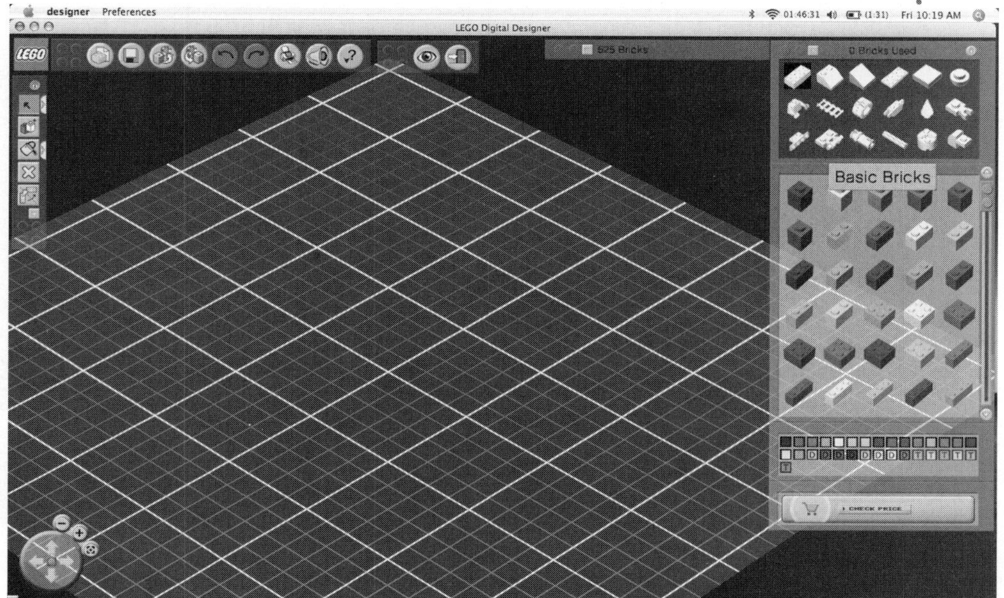
designer Preferences
01:46:31 (1:31) Fri 10:19 AM
LEGO Digital Designer
525 Bricks
0 Bricks Used
Basic Bricks
CHECK PRICE

3

4 Choose a brick category.

5 Individual bricks can be selected.

6 The brick is dragged to the base plate grid.

7 Bricks snap together.

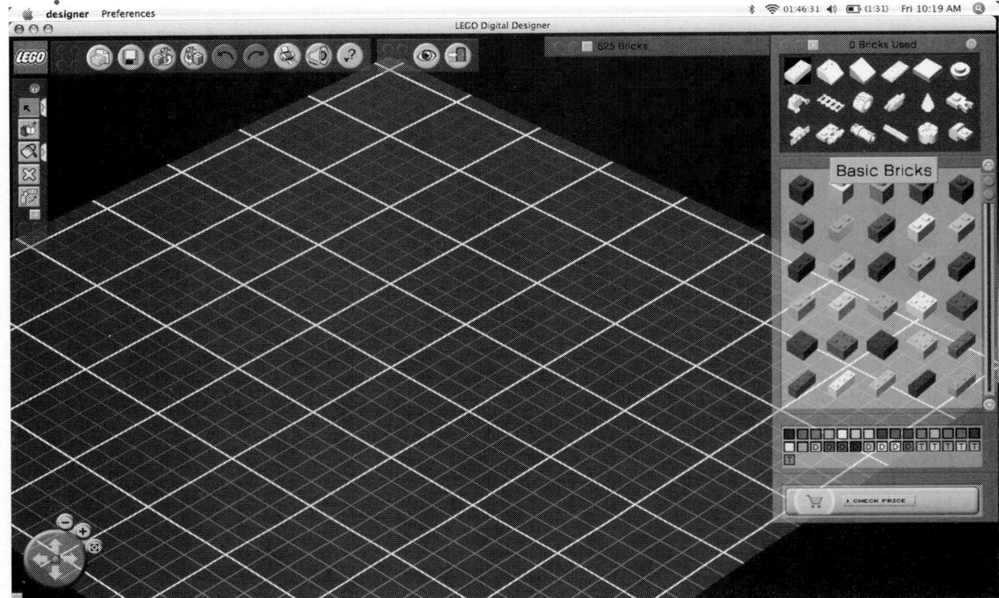

4

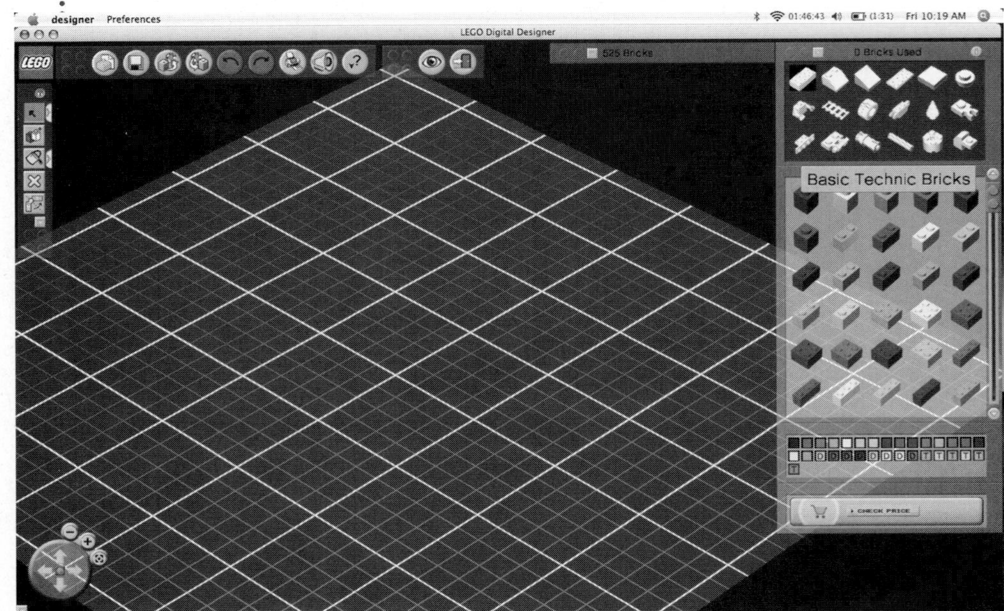

5

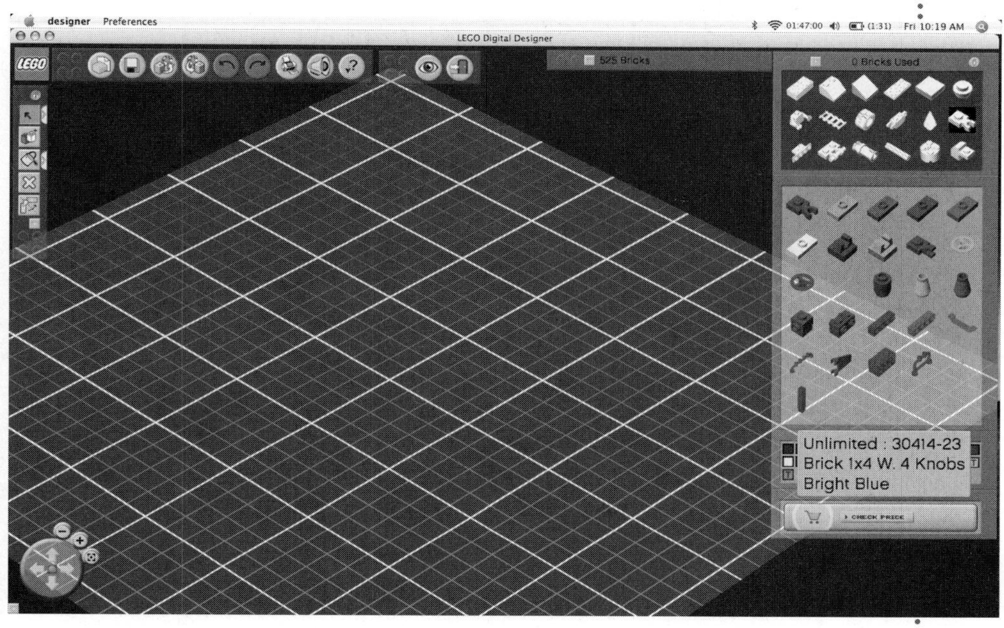
designer Preferences
LEGO Digital Designer
525 Bricks
0 Bricks Used
Unlimited : 30414-23
Brick 1x4 W. 4 Knobs
Bright Blue
CHECK PRICE

6

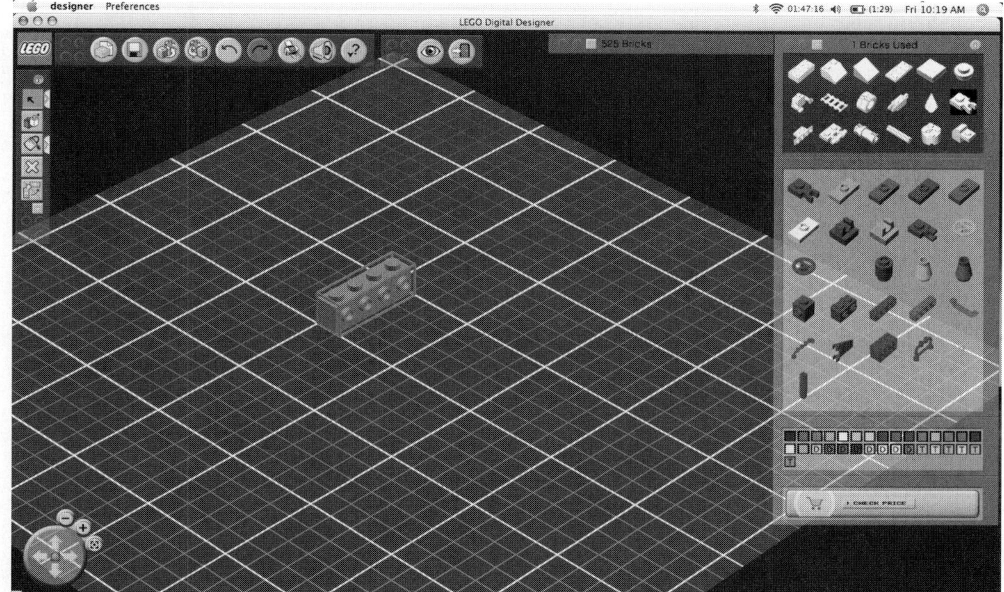
designer Preferences
LEGO Digital Designer
525 Bricks
1 Bricks Used
CHECK PRICE

7

8 Multiple bricks can be cloned for rapid assembly.

9 Complete structures can be snapped together. Be forewarned, LDD will allow you to "snap" bricks together that could never be joined in real life.

10 A completed project.

11 Use the Zoom command to get a better look at your new creation.

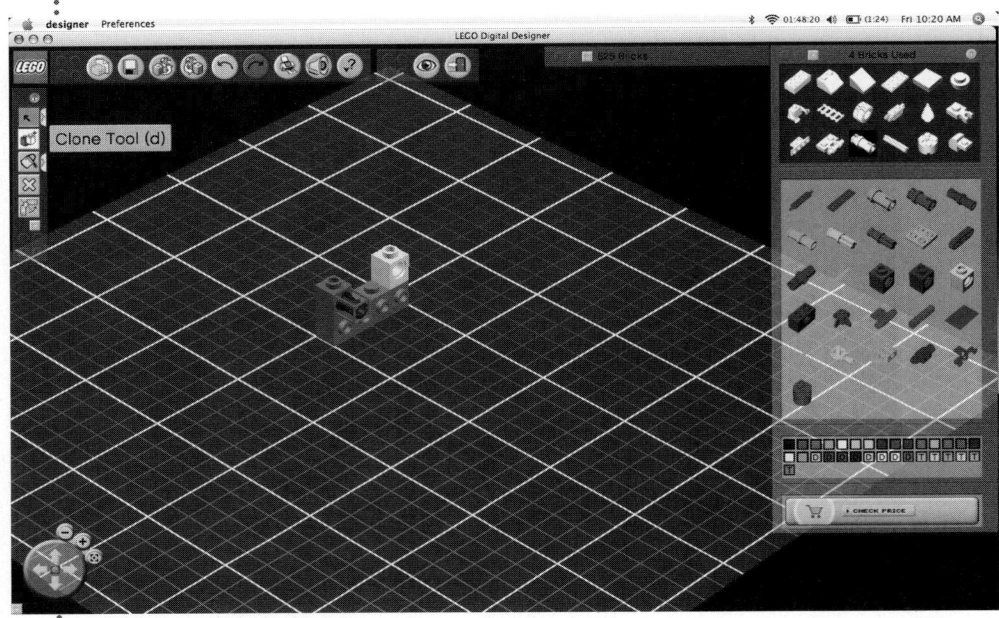

8

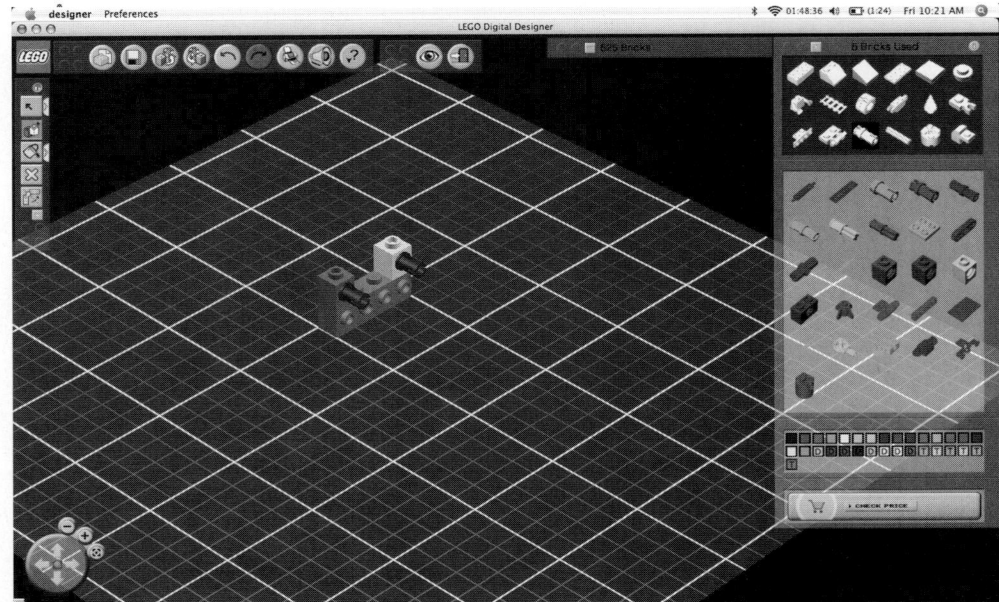

9

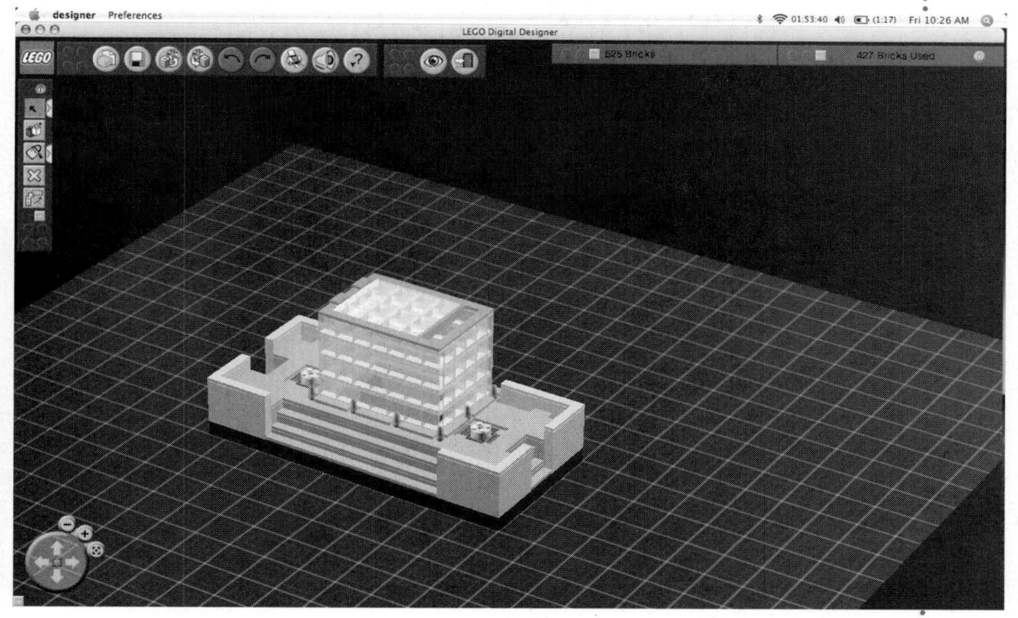
designer Preferences
01:53:40
(1:17)
Fri 10:26 AM
LEGO Digital Designer
LEGO
525 Bricks
427 Bricks Used

10

designer Preferences
01:53:54
(1:16)
Fri 10:26 AM
LEGO Digital Designer
LEGO
525 Bricks
427 Bricks Used

11

12 Pan the model for viewing different sides.

13 A separate window displays all of the steps used in building your model.

14 You can step through the assembly process as a guide for assembling your "real" bricks.

15 If you like what you've done, go buy it—online at the LEGO Shop.

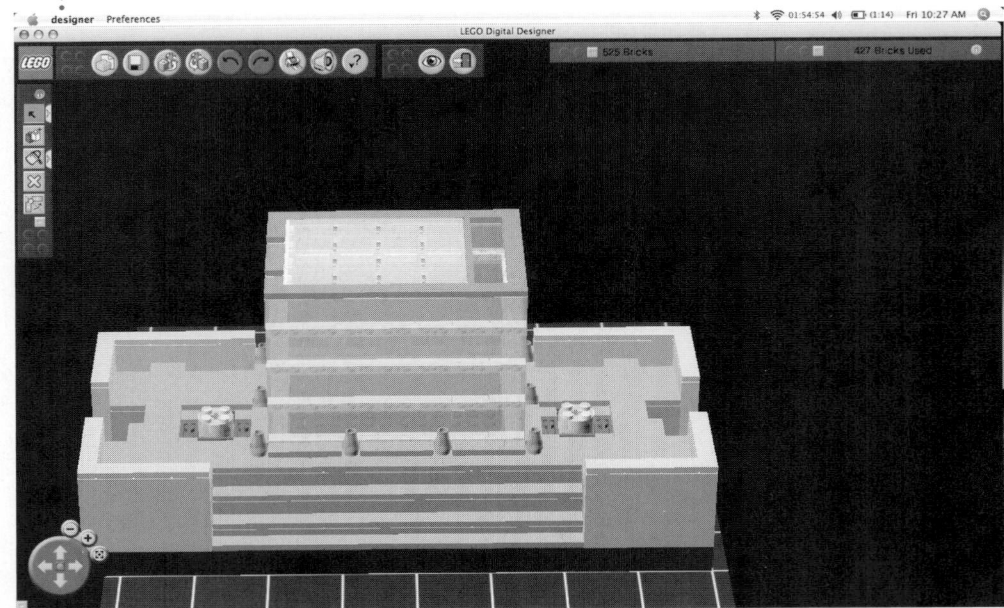

12

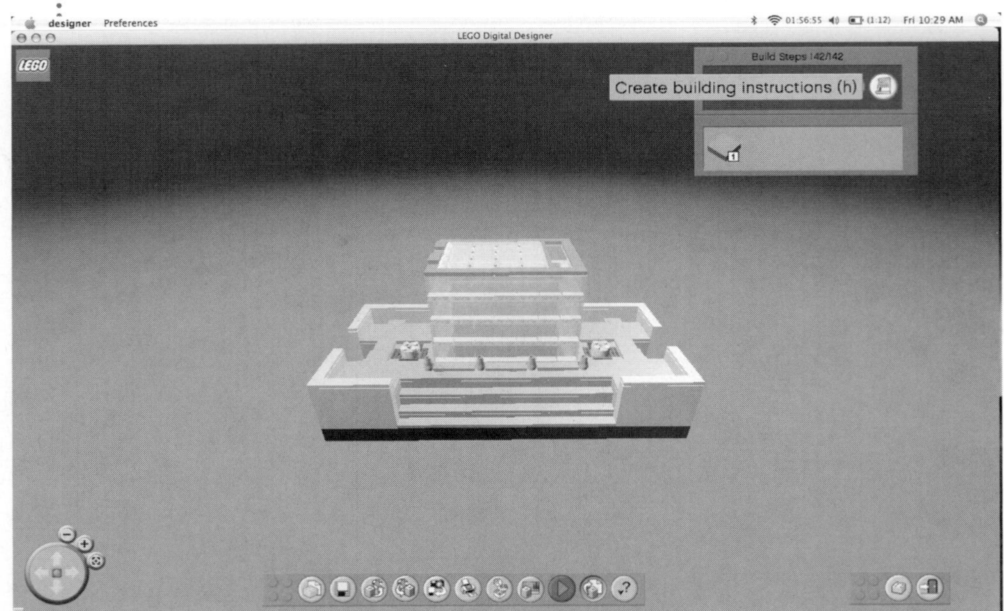

13

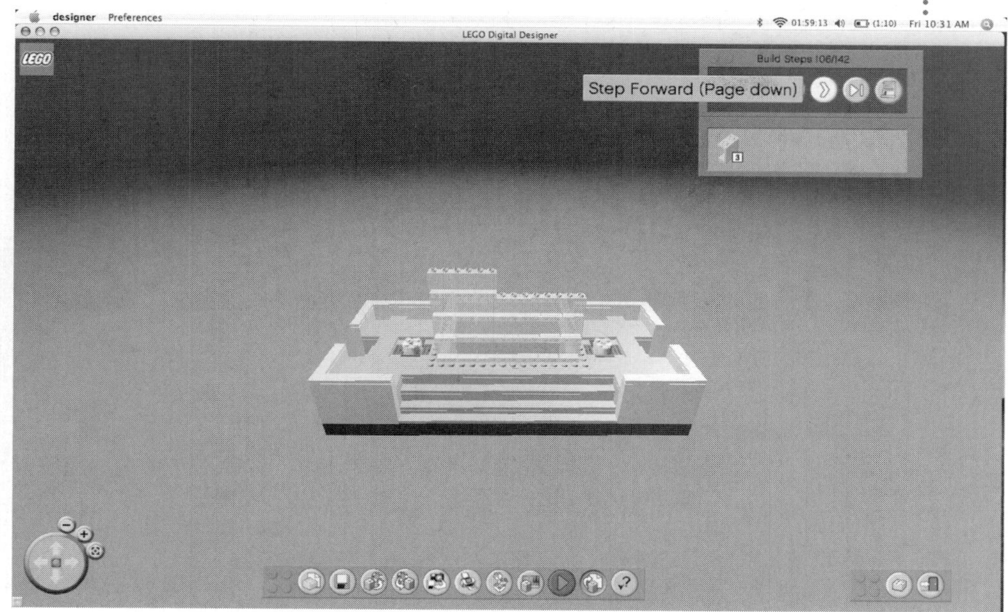
designer
Preferences
LEGO Digital Designer
Build Steps 106/142
Step Forward (Page down)

14

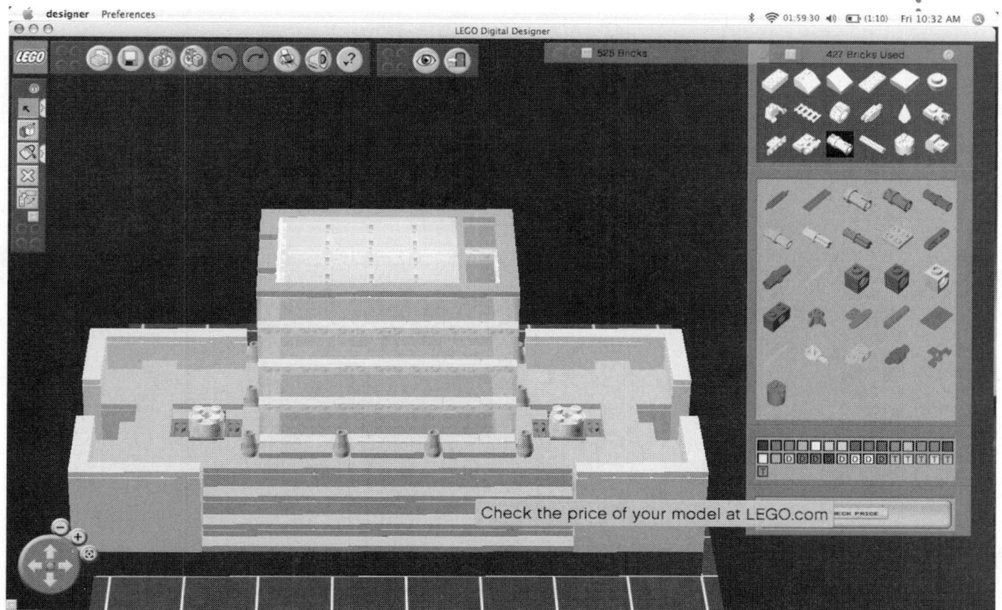
designer
Preferences
LEGO Digital Designer
525 Bricks
427 Bricks Used
Check the price of your model at LEGO.com

15

16 After the model has been uploaded to the LEGO Shop, you are provided a price and ordering instructions. Then within two weeks your own LEGO kit will arrive at your door.

16

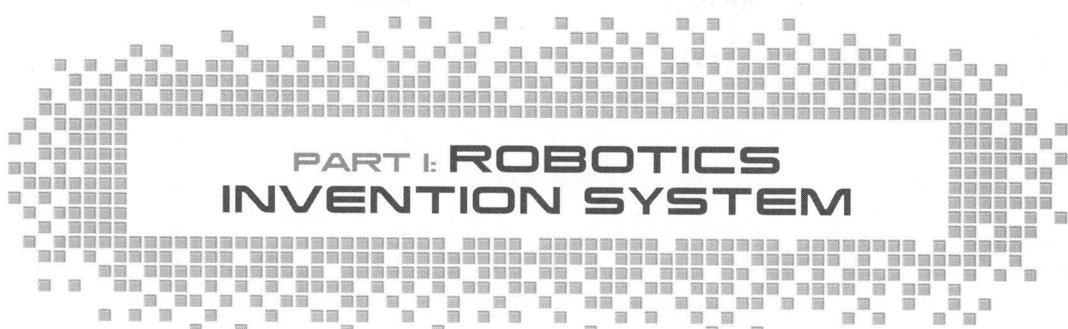

PART I: ROBOTICS INVENTION SYSTEM

1 YOUR FIRST ROBOT

FEW OF US possess either the artistic talent or the notoriety of Eric Sophie. Rather than designing robots that are merely regurgitations of the stock creations found in the LEGO MINDSTORMS *Constructopedia*, Sophie's robots are unique, exciting, and beautiful.

Born in 1970, Sophie quickly rose through the ranks of LEGO MINDSTORMS RIS robot builders to become internationally recognized for his comprehensive articulated creations. While his first robot to receive critical acclaim was his Praying Mantis in 1997, it took Sophie another five years before his talent caught the eye of the LEGO Group.

This Jamaica Queens, New York native combined 41 LEGO MINDSTORMS RIS motors with 12 RCX programmable bricks inside a gorgeous blue 50-inch-tall frame. After a year of design work, Sophie was finished with his vision, and the result was called Jamocklaquat (see Figure 1-1).

1-1 The 50-inch Jamocklaquat is controlled by 41 LEGO Technic motors and 12 RCX units. It took artist and LEGO enthusiast Eric Sophie about a year to complete the fully articulated robot. (Photograph courtesy of The LEGO Group, ©2003)

In July 2003, the LEGO Group recognized Eric Sophie's amazing contribution to the LEGO MINDSTORMS RIS product line. And Jamocklaquat was the guest of honor. In response to this official acknowledgment of his creativity, Sophie gave an eloquent endorsement of the LEGO MINDSTORMS kit.

"My hope is that this creation will reach people of all ages and open their eyes to the future. A future where we will learn to master this kind of technology. The LEGO elements put this technology in the hands of everyone," observed Sophie.

Today, Eric Sophie is extending his robot building skills with a remarkable "build it yourself" robot kit comprised of LEGO Technic elements. The jointly developed (i.e., co-developed by MC7 and Biomechanical Bricks) ZXR1 Builder's Frame is a barebones articulated robot frame that can be built from Sophie's 252-piece kit. Priced at $70, the ZXR1 Builder's Frame kit provides spine articulation, a special hip, and rotational friction for all major limb sections. Best of all, this kit also gives you the added benefit of learning firsthand how Sophie designs his robots.

THAT SPECIAL SOME-BOT-TY

Granted, we don't all possess the artistic talent of Eric Sophie, but most of us are just as proud of our robotic creations as he is of Jamocklaquat. So what was your first robot encounter?

Are you a robot kit builder? A Joinmax Digital Smart Dog? An OWI Kranius? A KTB mechatronics qfix Crash-Bobby? A Parallax Boe-Bot? A Heathkit HEROjr?

How about learning robotics by hacking? Maybe pre-built robots like WowWee Robotics Robosapien (see *The Official Robosapien Hacker's Guide*; McGraw-Hill, 2006) or Robosapien V2 helped lay the foundation for your introduction to advanced robot physics. Hey, you can't get an education in robotics much better than one gleaned from the spawn of the designer of Robosapien, Mark W. Tilden.

What about learning robotics to solve some of life's daily problems? You say that toys aren't your game, well how about a household robot like iRobot Corporation's Roomba (see *Take This Stuff and Hack It!*; McGraw-Hill, 2007)? It doesn't take much of an engineering feat to convert a floor vacuum system into a lawn mowing system. In fact, by just adding the remarkable Mind Control interface from Element Products to your Roomba you can experience all of the advantages of hacking a commercial robot without removing a single screw.

Oh, yes, I can hear you now.

Hey, what about LEGO robotics? If you're a longstanding LEGO aficionado, then the LEGO MINDSTORMS RIS was your first introduction to robot design and construction. Or, if you're new to LEGO MINDSTORMS, then the revolutionary NXT next generation robotics tool set represents your first interaction with robots. If you're one of these NXTers, then welcome to the most sophisticated, powerful, extensible, and easy way to build real "live" robots.

Sure, you built all of the example robot designs included with the LEGO MINDSTORMS kit, but now what? Well, most of us don't have the financial support for purchasing 12 LEGO MINDSTORMS RIS or NXT kits like Eric Sophie did for his Jamocklaquat robot. Luckily, you don't have to resort to this type of "extreme" technique for building your own robot creation.

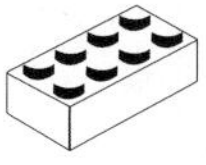

JOURNEY BACK TO THE DAYS OF THE ROBOTICS INVENTION SYSTEM

1 Make your selection.

2 Choose a user profile.

3 Create some freeform programming.

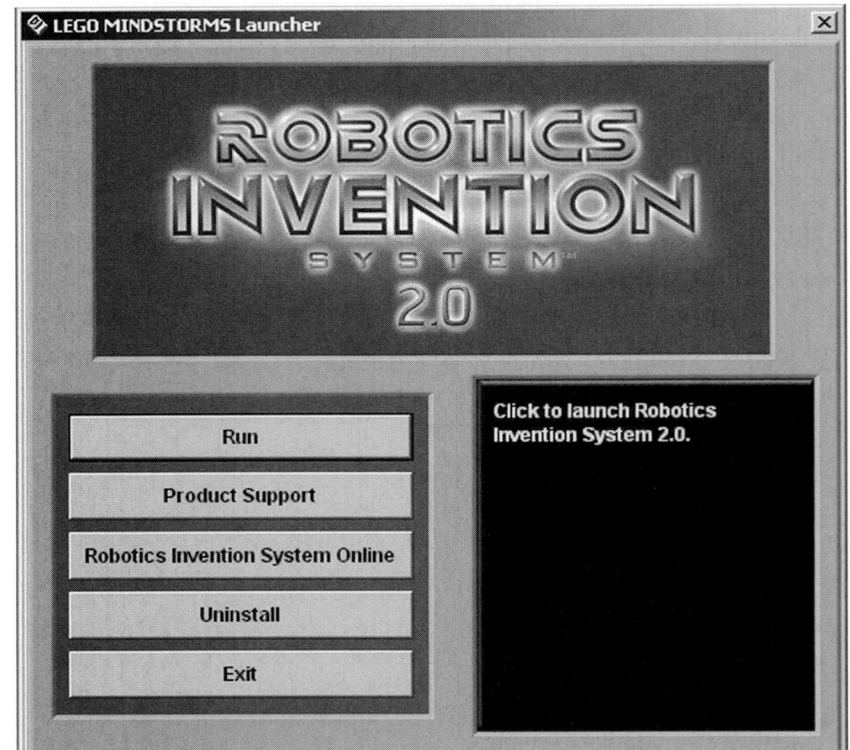

1

LOGIN
EXIT
Katherine
Katherine
Enter
New User
Delete User
2

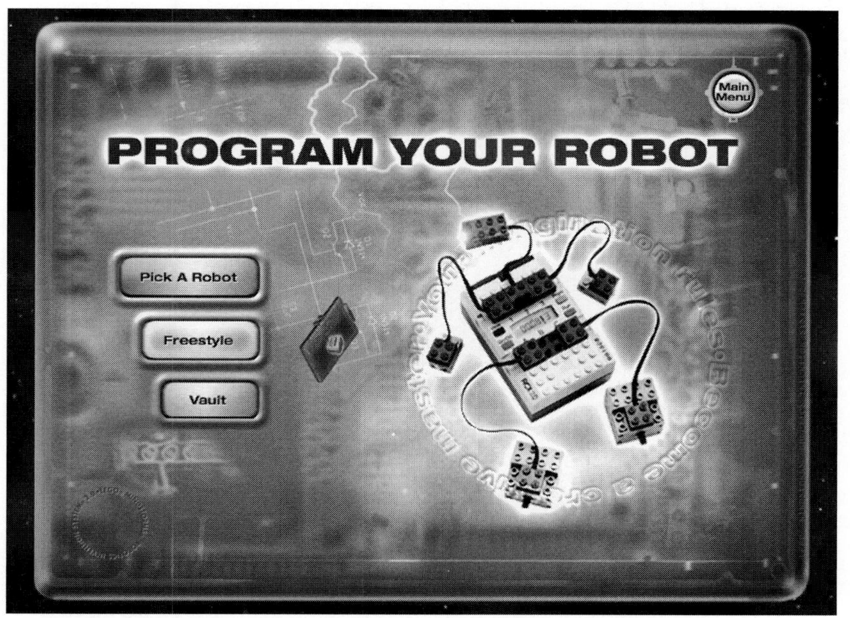
Main Menu
PROGRAM YOUR ROBOT
Pick A Robot
Freestyle
Vault
3

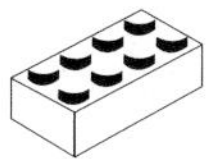

4 Drag and click your program blocks.

5 Download your program to the RCX Brick and have fun.

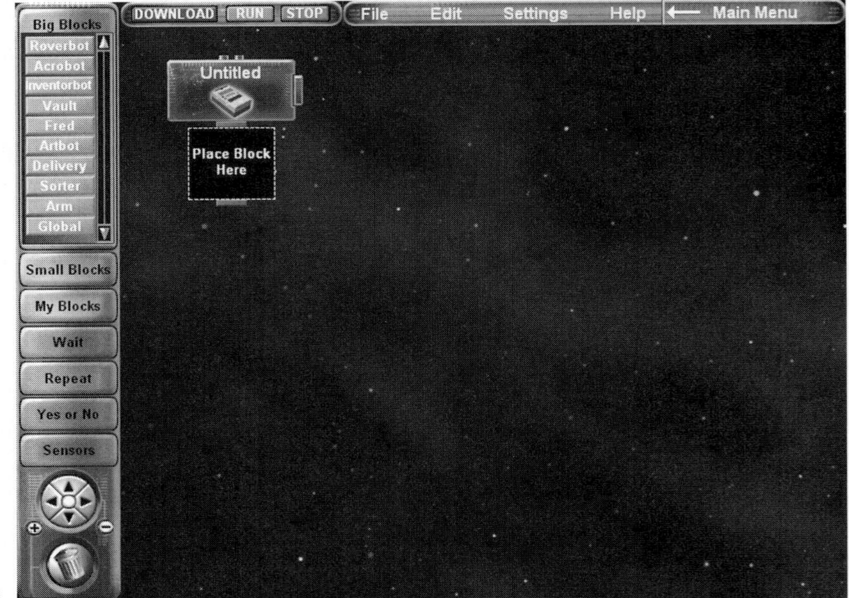

4

5

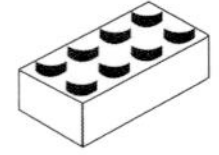

YOU HAVE SIX BRICKS AND 915 MILLION POSSIBILITIES, NOW GET TO WORK

Here is a little brain twister from the folks at the LEGO Group: How many possible combinations are there for combining six 2×4, 8-stud bricks? Oh, and all six bricks must be connected together. Well, before you start throwing away your evening trying to solve this little puzzle, maybe you should just find comfort in the answer provided by the LEGO Group. There are 915,103,765 possible combinations. Gulp; now that makes the LEGO MINDSTORMS NXT look really impressive doesn't it?

How to Join Six Bricks Together in Perfect Harmony

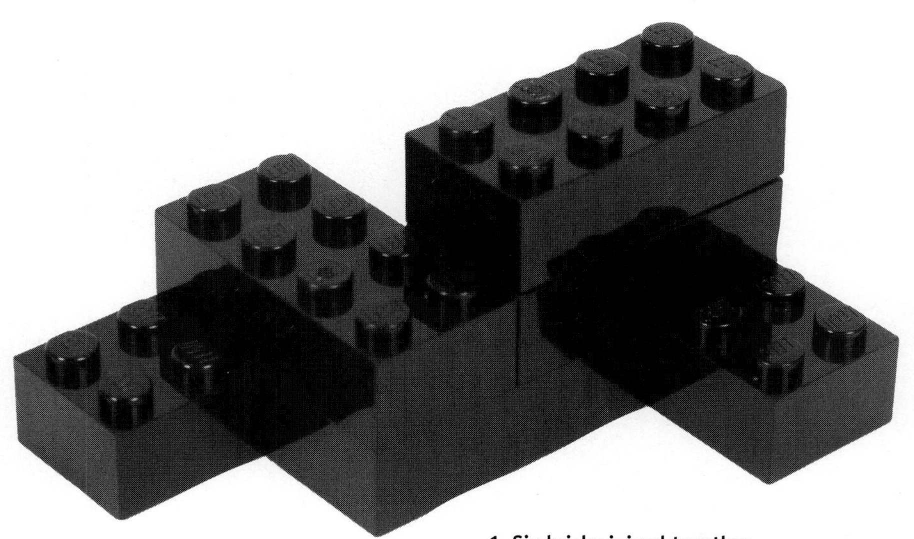

1 Six bricks joined together.

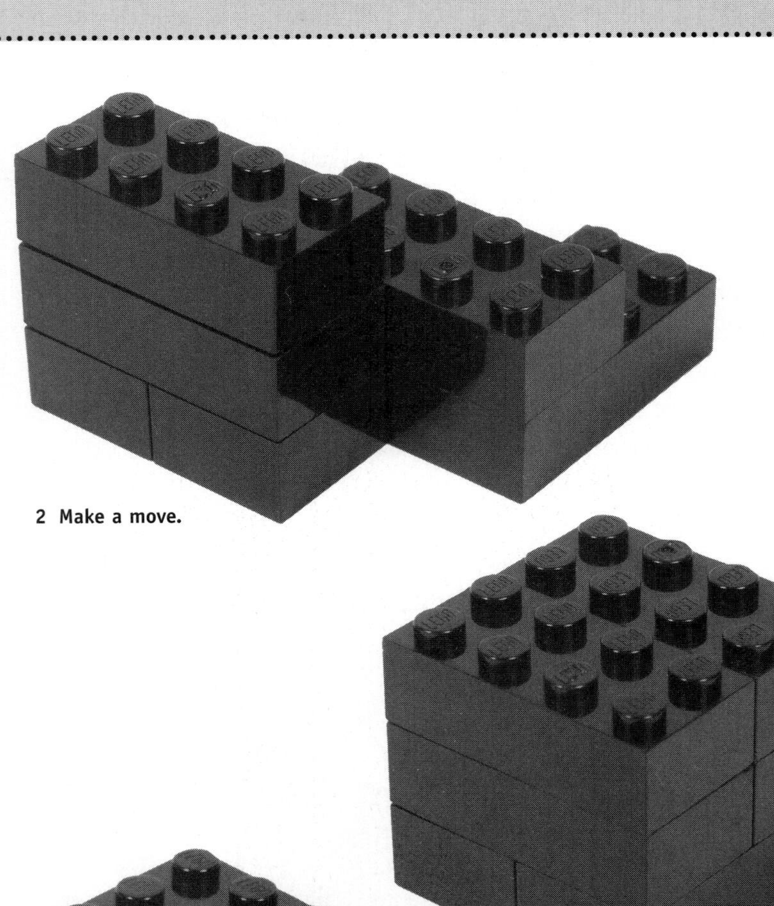

2 Make a move.

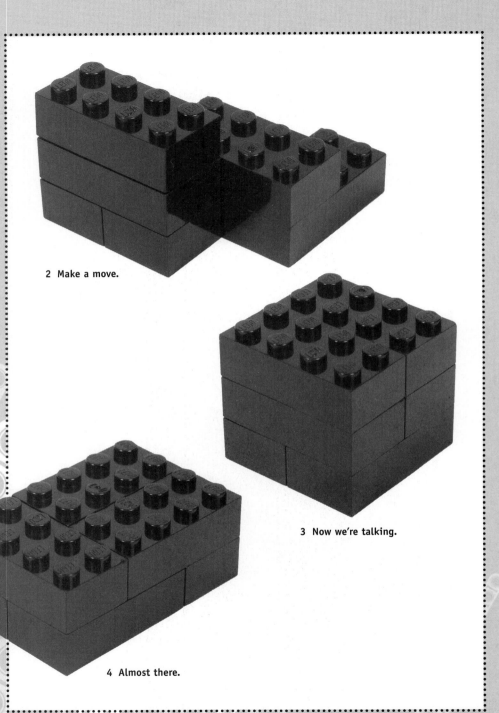

3 Now we're talking.

4 Almost there.

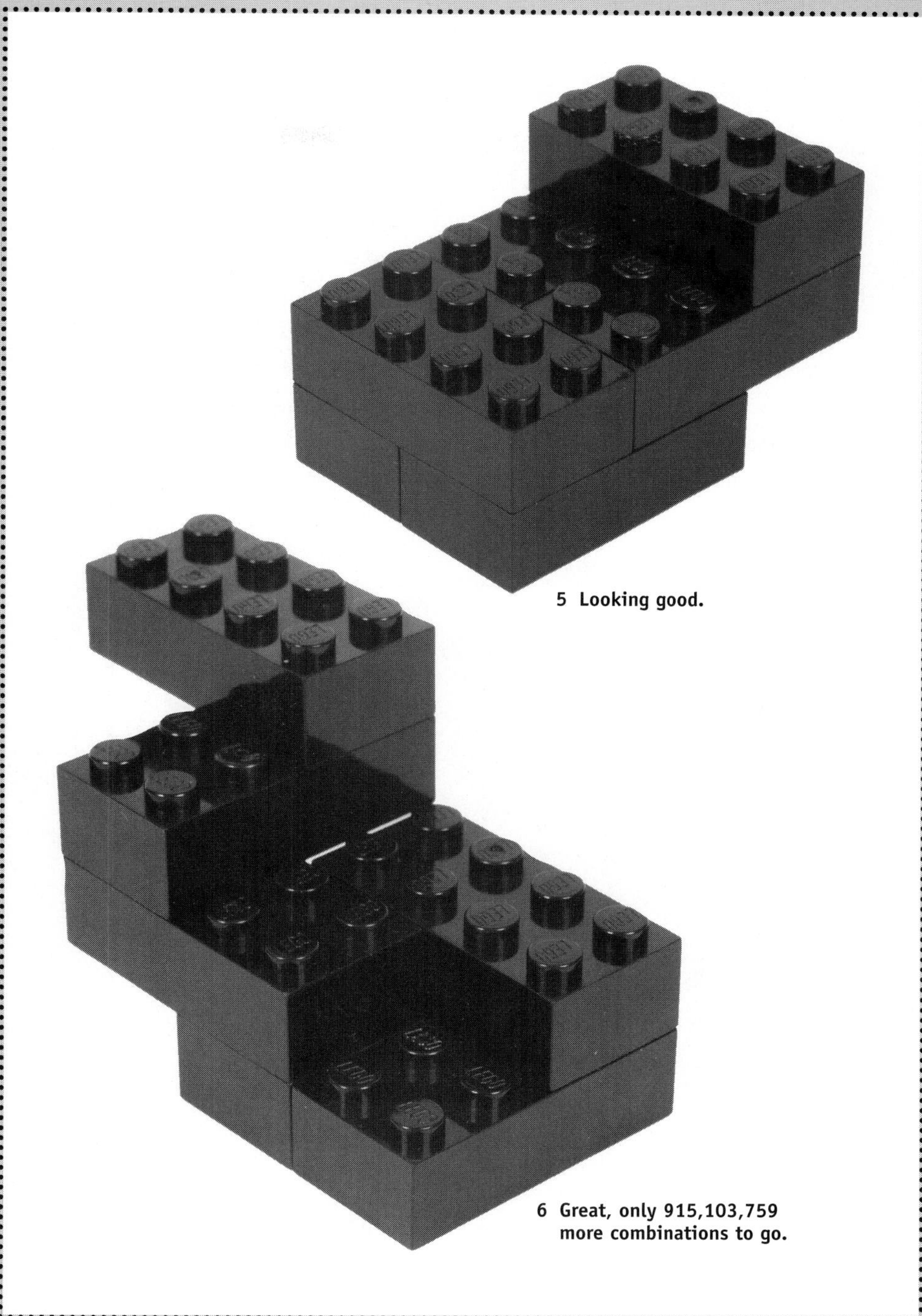

5 Looking good.

6 Great, only 915,103,759 more combinations to go.

DRIVEN TO STORE MORE

Leave it to LaCie to provide a hard drive solution that makes a bold, creative statement while supplying up to 500 gigabytes worth of storage. Known as the LaCie Brick, these hard drives are both stackable and colorful.

Designed by Ora-Ïto, the LaCie Brick is available as either a desktop edition (3.5-inch drives with capacities ranging from 160 to 500 gigabytes) (see Figure 1-2) or a mobile edition (2.5-inch drives with capacities ranging from 40 to 120 gigabytes) (see Figure 1-3). Priced between $99.99 and $399.99, LaCie Bricks also come in your choice of three colors: red, blue, or white.

So, who the heck is LaCie, anyway? A French-born company founded in 1989, LaCie quickly rose to fame in the hearts of all Mac owners for a line of

1-2 Storage that stacks up—LaCie Brick Desktop Hard Drive. (Photograph courtesy of LaCie)

superior SCSI interface external hard drives. Later, color-perfect monitors and support of PC hard drives helped entrench LaCie in the minds of all computer users.

OK, LaCie makes excellent hard drives, but what's with Ora-Ïto? Well, one of the special hallmarks of LaCie drives has been on the outside of their hard drives. LaCie regularly partners with some of the world's foremost designers for making the ugly technology of data storage pleasing to the eye, as well as to the pocketbook. The result has been incredible storage capacities housed inside the beautiful designs of Philippe Starck, Porsche Design GmbH, and, now, Ora-Ïto.

Now imagine the joy of corporate IT staffs everywhere as they build massive storage systems with LaCie Bricks. Just stack, pack, and serve.

1-3 Take your bricks with you—LaCie Mobile Edition. (Photograph courtesy of LaCie)

IT TAKES A VILLAGE, BUT A COUPLE BILLION MINIFIGURES CAN'T HURT

Just what is a minifigure, you ask? A minifigure, specifically a LEGO minifigure, is a diminutive humanoid being that is packaged with most LEGO theme sets. Licensed products like the LEGO Star Wars and LEGO Harry Potter sets contain starfighter drivers, maniacal masters of the universe, and assorted wizards and magicians.

One and all of these beings are LEGO minifigures.

Born in 1978, the original minifigure population consisted of seven different characters. (**NOTE**: Not to be confused with role-playing LEGO figures which originated in 1974.) These characters were the residents of the LEGO Town, Space, and Castle theme sets. It wasn't until the arrival of the first female minifigure, two months later, that the population really exploded.

Naturally, the appearance of this female character made the minifigures all smiles. In fact, LEGO minifigures remained a smiling mass until 1989; that's when those bad-boy pirate minifigures sailed into town.

Now minifigures could smile, laugh, smirk, frown, sneer, and grimace. Just like you and me; well, maybe just you, I'm always all smiles.

Still, LEGO minifigures were a stationary crowd. That is, until 2000, when LEGO Football introduced the spring-mounted element to the minifigure. This element was improved with LEGO Basketball theme characters. Now, minifigures could be seen hopping and bending all over the place—the masses had mobility.

Finally, in 2003, LEGO minifigures underwent a major makeover. Gone was the sickly yellow hide from the late 1970s; LEGO minifigures now bore more accurate skin tones, as well as professionally styled hair. There's nothing quite like a professionally coiffured minifigure for making a real fashion statement about your latest LEGO creation.

So what about size? Does it matter? Just how big is a LEGO minifigure? Although some texts have attempted to assign a scale height to minifigures, it just isn't so. The LEGO Darth Vader minifigure, for example, is the same height as the nurse from the City theme set who, in turn, carries the same dimensions as the LEGO Harry Potter. Of course, Harry's got that full head of hair.

This size ambiguity dissipated in 2005, however, with the release of the LEGO cast from Star Wars™ Episode III *Revenge of the Sith*. Suddenly, you had the diminutive Yoda™ and Astromech Droid ™ joining the ranks of Anakin Skywalker™ and company.

So in your LEGO village, everyone is darn near equal—the same height, the same looks, and the same disposition. If only real life was so bland. Nah, I enjoy my grouchy neighbor's unique viewpoint on life. Therefore, minifigures should stay permanently ensconced inside their LEGO enclaves.

Even so, the LEGO minifigure is a strong force with over 4 billion produced since 1978. If only we can empower them with the right to vote, then things would change. And I could be the first ruler of minifigures. Moo-ha-ha.

It's a Small, Wonderful Minifig World, Out There

1 LEGO minifigures or minifigs.

2 "Hey, we're all the same height."

3 "Where's your hair, Harry?"

4 Minifig pins make a great fashion statement.

5 Helmets truly make the minifigure.

6 Now that Harry's found his wig, nobody's going to take it, again.

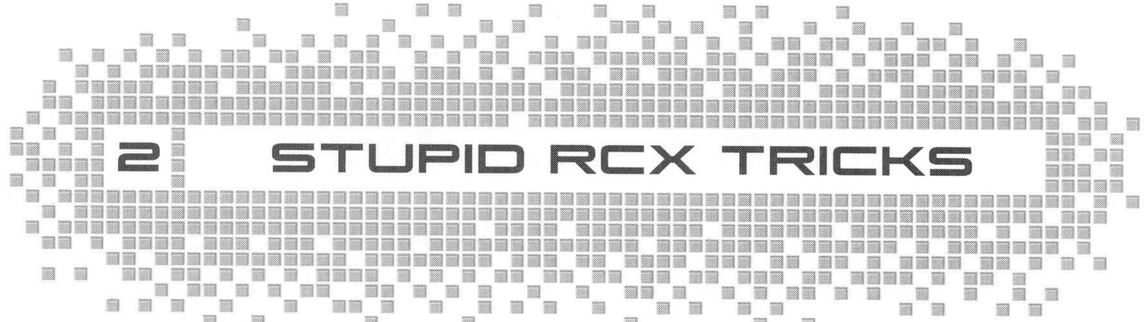

2 STUPID RCX TRICKS

BEFORE THERE WAS NXT, there was RCX. Functioning as the "brain" for the pre-NXT MINDSTORMS Robotics Invention System (RIS), the RCX was considered a "programmable" LEGO brick. Known as the "Robotics Command System," the RCX had three input ports, three output ports, four integrated control buttons, a built-in LCD screen, and an infrared (IR) transmitter. Able to fit in the palm of your hand, the RCX was more capable than a Timex-Sinclair ZX-81 computer. Whew, how's that for a throwback to the stone age of microcomputers?

Inside the RCX, an 8-bit Hitachi H8 microcomputer featuring an H8/300 central processing unit (CPU) served as the "rider" for driving your LEGO robot horse. A lonesome CPU isn't much good without some support, however.

Peripheral capability for the CPU was supplied by a warehouse full of on-chip support modules. In other words, the H8/300 CPU was enhanced with ROM, RAM, timers, a serial communication interface, an A/D converter, and I/O ports via these extra support modules. Now that's a microcomputer worth using.

But all of the digital capability in the world is worthless without power—electrical power. In the case of the RCX, you had to install six "AA" batteries inside a tray on the back of the RCX. Unfortunately, replacing these batteries inside a functioning RIS robot could become an exercise in frustration. Therefore, replacing the stock LEGO battery tray with an external rechargeable power supply became one of the first RCX hacks performed by most RIS owners.

HOW TO ADD YOUR OWN EXTERNAL BATTERY PACK TO THE RCX BRICK

1 The RCX Brick.

2 Remove the battery pack cover.

3 Remove the screws from the battery pack.

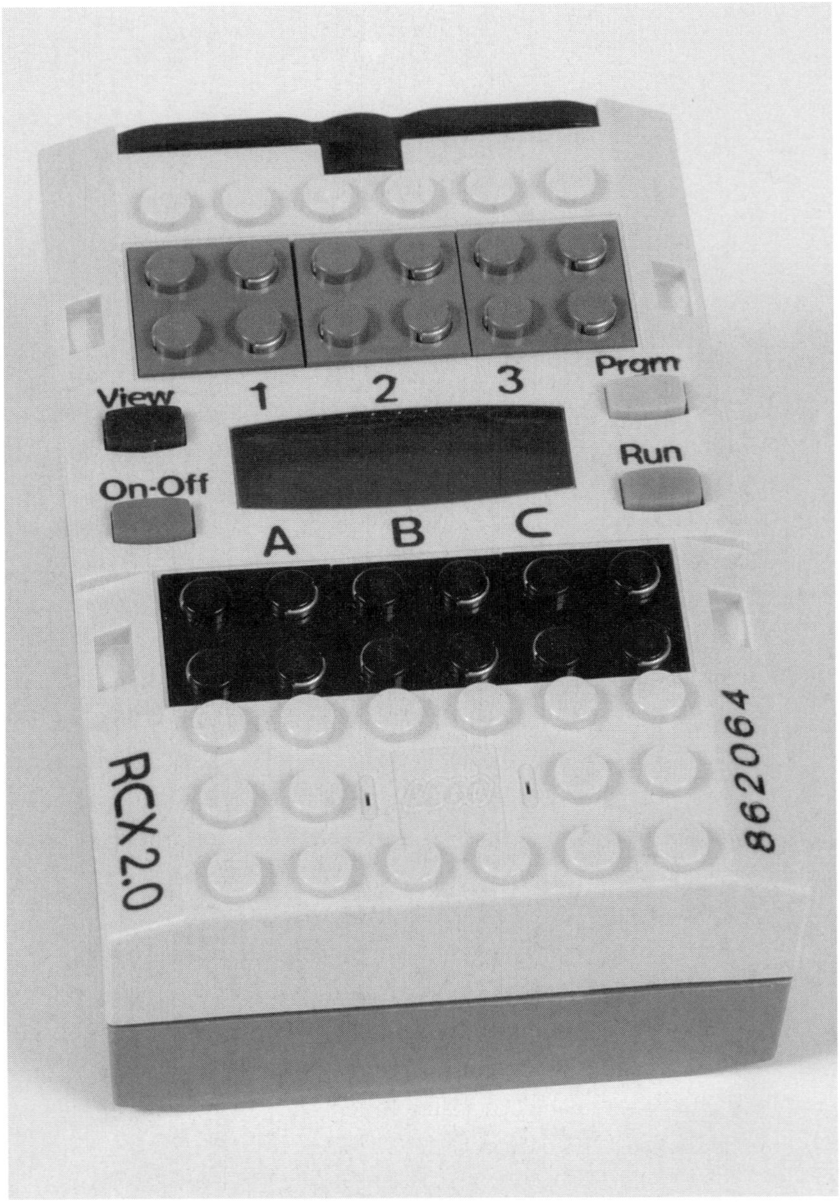

1

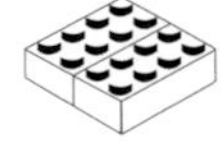

2

3

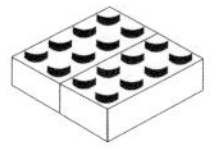

4 Separate the RCX cover.

5 Locate the top terminal.

6 This is the negative terminal.

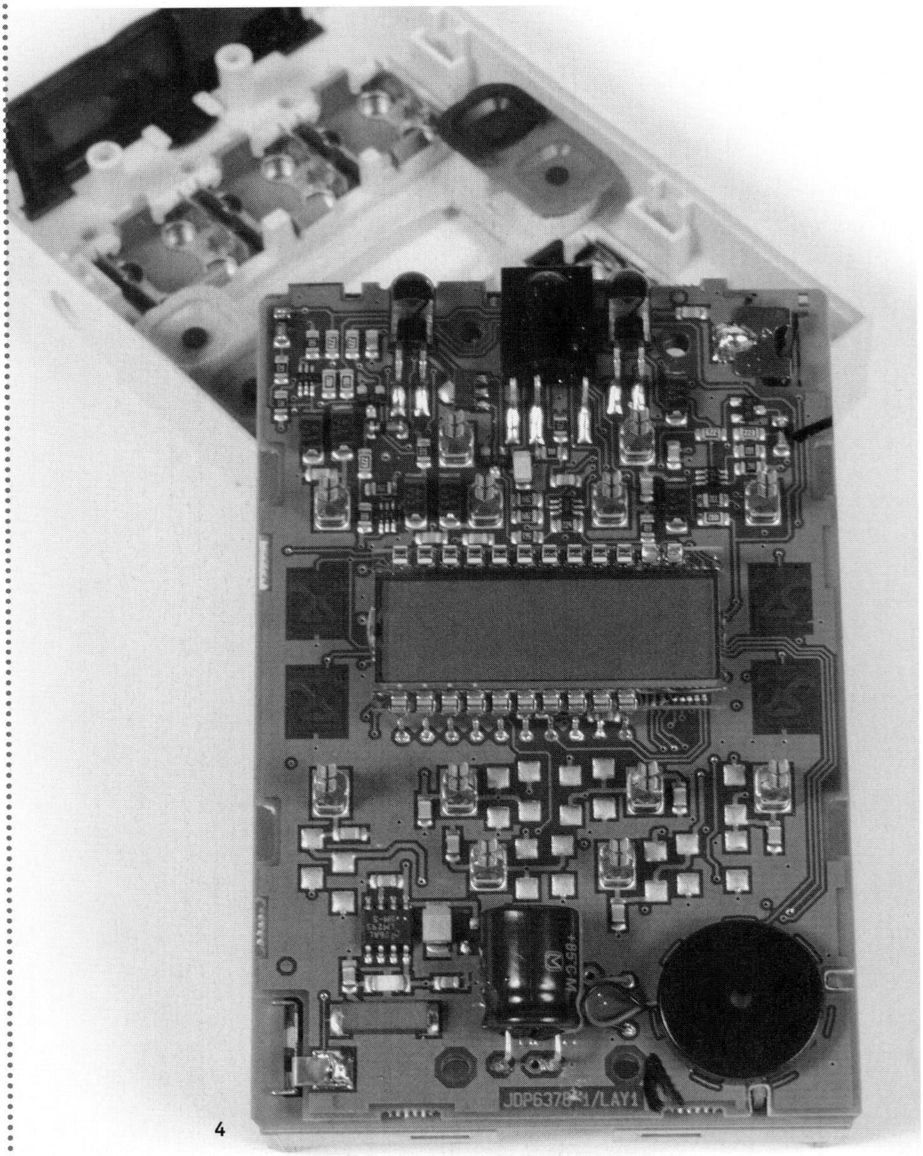

4

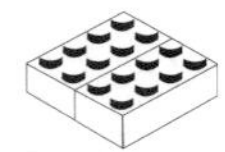

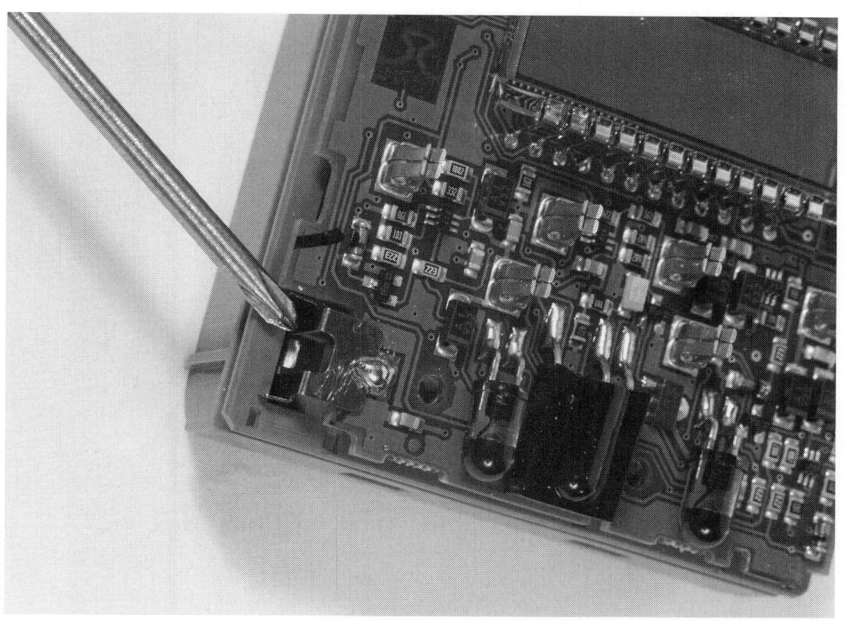

5

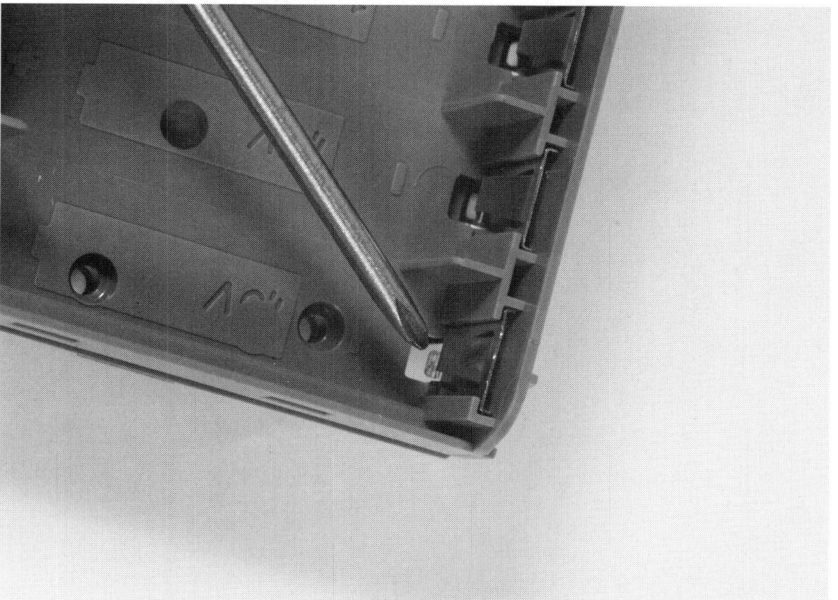

6

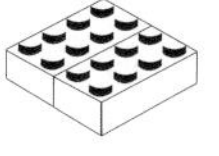

7 Locate the bottom terminal.

8 Verify that this is the positive terminal.

9 Route your wiring through this opening.

7

8

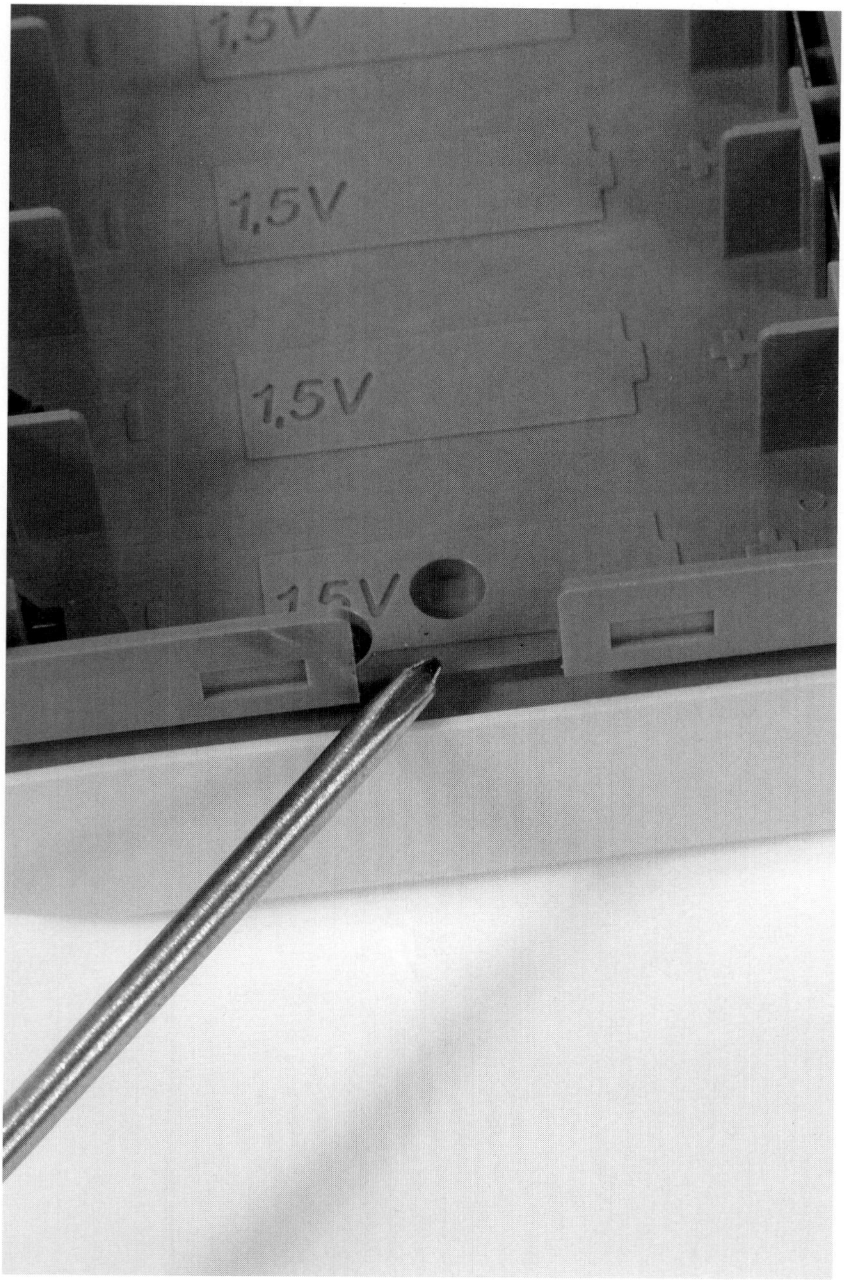

9

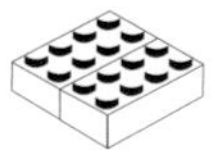

10 Use clips to attach your battery pack to each terminal.

11 Reassemble the RCX and close the battery pack cover.

10

11

ALL IN THE PALM OF YOUR HAND

Imagine the look on your co-worker's face when you pull your RCX out of your pocket to scroll through a list of phone numbers. Or, think how impressed your mates at the gym will be when you turn on your RCX music player and exercise to the tunes of your own melodies. Neither your PDA nor your iPod should be thrown away, yet, but you can help give your RCX some new life through some clever programming and a touch sensor. Here's how:

RCX ADDRESS BOOK. Take a stock RCX and mount a touch sensor on it as Sensor 2 (I use Sensor 2 purely for design aesthetics). Now write your code for holding your telephone numbers in as a series of "if" statements along with declared variables.

Because the RCX is limited to 16-bit precision integer variables, you will have to divide up your telephone numbers into a series of area code, prefix, and phone numbers. Also, some peculiarities with the display function can make the display difficult to read. In operation, the touch sensor will display a different phone number element for every "click" on the touch sensor (see Figure 2-1).

RCX MUSIC PLAYER. Just like the RCX Address Book trick, connect a touch sensor to Sensor 2 of your RCX (see Figure 2-2). In this case, however, use the BEEP block to compose your musical ditty. Then cut and paste the resulting code into your RCX Music Player program.

If you're looking for some extra credit, try these two tricks:

RCX REFRIGERATOR RAIDER ALARM. OK, remember that New Year's Resolution that you made for losing weight? Well, give yourself a break and use your RCX for blaring out an alarm every time the refrigerator door is opened.

Not to be confused with the LEGO MINDSTORMS RIS 2.0 advanced *Refrigerator Fred* robot, here's how to build a better RCX alarm:

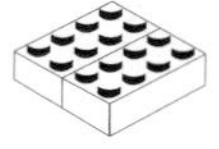

2-1 **RCX Address Book.**

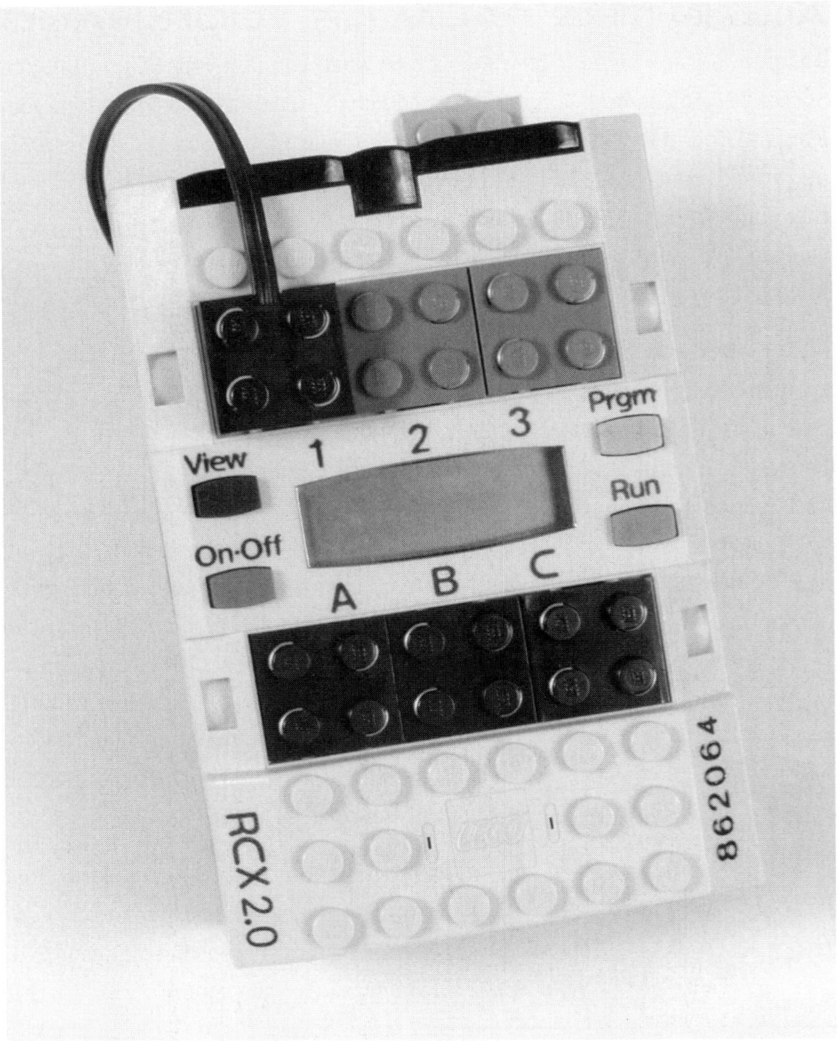

2-1

Connect the LEGO RIS light sensor to Sensor 2 of your RCX. Now program the light sensor to register a change when the light threshold is exceeded and BEEP the speaker. Oh, and don't forget to record how many times the frig door is opened. Then you'll have a number to help quantify that weight gain.

For extra, extra credit, add a LEGO motor along with a geared mechanism for waving an arm when the refrigerator light goes on. Not only will your RCX alarm record your raid, it will also greet you in a manner that is bound to embarrass you enough to go on a diet.

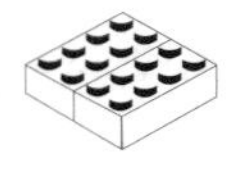

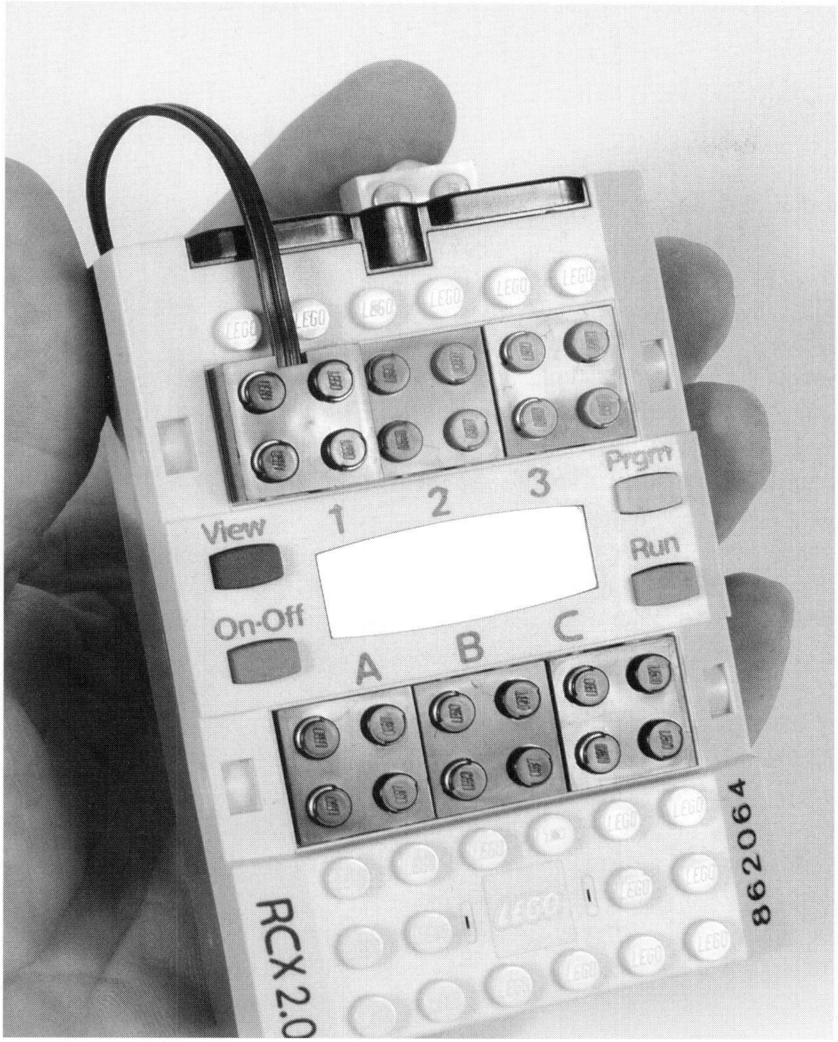

2-2 **RCX Music Player.**

2-2

QUICK DRAW MCGRAW RCX. Oh, a game for your RCX. Get your fancy fingers ready. Add a touch sensor to the Sensor 2 input port on your RCX. Then write a simple data logging program. This program issues a BEEP on the RCX speaker, then it begins polling the touch sensor (ouch, gosh that polling's gotta hurt) until a trigger has been logged. Your fast finger reaction time is then displayed on the LCD screen.

For extra, extra credit, try to record the s-l-o-w-e-s-t reaction time. This is a great way to demonstrate the memory storage limitations of the RCX, as well as proving to your friends that you are a person of incredible patience.

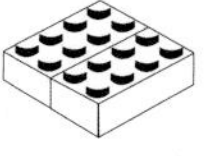

I'M BIONICLE® BILL, THE SAILOR

Can you imagine releasing a new toy in the aftermath of the September 11, 2001 tragedies in New York City, Pennsylvania, and Washington, DC? Especially, a toy that features frightening, armor-clad characters who violently confront their enemies? Well, the LEGO Group did just that with the opening of the BIONICLE universe in 2001.

And within a short five-year span, BIONICLE has taken on a life of its own. Rising from the ruins of America's worst nightmare and spawning a profitable lineup of incredible alien-like creatures, a publication line of popular magazines (including the gorgeously illustrated Farshtey/Sayger *Ignition* edition), a video game, a CGI-animated DVD movie, and dedicated Web site, the LEGO Group even cultivated a new term for this revolutionary theme set concept: Constraction.

Claimed to be a combination of "construction" toys and "action" figures, the BIONICLE Constraction category was joined by Knights' Kingdom in 2004. Although not as widely popular as BIONICLE, Knights' Kingdom has helped make the Constraction category of toys a legitimate element in the LEGO Group's brick empire.

Whether your passion is BIONICLE or Knights' Kingdom, the storyline is basically the same: the LEGO Group plants the most minimal amount of a concept. For example, in the initial year of BIONICLE this story concept was titled, "Quest for the Masks." Supported by a handful of construction, err, I mean, Constraction kits and an online community, BIONICLE users are invited to "grow" or develop the story and action themselves.

The result is a unique, fully blossomed interaction between people, storytelling, and role-playing. Not bad, for a gutsy marketing venture in the twilight of our darkest hour.

In a universe that is filled with strange names like Rahkshi, Vahki, Metru Nui, and Piraka, the BIONICLE line continues to evolve. For those of you keeping score at home, following the 2001 debut, the BIONICLE storylines changed to:

- 2002: The Bohrok Swarms
- 2003: The Bohrok-Kal Strike and the Mask of Light
- 2004: Metru Nui
- 2005: Web of Shadows
- 2006: Piraka: The Gang is on the Loose

Likewise, the LEGO Group released massive battle sets beginning in 2004. The Battle for Metru Nui included Tower of Toa (8758), Visorak Battle Ram (8757), and Battle of Metru Nui (8759). Each set included some great LEGO elements along with new "mini" BIONICLE critters. Along with the standard BIONICLE kits, these battle sets make great sources for parts to hack onto your NXT robots.

But if you're new to BIONICLE, then the best bet for supplementing your NXT construction elements is with the Special Edition BIONICLE Ultimate Creatures Accessory Set (6638). Released in early 2006, this $19.99 "jar" contains over 300 LEGO elements which enable you to build at least one BIONICLE creation.

Inside each of these canisters you will find over 2 1/2 pounds worth of Kanohi masks, tools, Rahkshi heads and spines, armor, weapons, feet, and some random bodies. In fact, you *should* be able to construct at least one Vahki or Rahkshi or two Matorans.

Why is my list of contents so vague? Well, it appears that the LEGO Group is randomly filling each of these Special Edition sets with an assortment of pieces and parts from older BIONICLE lineups. That's why the set is labeled, in a typically un-LEGO-like fashion, as containing "300+" pieces.

Ironically, inside the three Special Edition BIONICLE jars that I examined, I couldn't find any BIONICLE Eye (41669) pieces. Therefore, you will have to develop some creative creature creations that don't rely on the BIONICLE Eye for assembly.

For example, one of these jars consisted of parts from Panrahk (8587), Guurahk (8590), and Vorahk (8591) while another one held elements from Kurahk (8588), Lerahk (8589), and Turahk (8592) sets. Regardless, either of these Special Edition sets could be used for building one Rahkshi Kaita.

Inside yet another one of these jars I found a mother lode of Matoran (872x series) parts including power carvers used by Velika (8721), chargers of Dalu (8726), and Garan's (8724) pulse bolt generators. Remarkably, this same jar contained enough parts to build a complete Velika *and* a battle-ready Dalu. Very sweet.

Likewise, using any of the elements from a Special Edition BIONICLE Ultimate Creatures Accessory Set is one of the best ways to hack your NXT creations into something special. Even if you *are* too old to pronounce any of the BIONICLE names.

HOW TO MAKE A BIONICLE SPECTACLE

1 A bountiful Bionicle bonanza.

2 Create your own Bionicle creatures.

1

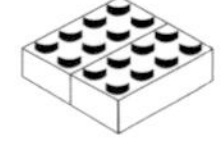

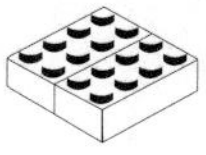

2

3 Unless you build behemoth Bionicles, you will need to use micromotors for animating your creations.

4 This room alarm system is doubling as a Bionicle.

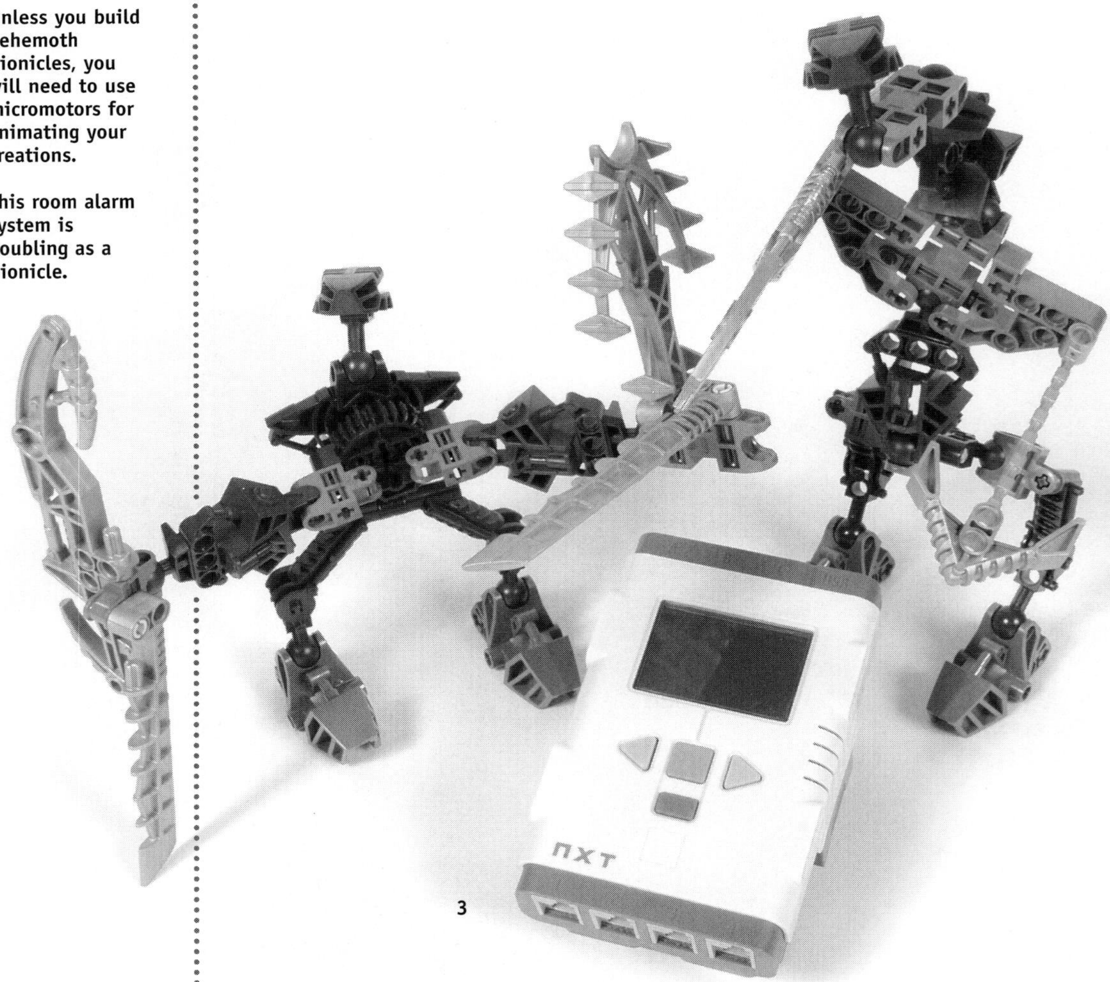

3

4

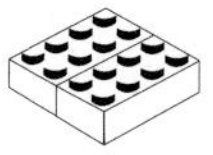

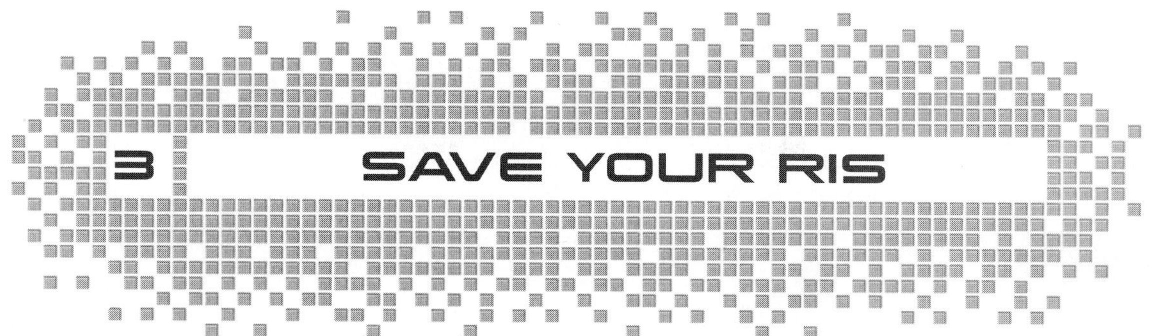

3 SAVE YOUR RIS

PACKED WITH 718 LEGO Technic elements, the LEGO MINDSTORMS RIS 2.0 (3804) represented a high watermark in amateur robotics. Supplied with two high-efficiency geared electric motors, two touch sensors, one light sensor, and the venerable microcomputer-on-a-brick RCX programmable brick, the RIS could make anyone into a competent robot designer.

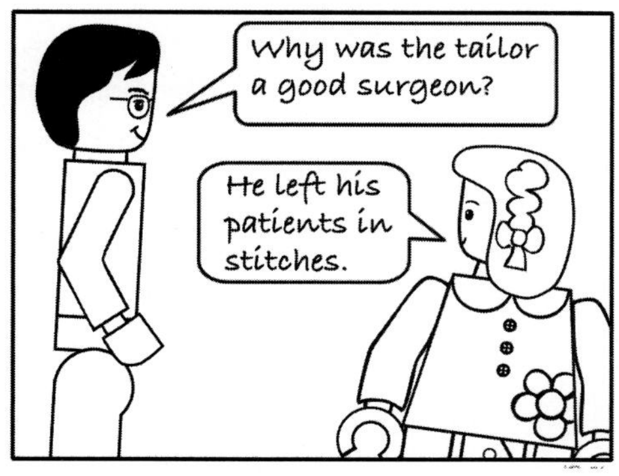

Even the most remedially challenged robo-designer could knock out a fairly functional robot within 30 minutes of opening up the RIS box. Featuring a 100+ page instruction book, titled *Constructopedia*, that contained elaborate designs for three robots (with at least another three variations of each design), the RIS was able to add "life" to each robot creation with a unique programming language integrated design environment (IDE).

Organized like a series of, well, bricks, the RIS IDE was an intuitive click-and-drop programming language that could become more powerful as the user's capabilities matured. Just build your program from a series of RIS IDE bricks, download it to the RCX via an infrared transmitter, and test your robot. It couldn't get much simpler.

Some intrepid experimenters created other programming languages for controlling the RCX, but most users were content with the standard RIS IDE. That is, unless you were a Mac OS RIS user.

The RIS IDE was a PC-only programming environment. If you took your Mac OS ire to LEGO customer support you were provided with the incompetent recommendation to use Virtual PC for Mac with the RIS software. That type of reply was neither an option nor was it acceptable to Mac users. Yes, there was the ROBOLAB software which is Mac OS compatible. Unfortunately, a site license cost $235 and many interested school districts were unable to justify this "extra" expense.

All of these faults and limitations were addressed with the NXT, making robot designers everywhere very happy.

That concludes our quick jog down memory lane.

So, why should you save your old RIS? Aside from the 711 LEGO elements (that's the total RIS parts list minus the RCX-specific elements) which can be combined seamlessly with your NXT set, you can also integrate the RCX into your NXT robot designs for added dual "brain" functionality. Hey, two heads are better than one, right? Well, almost.

Don't expect the NXT to play well with the RCX right away. You will have to do some simple modifications to the RIS sensors and motors to make them compatible with the NXT set.

Likewise, if you'd like to reproduce some of those exotic RCX "extras" like the Micromotor and the LEGO Lamp Brick, then you're going to have to do some creative hacking. That is, both LEGO brick hacking *and* electronics hacking. No it's not impossible, just a time-consuming task that does require patience, dexterity, and some finesse. But I'm confident that you can handle it. Here, hold my hand.

Oh, and probably the best reason for saving your old RIS set is that it included the only real "tool" ever released by the LEGO Group—the Brick Separator. Sure, this doorstop-shaped device only fetches about $1.69 on the LEGO marketplace, but the savings in pedicures that this little tool will afford makes it, well, priceless.

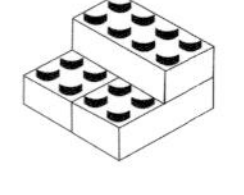

KEEP YOUR OLD RIS KIT

1

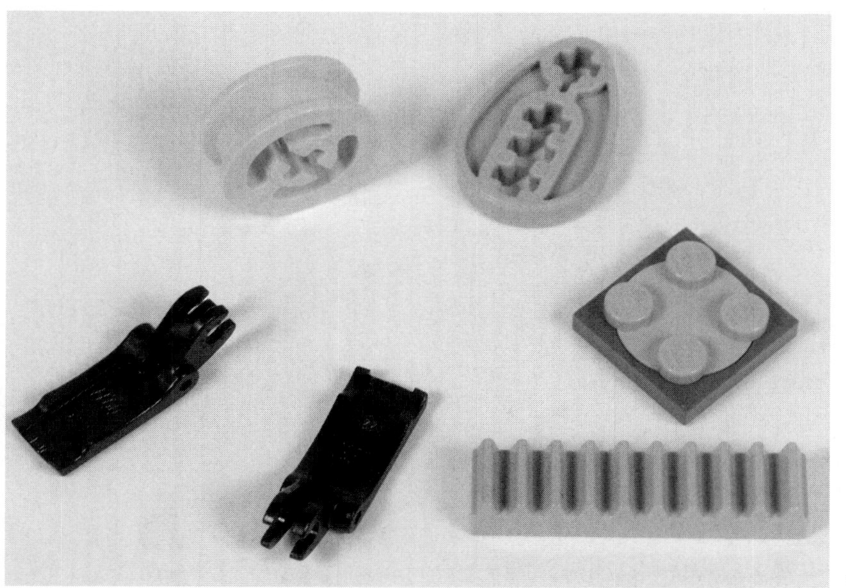

2

1 This tubing is great for NXT projects that move water.

2 Skid pads and turntables for moving NXT parts.

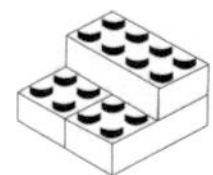

3 A Clutch Gear and a steering wheel for getting from point A to point B.

4 Gee, wings; need I say more.

5 Great big wheels and tires are perfect for driving through those robot obstacles courses.

6 A simple worm drive. And, no, this gearbox is not hooked into the RCX...yet.

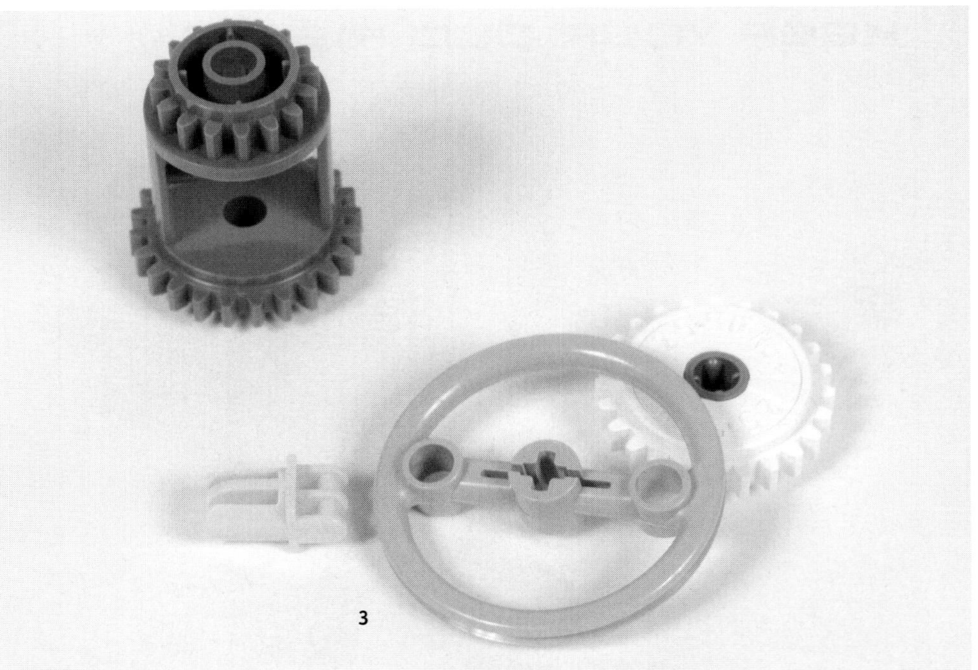

3

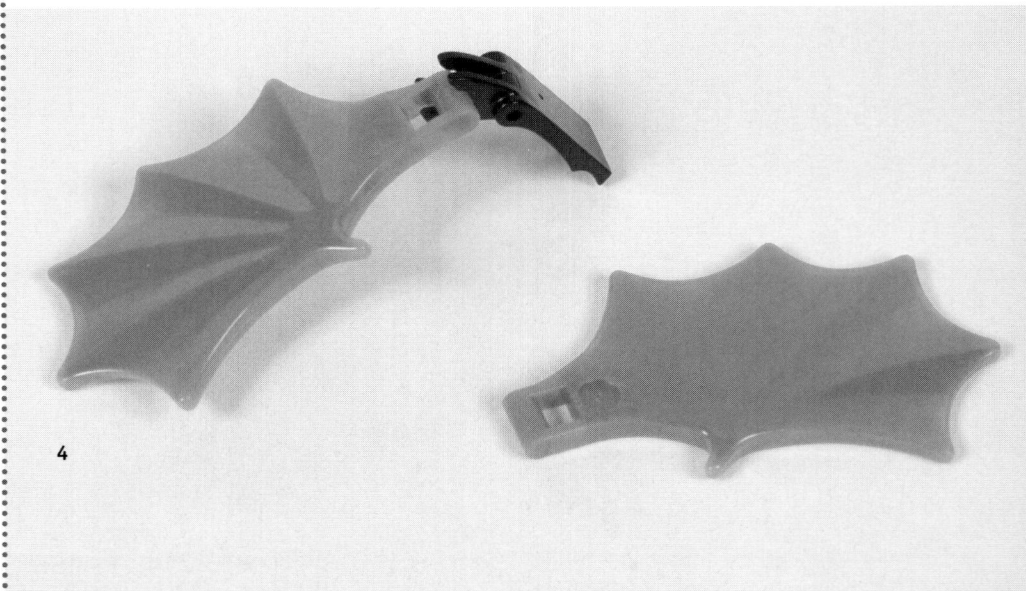

4

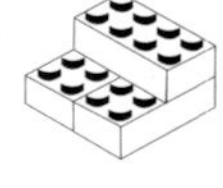

5

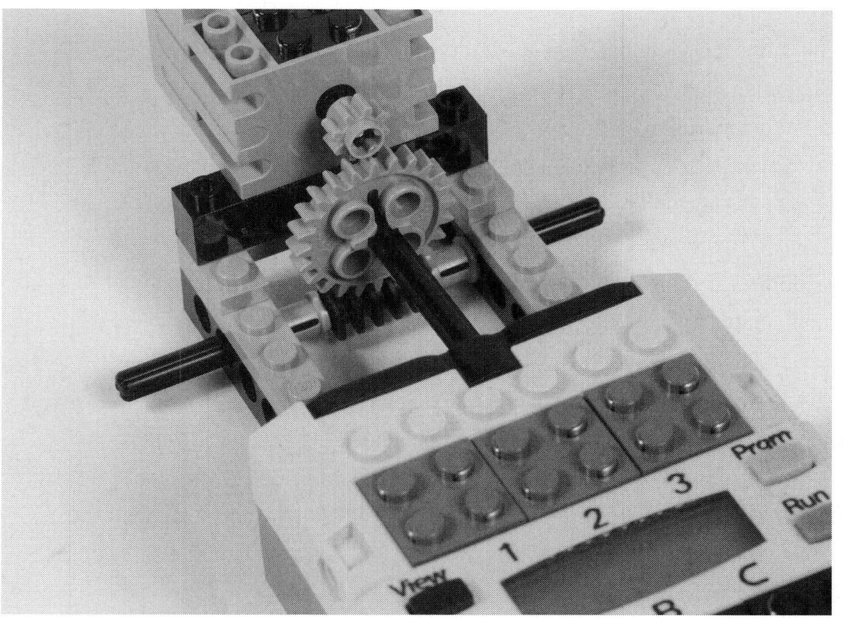

6

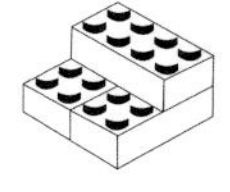

SOME MOTIVE FOR GOING LOCO WITH LEGO

Think that you can handle a logic puzzle? OK, here ya go.

Can you believe that mathematics and logic used to be really, I mean *really, big* news? So big was math news that newspapers would actually run mathematical and logic puzzles in them.

Yeah, that's right, like the horoscopes that we read in newspapers today, puzzles were a regular pastime in turn-of-the-century dailies. Oh, that's the twentieth century, not this newfangled modern century that has just barely begun.

Now get this: people who dabbled in these puzzles were actually called "amateur" or "lay" or "recreational" mathematicians. Gosh, in today's world of entertainers, athletes, and celebrities it's tough to think that folks really pursued an avocation in mathematics. But they did and sometimes these amateurs made a splash on the society page as well.

Just like the headlines about a love-scorned starlet, these lay mathematicians would occasionally cross calculators and the sparks would fly. Oh no, math fight!

One of the biggest brouhahas that arose during this math puzzle heyday was the one between Sam Loyd and Henry E. Dudeney. Loyd the American and Dudeney the Englishman began their relationship as active pen pals.

Collaborating on puzzles that they both published in newspapers, magazines, and books, these two lay mathematicians were fast friends and public darlings. In fact, Dudeney was such a massive celeb across the pond that he purportedly adopted a nom de plume, authoring many of his early puzzles under the name "Sphinx."

Sometime around the early 1900s, Dudeney accused Loyd of stealing his original puzzles and then publishing them as if they were the American's own work. Accusations and denials sailed back and forth across the Atlantic Ocean, but the damage was done and this amazing puzzle-publishing duo was done.

Sadly, Loyd died in 1911 before any form of reconciliation could be made between these two old friends. Lucky for us, however, these two recreational mathematicians left us a terrific little logic puzzle about railroads.

Well, there you go again, who cares about railroads today? Well, if you look beyond the train aspect of this puzzle you can get a very good education in database sorting. Say what? Yup, in fact, a group of scientists wrote a journal article about this concept using the original Loyd puzzle.

According to the abstract from this journal article about "reversing trains" by Nancy Amato, Manuel Blum, Sandra Irani, and Ronitt Rubinfeld:

> There is a train, locomotive and n cars, that must be entirely reversed using only a short spur line attached to the main track. The efficiency of a solution is determined by summing, for all cars, the total distance moved by each car, where distance is measured in car lengths. We present an O(n^2 logn) algorithm for accomplishing this task.

The authors of this article cite Sam Loyd as the author of this puzzle. In fact, this puzzle may be one of those that Dudeney claimed that Loyd pilfered from him. You be the judge. Here is Dudeney's original puzzle as reprinted from *Amusements in Mathematics* (devoid of the original illustration):

223.—A RAILWAY MUDDLE.

> The plan represents a portion of the line of the London, Clodville, and Mudford Railway Company. It is a single line with a loop. There is only room for eight wagons, or seven wagons and an engine, between B and C on either the left line or the right line of the loop. It happened that two goods trains (each consisting of an engine and sixteen wagons) got into the position shown in the illustration. It looked like a hopeless deadlock, and each engine-driver wanted the other to go back to the next station and take off nine wagons. But an ingenious stoker undertook to pass the trains and send them on their respective journeys with their engines properly in front. He also contrived to reverse the engines the fewest times possible. Could you have performed the feat? And how many times would you require to reverse

> the engines? A "reversal" means a change of direction, backward or forward. No rope-shunting, fly-shunting, or other trick is allowed. All the work must be done legitimately by the two engines. It is a simple but interesting puzzle if attempted with counters.

Now, here is the Sam Loyd version as reprinted from *Sam Loyd's Cyclopedia of 5000 Puzzles, Tricks, and Conundrums with Answers* (again, devoid of the original illustration; also reprinted in *Mathematical Puzzles of Sam Loyd*):

> PRIMITIVE RAIL ROADING
>
> Swing to the widespread interest taken in a simple little Rail Road Switch Problem which I sprung upon my friends some time ago, as well as in response to the request from many for another practical lesson in railroading, I present one which is an offshoot from the first, and illustrates the difference between side tracking a train or passing it through a Y branch, which reverses the direction of the trains. In this specimen of primitive railroading we have an engine and four cars meeting an engine with three cars, and the problem, as in the previous one, is to ascertain the most expeditious way of passing the two trains by means of the switch or sidetrack, which is only large enough to hold one engine or one car at a time. No ropes, poles or flying switches are to be used, and it is understood that a car cannot be connected to the front of an engine. It shows the primitive way of passing trains before the advent of modern methods, and the puzzle is to tell just how many times it is necessary to back or reverse the directions of the engines to accomplish the feat, each reversal of an engine being counted as a move in the solution.

Now before you CSI this thing to death looking for a smoking gun, let me point out that Dudeney's date of publication for his puzzle book was 1917, while Loyd's puzzle book was posthumously published in 1914—three years after his death. Oh, eerie, no?

Please don't cry foul and accuse me of masterminding some sort of bogus conundrum (which would be no small feat in itself). One important bit of history that you might want to consider is that Dudeney did publish a similar puzzle much earlier in his career. According to Donald E. Knuth, Dudeney published a puzzle titled *The Mudville railway muddle* in 1897 (Issue 10, Number 10) in *The Weekly Dispatch*. [Note: Dudeney also published a puzzle in 1911 for his column "Perplexities" in Volume 41 of *The Strand Magazine* titled, *43. A railway muddle: Get long trains past each other.*]

Be advised that there have been several "modernizations" of this same type of railroad-inspired sorting puzzle. For example, Martin Gardner in his *My Best Mathematical and Logic Puzzles* presents a slightly different spin on the original concept.

In Gardner's version, there is one engine, two cars, two switches, and a tunnel. Sounds like the beginning for a bad Vaudeville joke, doesn't it? At any rate, Gardner cites in the solution to his railroad puzzle that "many different train-switching puzzles" have been previously published in the books of both Loyd and Dudeney.

So who came first? That's for you to sleuth out (see Figure 3-1).

Oh, alright, now I get it, you're asking what the heck does this have to do with LEGO bricks, let alone the RIS. Well, I'm glad that you asked that.

Are you familiar with the LEGO Group's Trains construction sets? This remarkable line of railway kits featured locomotives, track, rolling stock, structures, and, of course, minifigures (see Figure 3-2). By using just a handful of these kits, you can reconstruct any of these famed train-switching puzzles.

Here's a modest shopping list for solving one version of this age-old puzzle:

- Burlington Northern Santa Fe Locomotive (10133)
- 9-volt Train Motor (10153)
- 9-volt Speed Regulator (4548)
- (2) Santa Fe Cars Set I (10025)
- Straight Rails (4515)
- Switching Rails (4531)

3-1 Solutions to the railway problem(s).

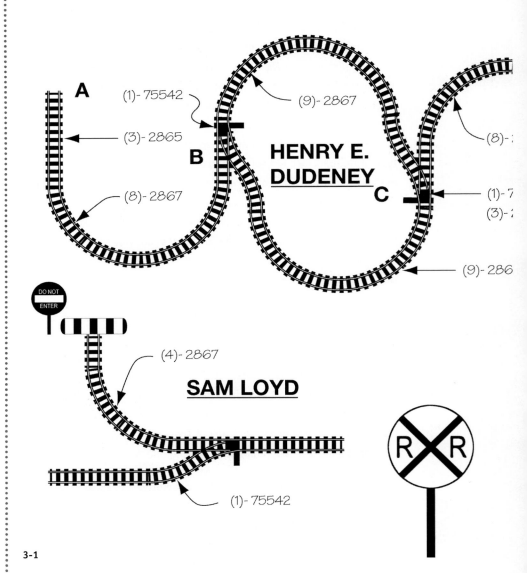

A

(1)- 75542

(9)- 2867

(3)- 2865

(8)-

B

HENRY E. DUDENEY

C

(1)- 7

(3)-

(8)- 2867

(9)- 286

DO NOT ENTER

(4)- 2867

SAM LOYD

R R

(1)- 75542

3-1

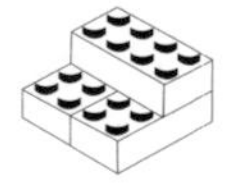

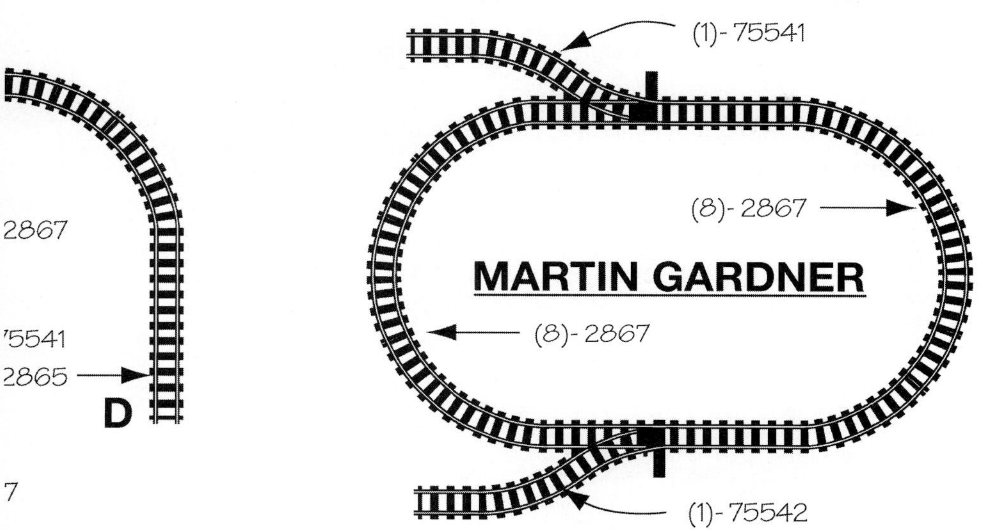

LEGO PARTS

#2867

#2865

#75541
Switch Left **#75542**

MATERIALS LIST

	#2867	#2865	#75541	#75542
Sam Loyd	5	7	0	1
Henry E. Dudeney	34	6	1	1
Martin Gardner	18	6	1	1

A Puzzle in Railroad Logic

Sheet 1 of 1 | kp

Notes
Solve these train-switching puzzles with LEGO Trains.

SAVE YOUR RIS

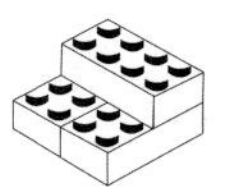

Once you've solved the puzzle, you can return to the previously mentioned journal article by Amato et. al., and form an algorithm that mimics your switching solution. You are now ready to attempt to implement this database sorting technique into your RCX and design a robot-based train-switching solution.

Once solved, you will have an enigma wrapped inside a conundrum, or something like that, and the Sphinx will be smiling.

3-2 A suitable LEGO locomotive.

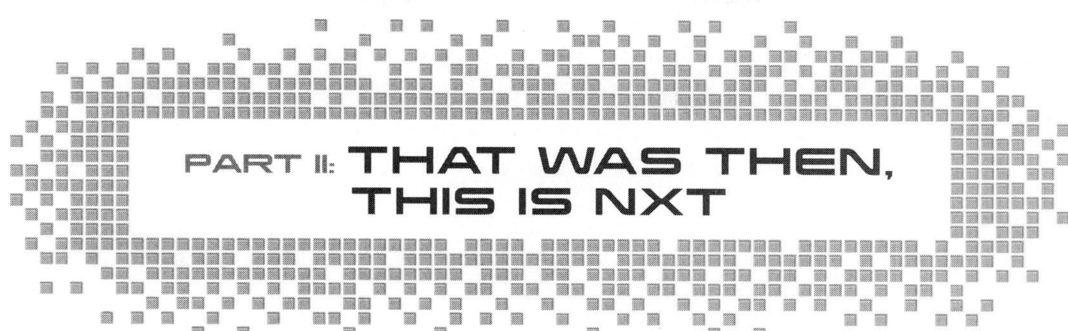

PART II: THAT WAS THEN, THIS IS NXT

4 AS SMART AS A BRICK

REMEMBER THAT EPISODE of *Star Trek* (recently dubbed "The Original Series" as a means for differentiating the various flavors of *Star Trek* that have emerged since the demise of the late 1960s teleplay) where a female intruder boards the U.S.S. Enterprise, knocks out the entire crew, and absconds with Spock's brain?

In that 1968 episode (although it was produced as Episode 62, it was aired as Episode 61 in 1968), which was titled, ironically enough, "Spock's Brain," viewers enjoyed two of the greatest script lines ever written for a televised series. The first of these chestnuts is uttered by one of the male inhabitants of the planet where the female intruder who entered the starship and stole Spock's brain lives.

This character, called a Morg, tells Captain James T. Kirk that the women of his planet are the "... givers of pain and delight." WOW, what a great line. And if you recall what this fellow looked like, you can certainly argue that he must have been a frequent recipient of the former.

Learning that this celestial body is inhabited by "... givers of pain and delight," Kirk makes a beeline to the center of the planet, locates the female intruder, and browbeats her into the second great line from this episode. His incessant interrogation regarding the whereabouts of Spock's brain torments this woman, called an Eymorg, into asking, "Brain, brain, what is brain?" Bravo. Thank you, James T. Kirk.

Well, as you build your NXT robots you might find yourself wondering out loud, "Brick, brain, what is a programmable brick brain?"

The brain in your LEGO MINDSTORMS NXT robot is neither Morg nor Eymorg. It is the NXT programmable brick. And the NXT Brick is the new programmable brick on the block; out with the Hitachi microcontroller of the RCX and in with a pair of Atmel Corporation microcontrollers for the NXT Brick.

Yes, you read right, there are two processors inside the NXT Brick. These dual Atmel Corporation processors are:

- Atmel 32-bit ARM® processor, AT91SAM7S256
- Atmel 8-bit AVR® processor, ATmega48

If we refer to the Atmel documentation for these two microcontrollers, we learn that the AT91SAM7S256 is part of an Atmel series of low pin–count Flash microcontrollers based on the 32-bit ARM7DMI® ARM Thumb® RISC processor (see Figure 4-1). The AT91SAM7S256 has a maximum clock speed of 55 mega-

4-1 The main brain of the NXT Brick, the Atmel AT91SAM7S256. (Photograph courtesy of Atmel Corporation)

4-1

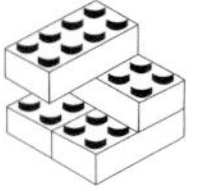

hertz, 64 kilobytes of high-speed, on-chip SRAM, 256 kilobytes of integrated high-speed Flash memory, 11 peripheral DMA channels, one USB 2.0 (12 megabits per second) device port, three 16-bit timers, four PWM controllers, 32 I/O pins, and requires a 3.0- to 3.6-volt power supply.

When you look at this laundry list of features, there are several key capabilities that need to be highlighted:

- RISC—a Reduced Instruction Set Computer processor based on ARMv4T Von Neumann architecture. The ARM7DMI processor provides .9 millions of instructions per second (MIPS) per megahertz. Therefore, you should expect a capability of 49.5 MIPS at a maximum clock speed of 55 megahertz.
- ARM—an Advanced RISC Machine with 32-bit instruction set.
- Thumb—this extension to ARM is a critical subset of the ARM 32-bit instruction set that has been encoded into a 16-bit instruction set.
- Flash memory—can be programmed in-system either via the JTAG-ICE interface or through a parallel interface on a production programmer prior to mounting.
- USB device—a port that is ideal for peripheral applications requiring high-speed wired connectivity to a PC or cellular phone.

One final mention, the ARM7DMI can execute both the high-performance 32-bit ARM and high-density 16-bit Thumb® instruction sets. These instruction sets are toggled by the processor as an ARM state or a Thumb state. In the ARM state, the 32-bit instructions are executed conditionally, while the 16-bit Thumb state instructions are a re-encoded subset of the ARM instruction set.

TWO BRAINS MUST BE BETTER THAN 32 BITS

Helping the AT91SAM7S256 make ends meet, the high-performance, low-power AVR® 8-bit ATmega48 microcontroller acts as a co-processor, handling some of the more mundane system tasks for the NXT Brick (see Figure 4-2).

Inside the ATmega48 there is 4-kilobyte reprogrammable Flash Program Memory, 512-byte SRAM, 256-byte EEPROM, an 8-channel, 10-bit A/D converter (TQFP/MLF), 23 I/O pins, and a voltage-determined maximum clock

4-2 Portable devices thrive on low-power AVR microcontrollers like the ATmega48 which lives inside the NXT Brick. (Photograph courtesy of Atmel Corporation)

speed. This latter feature is characterized as a maximum clock speed of 10 megahertz with an operating voltage range of 2.7 to 5.5 volts and a higher 20-megahertz speed grade for a voltage range of 4.5 to 5.5 volts.

Although the ATmega48 is much smaller than the main AT91SAM7S256 processor, this 8-bit microcontroller is able to achieve throughputs nearing 1 MIPS per megahertz by executing its RISC instruction set in a single clock cycle.

Now according to the LEGO Group, the general specifications for these two microcontrollers as they are implemented in the NXT Brick are:

MAIN PROCESSOR: Atmel 32-bit ARM7 processor, AT91SAM7S256

Featuring:

- 256-kilobyte Flash
- 64-kilobyte RAM
- 48 megahertz

CO-PROCESSOR: Atmel 8-bit AVR processor, ATmega48

Featuring:

- 4-kilobyte Flash
- 512-byte RAM
- 4 megahertz

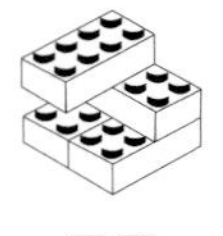

BORED? PLAY WITH THE AT91SAM7S-EK EVAL BOARD

Rather than ripping your NXT Brick apart for hacking the AT91SAM7S256, order the AT91SAM7S-EK Evaluation Board from Atmel, instead (see Figure 4-3). This board comes populated with the AT91SAM7S256 microcontroller along with several key components:

- USB port
- two serial communications ports
- four buffered analog inputs
- four user-defined LEDs with pushbuttons
- expansion connector
- prototyping area

This board also comes with a DVD-ROM containing all documentation, as well as AT91 software, C code samples, and assembly listings.

4-3 You can play around with the NXT Brick's brain using this Atmel evaluation board. Be forewarned, that's how Dr. Frankenstein got started. (Photograph courtesy of Atmel Corporation)

IT'S NOT HARALD MELVILLE AND THE BLUETOOTHS

What's with this name—Bluetooth? According to the Bluetooth SIG Web site (www.bluetooth.org), a tenth-century Danish King name Harald Blatand or Harold Bluetooth was selected as the code name for this new wireless technology by the Bluetooth Trade Association. King Bluetooth was a unifier in Scandinavia and the members of the Trade Association felt that his name was apt for inspiring collaboration between diverse industries.

You can also sense this Scandinavian origin in the Bluetooth wireless technology logo (see Figure 4-4). If you look closely you can see the runic letters "H" and "B" superimposed over each other. Oh, it might help you to know that a runic "H" looks a lot like an asterisk "*."

4-4 The Bluetooth® word mark and logos are owned by the Bluetooth SIG, Inc.

Generally speaking, the following functional assignments are performed by these two microcontrollers:

- Atmel 32-bit ARM7 processor, AT91SAM7S256
- Bluetooth communication
- USB connectivity
- LCD screen management
- Digital output port management
- Digital input port management
- Atmel 8-bit AVR processor, ATmega48
- NXT Brick button interface
- Analog-to-digital output port conversion
- Analog-to-digital input port conversion
- Digital AT91SAM7S256 I^2C communication

OPEN HAILING FREQUENCIES

So, what about this Bluetooth thing? The official statement from the Bluetooth Special Interest Group (SIG) states that:

> Bluetooth wireless technology is set to revolutionize the personal connectivity market by providing freedom from wired connections—enabling links and providing connectivity between mobile computers, mobile phones, portable handheld devices and much more. Bluetooth wireless technology redefines the very way we experience connectivity. The Bluetooth SIG, comprised of leaders in the telecommunications, computing, consumer electronic, network and other industries, is driving the development of the technology and bringing it to market. The Bluetooth SIG includes Promoter companies 3Com, Agere, Ericsson, IBM, Intel, Microsoft, Motorola, Nokia and Toshiba, and more than 2,000 Associate and Adopter companies.

Technically speaking, the Bluetooth specification is a low-cost, low-power radio standard that is used for connecting devices. Based on a nifty time-sharing architecture featuring frequency-hopping and tiny packet sizes, Bluetooth uses the 2.4-gigahertz radio band with a transmission/reception range of approximately 30 feet.

Personal computers, like the Apple Computer PowerBook G4 portables, offer Bluetooth 2.0+ Enhanced Data Rate (EDR). Bluetooth 2.0+EDR, while backwards-compatible with Bluetooth 1.x, is up to three times faster than the older standard.

Thoughtfully, the BlueCore 4 Bluetooth chip inside the NXT Brick is fully compliant with Bluetooth 2.0+ EDR (see Figure 4-5). Therefore, your personal computer, PDA, and cell phone should all be able to communicate with your robot.

While you might not have heard of them before, Cambridge Silicon Radio (CSR; csr.com) is one of the big players in the

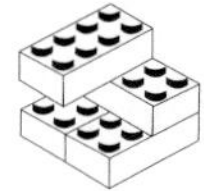

4-5 CSR's BlueCore4—3-times faster for Bluetooth Enhanced Data Rate. (Photograph courtesy of Cambridge Silicon Radio)

single-chip radio device market of short-range wireless communication. The CSR main offering is BlueCore. BlueCore is a fully integrated 2.4-gigahertz radio, baseband and microcontroller used in over 60 percent of all qualified Bluetooth v1.1-, v1.2-, and v2.0-enabled products from Apple, Dell, IBM, Motorola, NEC, Nokia, Panasonic, RIM, Sharp, Sony, Toshiba, and now the LEGO Group (see Figure 4-6).

CONNECT THE DOTS

There are two major hacks that you should do to your NXT Brick before you build any robots. In reality, neither of these hacks should be necessary, but the painless operation of your NXT Brick demands that these two glaring hardware oversights be corrected ASAP.

First, the power management system on the NXT Brick requires the insertion of six "AA" batteries into a tray held on the back of the programmable brick. Sound familiar? It should; that's the same ridiculous power plant that was implemented so poorly on the RCX.

So here we are primed for the "next step" in robotics and have to take a step backwards. Granted, the educational versions of the LEGO MINDSTORMS NXT kit are equipped with a rechargeable Lithium ion battery pack, but the retail version is not. Alternatively, you could purchase the NXT Rechargeable

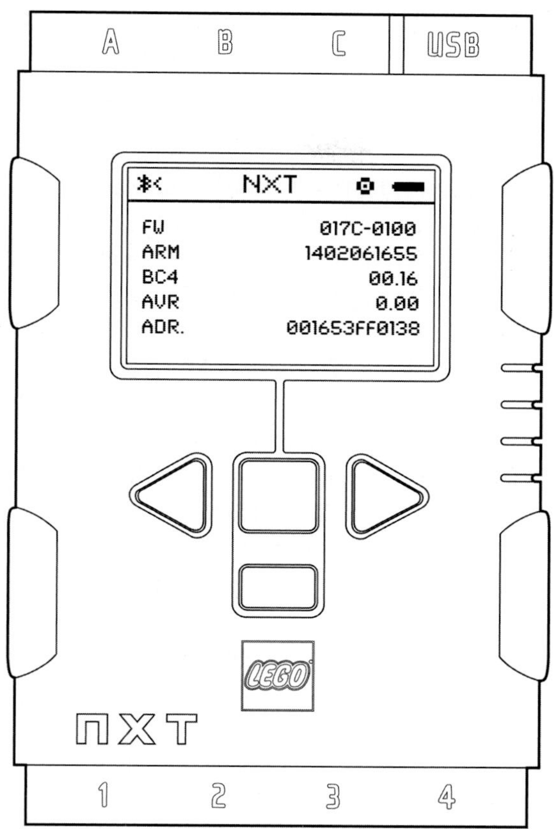

4-6 Check your NXT Brick's firmware with its built-in software.

Battery (W979798) which costs $48. Unfortunately, get ready to be zinged again: In order to charge the NXT Rechargeable Battery, you must use the AC Adapter (W979833). This little "extra" item will set you back an additional $23. This is crazy—$71 for a rechargeable battery. The LEGO Group is truly the "givers of pain and delight."

Well, you can easily add your own hacked rechargeable battery pack for considerably less money. Even better, this hack will only take about one hour to complete. See *Katherine's Best Hacker Projects* for a solar-powered solution.

Once you have your NXT Brick power management system all squared away it's time to hack a set of new six-wire connectors. Specifically, you will want to hack industry-standard connectors as replacements for the nasty proprietary connectors that are used on the programmable brick and all of the sensors and motors.

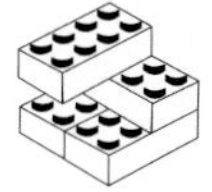

LEFT BRAIN FOR THE RIGHT BRICK

Spawned by the same "brain" that paved the way for the NXT Brick, legendary LEGO professor Mitchel Resnick has helped found the Playful Invention Company (PICO). In the words of Resnick himself, PICO will enable kids to "build on their interests in art and music, while learning important science and engineering ideas in the process."

The first commercial product from PICO was the PicoCricket™ Kit.

Released in July 2006 (ironically coinciding with the initial launch of the LEGO MINDSTORMS NXT kit), the PicoCricket Kit carried a retail price of $250. Inside this kit was a small programmable device known as the PicoCricket which was about the size of a rectangular Altoids® tin. Sound familiar? Devoid of the "engineering" pretense that earmarks the NXT product, the PicoCricket can just as easily be interfaced with lights, motors, and sensors as it can be used with pipe cleaners, bottle caps, and Play-Doh.®

Backed with a simple programming interface, designers can quickly program a creation to spin, light up, and play music.

I can hear you now; "isn't this kit a competitive product to the NXT kit?" Well, you're almost right. The PicoCricket was actually developed with support from the LEGO Group. In fact, Resnick and company see the PicoCricket as an extension of the NXT with an emphasis on artistic expression rather than building engineering skills. As a result the PicoCricket Kit enables designers to create and program artistic inventions like musical sculptures and interactive jewelry rather than building autonomous robots (see Figure 4-7).

It's like Clikits™, but with a programmable brain, sensors, motors, and pom-poms. Oh, imagine the possibilities.

4-7 The PicoCricket is a playful alternative to the NXT Brick. (Photograph courtesy of Playful Invention Company)

Basically, you will need 14 replacement connectors. Rather than attempt to fabricate your own plugs for fitting into the NXT Brick's proprietary connectors, a far better solution is to convert the LEGO Group's six-wire connectors into RJ12 (or RJ11, RJ14, and RJ25) connectors.

First of all, let's make sure that all of us are on the same sheet of music. The prefix in these replacement connectors, RJ, means "registered jack." In this context, RJxx connectors represent a family of general designs that are commonly used for telephony and data transmissions. Therefore, RJ12, RJ11, RJ14, and RJ25 are all common modular connectors (i.e., both plugs and jacks) capable of holding six wires.

While these common registered jacks are all capable of holding six wires, a set of standards state that the RJxx nomenclature should be reserved for specific wiring or conductor values. According to these standards, RJ11 connectors should only hold two conductors or wires. Likewise, telephony standards for RJ14 connectors stipulate a four-conductor wiring configuration. The remaining six-wire connectors, RJ12 and RJ25, are declared as six-conductor connectors.

HACKERS OF THE WORLD UNITE

While some manufacturers actually go to great lengths to attempt to thwart hackers (these extreme countermeasures rarely succeed), the LEGO Group is embracing the hacker mindset with the release of its NXT firmware as open source code. In a May 1, 2006 announcement, Søren Lund, the director of LEGO MINDSTORMS NXT, claimed that the LEGO Group's "ongoing commitment to enabling our fan base to personalize and enhance their MINDSTORMS experience has reached a new level with our decision to release the firmware for the NXT Brick as open source."

Along with the release of the NXT Brick firmware as open source code, Lund also announced that three developer kits (i.e., software, hardware, and Bluetooth developer kits) would be available to the "growing global audience of robotics enthusiasts." These kits contain the driver interface specifications, wiring pin-outs for the proprietary NXT six-wire connectors, and details about the embedded Bluetooth protocol.

In other words, the RJ11 standard dictates a two-wire connection, RJ14 connectors are declared a four-wire configuration, and RJ12 or RJ25 can utilize all six wires. There you go, any of these six-wire modular connectors could be used, but if you wish to adhere to modern telephony wiring standards, then either an RJ12 or RJ25 jack should be substituted for each NXT six-wire connector.

You can accomplish this conversion one of two ways: easily or really hard. In the really hard conversion, you must open the NXT Brick, unsolder each of the proprietary connectors, and solder new modular RJ12 jacks in replacement for each removed connector. Ugh.

A much easier conversion hack can be performed by snipping each of the supplied NXT connector cables in half and crimping either an RJ12 or RJ25 jack onto the cable's cut end. Actually, I prefer to cut my NXT cables to a length of about 2 1/2 inches long. Then I crimp an RJ12 jack onto the cut end. Now I have a short and very manageable conversion for the proprietary NXT connectors.

Likewise, I can use flat six-conductor wire with an RJ12 plug crimped onto each end for interfacing the NXT Brick with its sensors and motors. This technique enables me to create cables with custom colors and custom lengths. Very, trés chic.

This is a simple, low-cost solution to a potential proprietary connector nightmare. You'll thank me for this hack when you try to add other homebrew motors or sensors to your NXT robot.

Aren't you the least bit curious why the LEGO Group elected to use proprietary connectors on the NXT Brick? According to a confidential statement from the LEGO Group, this bizarre connector was chosen due to "toy safety requirements." Specifically, to prevent the NXT Brick from being connected to "other high voltage connectors." Oh, really?

So the mouse ports on my daughter's Barbie® computers (e.g., RJ45) are not compliant with toy safety requirements? "One to beam up, Mr. Scott."

INSIDE THE LEGO MINDSTORMS NXT BRICK

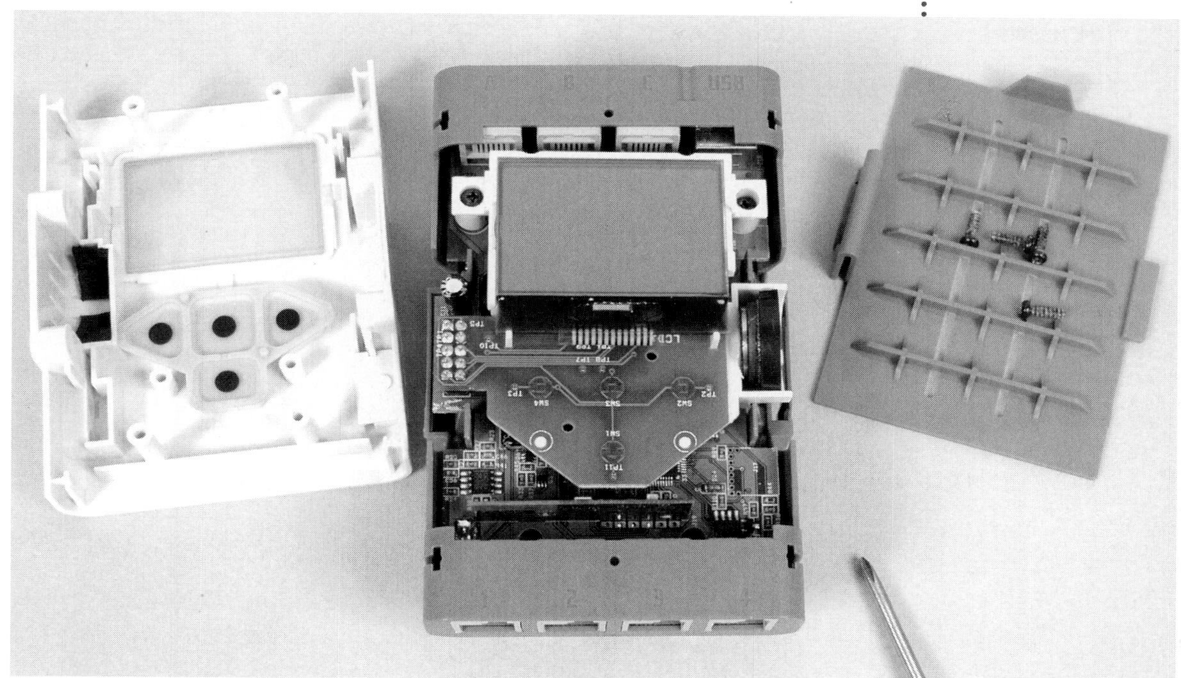

1 A freshly opened NXT Brick with the LCD (upper portion) and the keypad (lower portion) still covering the main printed circuit board (PCB; in the middle of the photograph).

2 The main PCB. The speaker is barely visible along the upper edge of the photograph. Both Atmel microcontrollers are clearly visible.

3 The BlueCore Bluetooth daughter board attached to the main PCB. The four input ports are visible behind the Bluetooth board.

HOW TO HACK YOUR OWN NXT SENSOR CONNECTORS

1

2

1 This is how you are "supposed" to connect a sensor to the NXT Brick.

2 This is a special "proprietary" six-connector plug that is used for connecting sensors and motors to the NXT Brick.

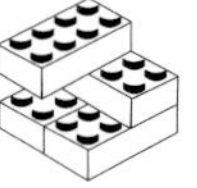

3 LEGO plug (left) versus a standard phone system RJ12 plug (right).

4 The standard RJ12 plug will not fit inside a sensor without modification.

5 Gather some modular RJ12 plugs and some six-conductor wire.

6 Carefully remove the retainer tab clip from the RJ12 plugs.

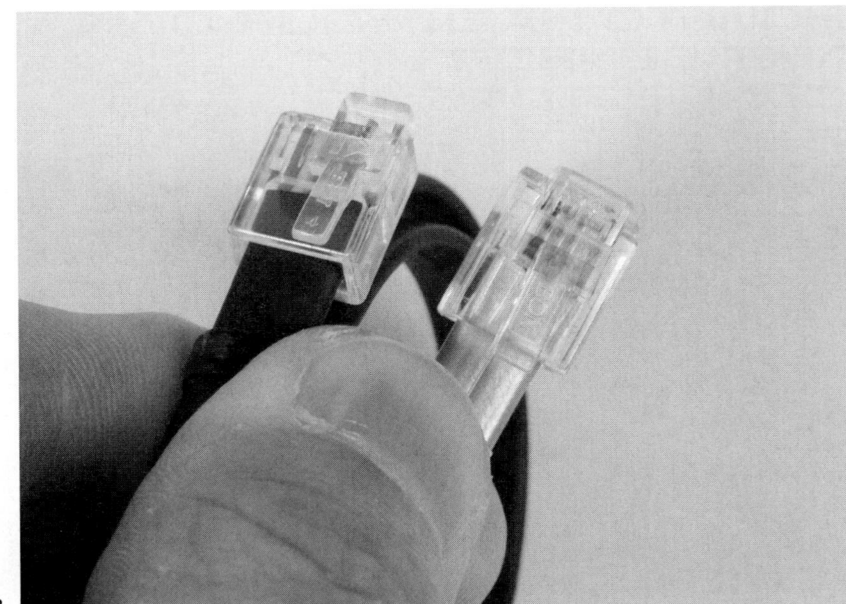

3

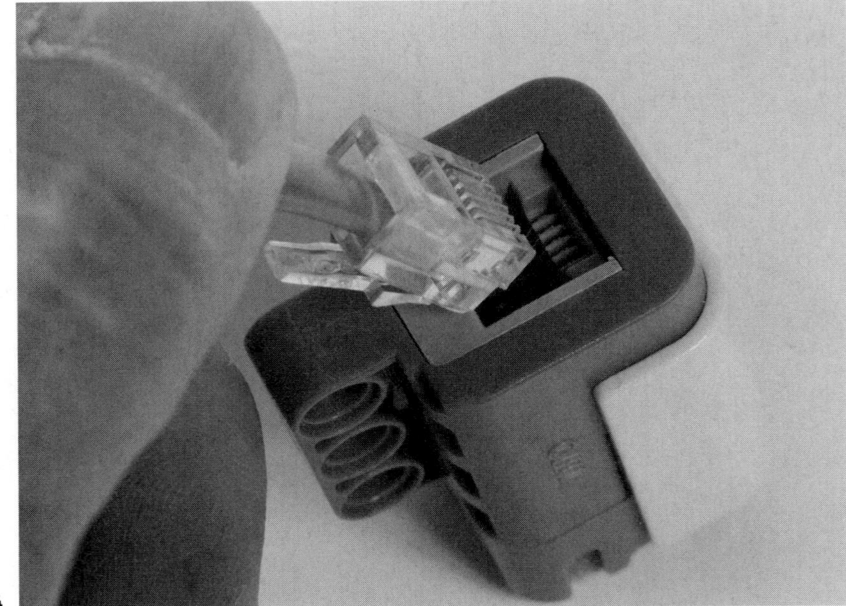

4

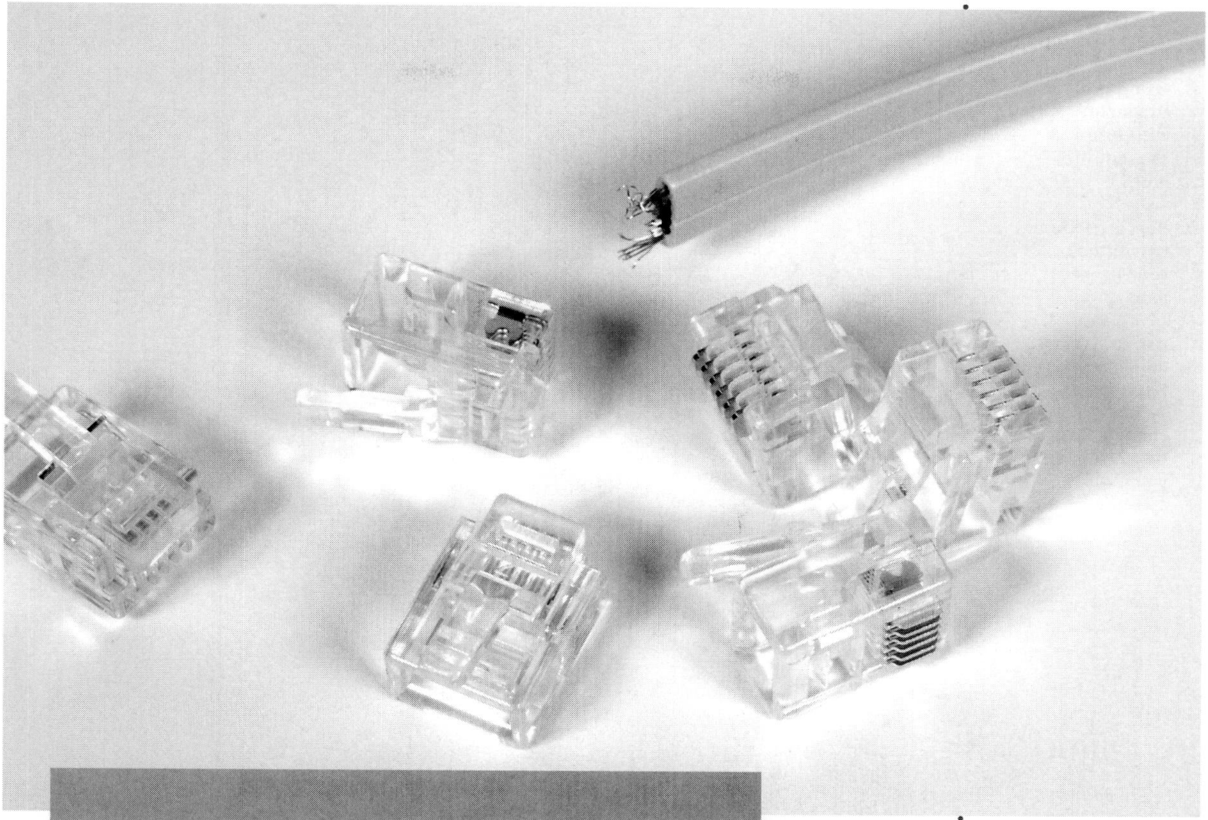

5

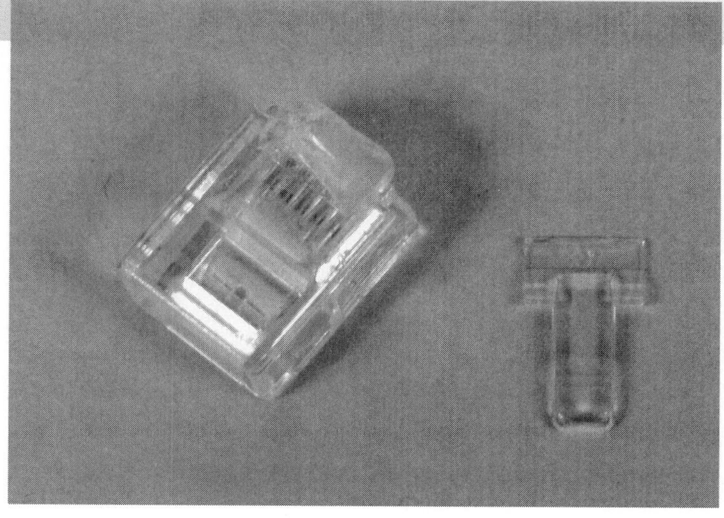

6

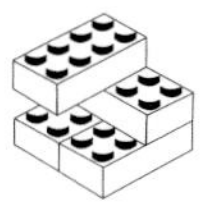

7 Sand the RJ12 plug.

8 Crimp six-conductor wire into the modified plug.

9 Now a friction fit will hold your hacked plug in place.

10 Use low-cost modular telephone 2-line connectors for wiring your NXT creation.

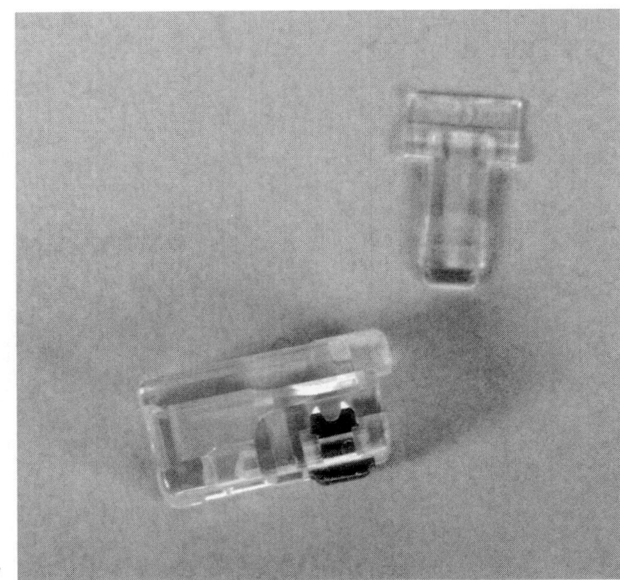

7

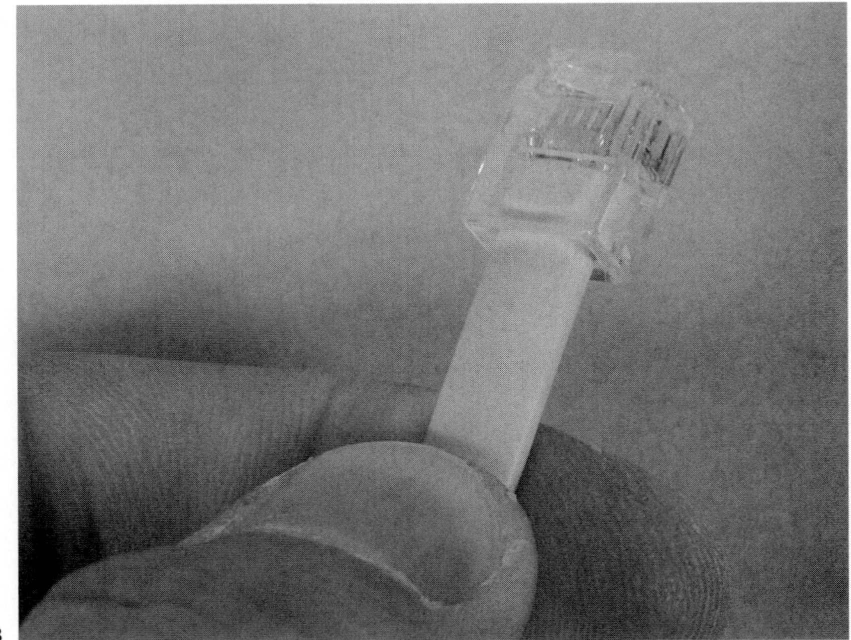

8

9

10

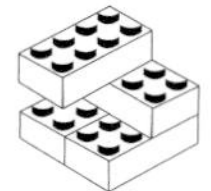

11 **Remember, these hacked plugs slid into the NXT Brick, sensors, and motors with a friction fit. Rough handling will dislodge your hacking effort. So, be careful.**

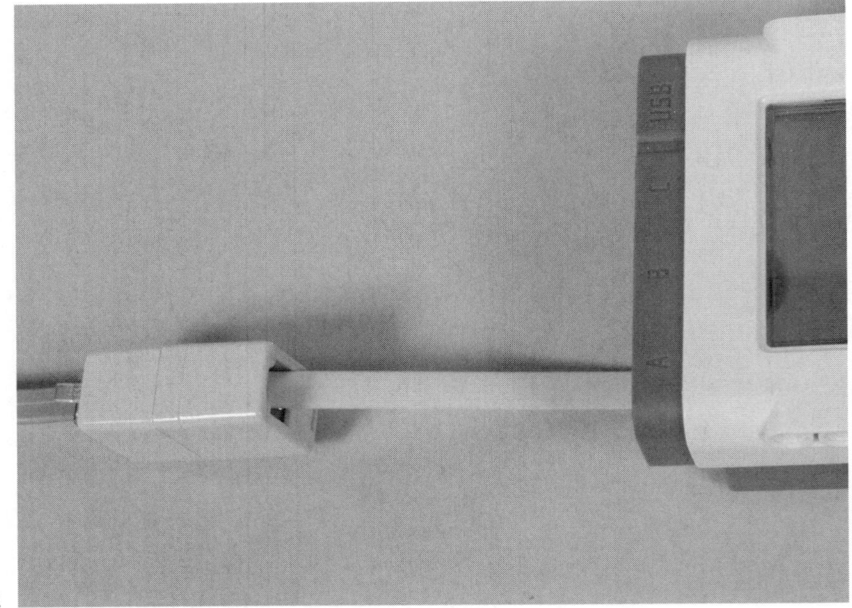

11

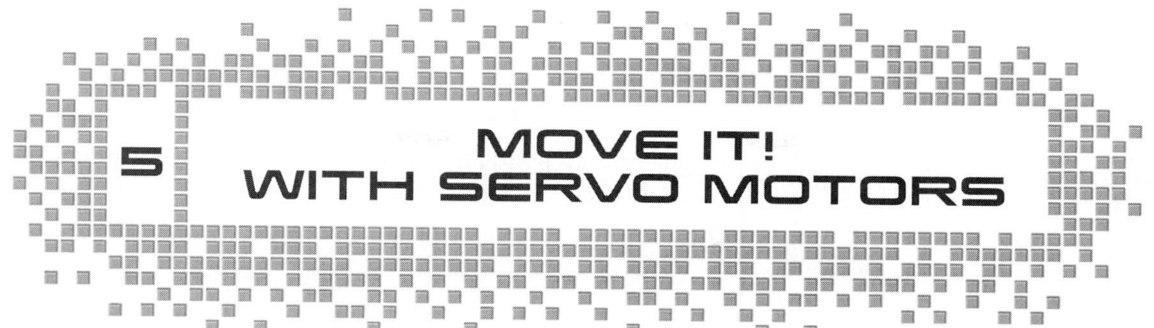

5 MOVE IT! WITH SERVO MOTORS

YOU CAN'T FIGURE out where you are unless you know where you're going. Right? In fact, global positioning system (GPS) manufacturers like Garmin® International have built a business based on this exact hypothesis.

But how does this GPS thing work?

What began as a satellite-based U.S. Department of Defense navigation network called NAVSTAR is today known as GPS. Initially launched in 1978, the entire GPS network was completed with its current population of 24 satellites in 1994. Unlike your cell phone network, however, GPS operates as a free system that can be accessed anytime and anywhere.

Why was the tailor a bad cook?
He kept putting flies in the meals.
KP·06

Each of these GPS satellites orbits the Earth at an altitude of 12,000 miles, twice a day. While the satellite transmitter is actually powered by a photovoltaic array (aka, solar cells), tweaks in orbit are accomplished with a set of rocket engines. These adjustments in orbit prevent the satellites from crashing into each other, ensuring that each transmitter operates properly throughout its expected 10-year lifespan.

Spitting out two signals, known as L1 and L2, that have a transmission strength of about 50 watts, each satellite broadcasts enough information to help you find yourself—literally. Although unable to accurately transmit into buildings, GPS receivers like those from Garmin can help you track yourself down to within 15 meters anywhere in the great outdoors.

What song would Bob Hope sing if he played a vampire in a movie?

"Fangs for the Memory."

In order to plot your current latitude and longitude position (i.e., 2D position), a GPS receiver needs the signals from three different satellites. More advanced 3D positional measurements (e.g., latitude, longitude, and altitude) require the reception of signals from at least four satellites. Once this data is available, advanced GPS receivers can massage the data into calculations for your speed, heading, and cumulative distance traveled.

These 2D and 3D positional placements are determined by triangulation of a signal's transmission and reception time as calculated from several different satellites. The difference in these two times provides a distance measurement for the movement of each respective satellite. Armed with the distances for at least three different satellites, the GPS receiver triangulates the distances and plots your calculated position on a map.

You are here.

WHAT'S THE FREQUENCY, KENNETH?

Although each GPS satellite transmits two signals, known as L1 and L2, only the L1 signal is used by commercial GPS receivers. Operating at a UHF frequency of 1575.42 megahertz, the L1 signals consists of three pieces of data:

- Satellite ID Data—identifies which satellite your GPS receiver is using for positional information.
- Ephemeris Data—provides orbital information for every GPS satellite.
- Almanac Data—reveals satellite status, current date, and time.

Oh, and if that 15-meter positional accuracy worries you, then consider a Garmin GPS receiver that uses the Wide Area Augmentation System (WAAS). By using WAAS, you can find yourself within 3 meters of your real position.

DRIVING RIS LAZY

In robotics, the easiest way to find out where you are is to count the number of steps it took for you to get from there to here. In order to control the older RIS motors, LEGO programmers had to resort to some code tricks for tracking a robot's movements. The keys to these tricks were speed and direction.

- **SPEED**. In this context, "speed" doesn't refer to revolutions per minute (RPM) for the motor, rather it refers to the frequency with which the RIS motor is switched on and off. This frequency is called *pulse width modulation*.
- **DIRECTION**. In this context, "direction" does not refer to the direction of travel for the robot, rather it refers to the movement of the motor's shaft. This motor action controls the forward and reverse direction of the motor.

All of this code trickery disappeared with the arrival of the NXT interactive servo motors. Each of the three motors packaged with the LEGO MINDSTORMS NXT kit contains a brand new rotation sensor that is embedded inside each motor's gearbox. This sensor, known as a tachometer, enables you to program precise movements into each motor.

Two key benefits from installing this tachometer inside each motor are duration and steering. These benefits were not available on the older RIS motors.

1. **DURATION**. The tachometer precisely measures the internal movements of the motor shaft. This precision enables the user to control each motor via either rotations or in clearly defined 1-degree units of rotation. For example, you can command the motor to either spin one-fourth of a rotation or turn to 90-degrees. Either way, the results are the same and this type of control precision provides accurate, predictable motor duration.
2. **STEERING**. Have you ever noticed that no matter how well you programmed your RIS robots, they would slowly fade to either the left or right? This type of motor fading results from the loss of synchronization. The NXT interactive servo motors can be synchronized with each

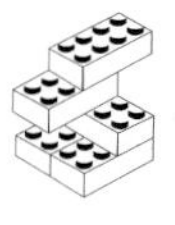

other through the steering property. Only two NXT servo motors (the motors attached to output ports B and C; the motor attached to output port A is reserved for other non-mobility-related functions—like swiveling a head) can be synchronized with this property.

Just how does the tachometer make these behind-the-scenes miracles work? First, there is a pulse. A pulse happens at every two degrees of motor shaft travel. The tachometer detects these pulses. When the tachometer senses a pulse, it sends two signals—there is a signal at the positive edge of each pulse and another signal at the negative edge of each pulse.

In a piece of some pretty nifty engineering, the LEGO Group combined the tachometer's ability to measure 180 pulses for each rotation of the motor's shaft along with the two signals per pulse. The result is 360 counts or beats per revolution of the motor's shaft. Armed with this 1 degree of rotation equals one tachometer count correlation, the Atmel microcontroller is able to maintain precise, synchronized motor control.

The program element that enables all of this motor control is the Move Block. Just drop the Move Block along with some control blocks on your programming work area and you can precisely control your robot's movements.

HOW TO HACK VARIOUS MOTORS INTO YOUR NXT BRICK

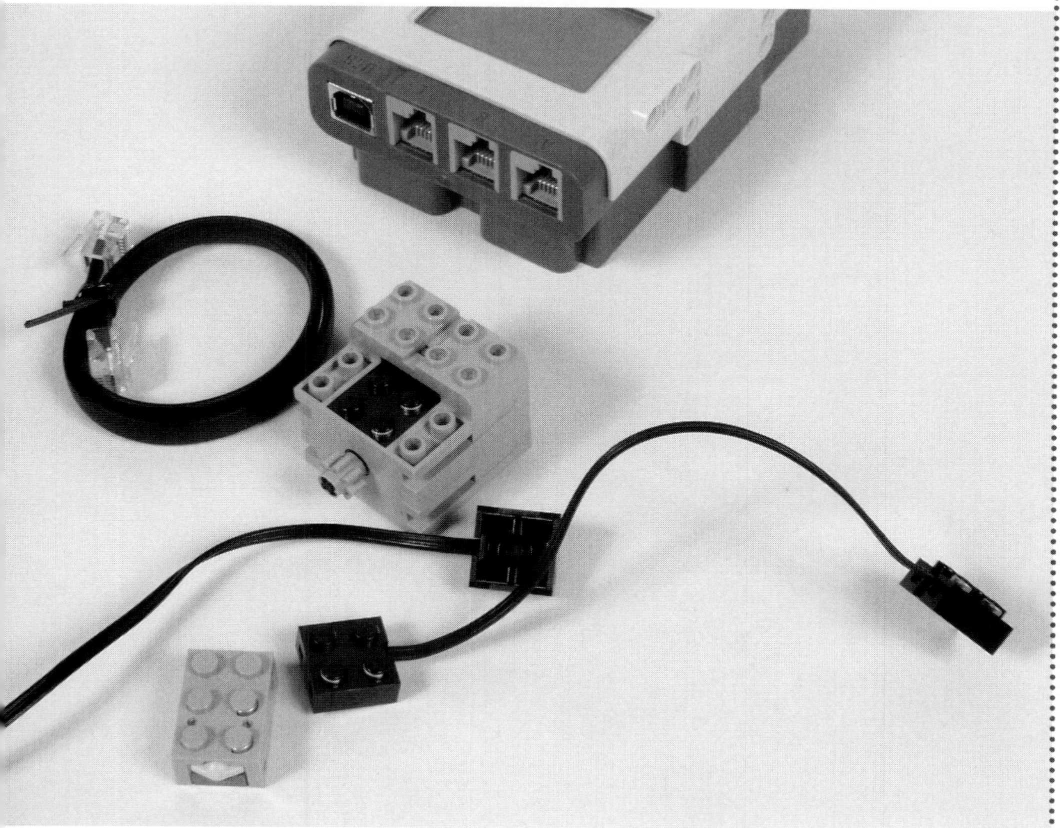

1

1 You can use your older RIS motors with the NXT Brick. LEGO sells a converting connector for this purpose.

2 You can wire your own converting connector, however, by soldering wire onto these opposing brick terminals.

3 Attach your soldered converting connector to an RIS motor.

4 You can verify your connections without jeopardizing your NXT Brick. Just hook a multimeter to the wires of your soldered converting connector. Twist the RIS motor's shaft. You should see a flicker of current displayed on your multimeter.

5 Pager motors work great with a hacked NXT Brick.

2

3

4

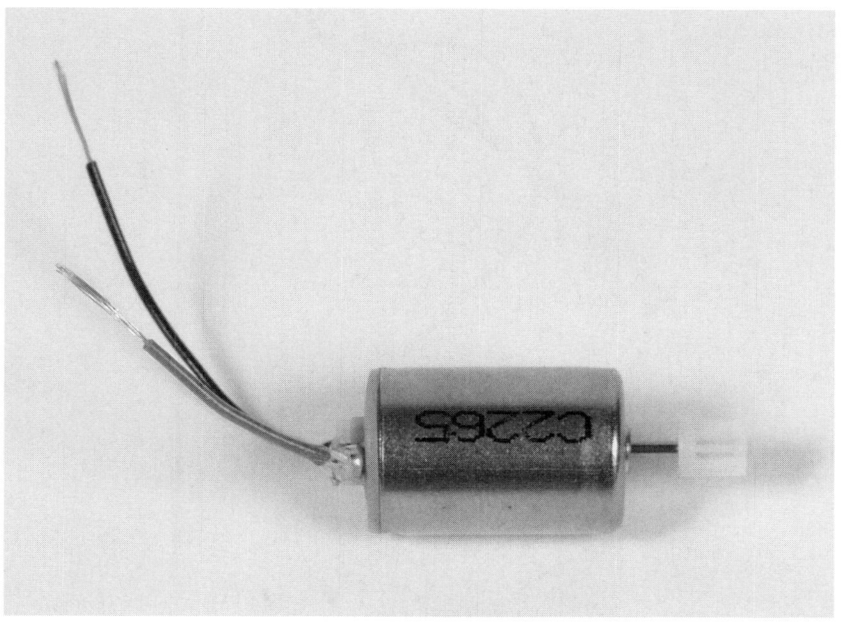

5

6 Test your motors with clips prior to soldering them in place.

7 Geared servo motors from toys make an even better hack.

8 Wire a couple of motors together and see if they run.

9 More complex motors can be driven with an PNP/NPN transitor combination and an electrolytic capacitor.

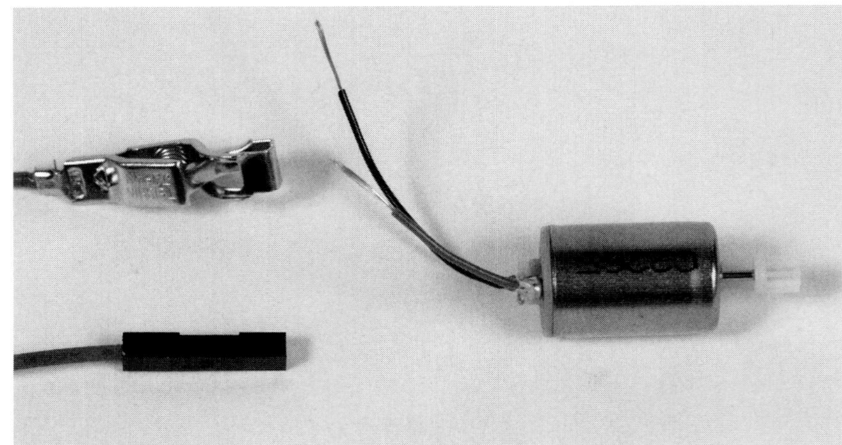

6

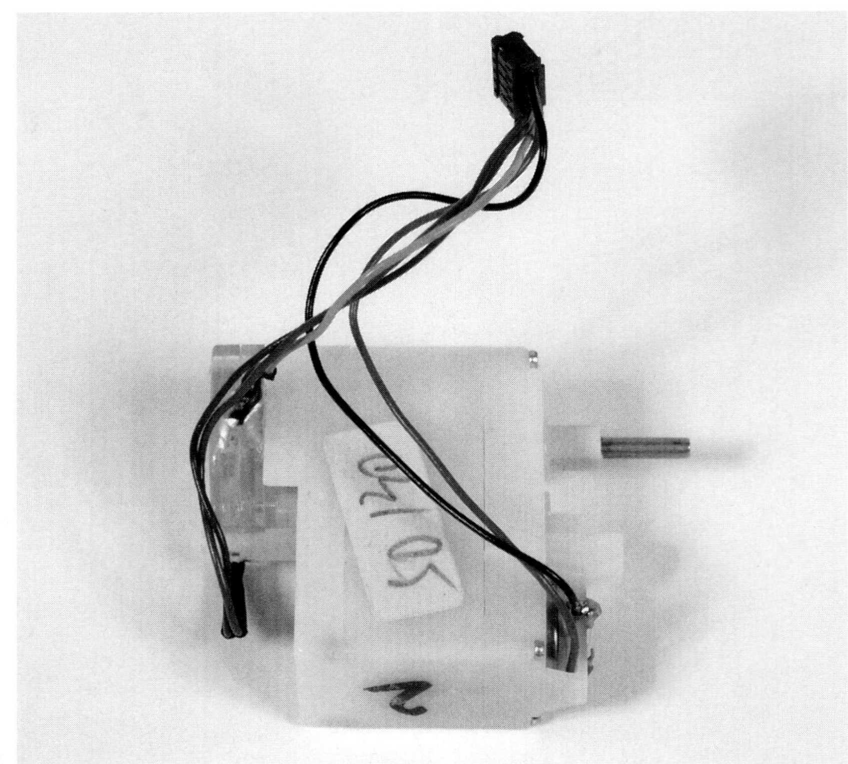

7

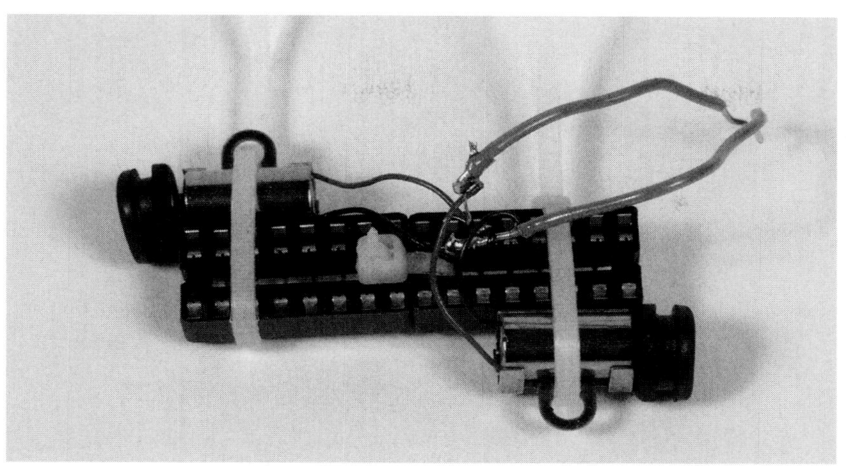

8

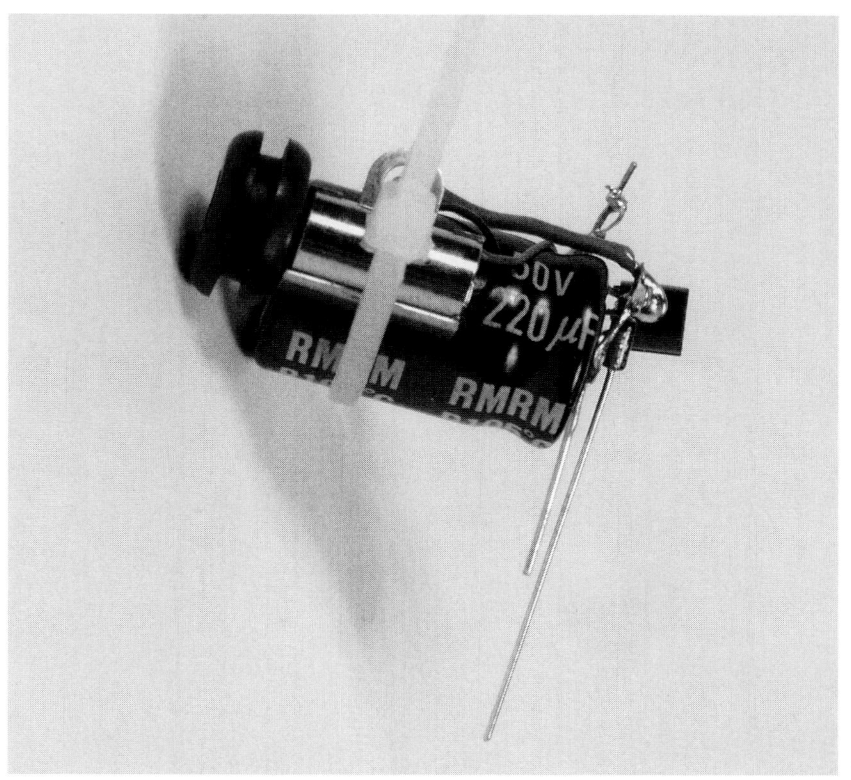

9

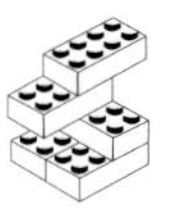

10 **Help your motor seek out light with a photocell or a photodiode.**

11 **Add some photovoltaic cells for generating juice for your motor monster.**

12 **Even newer TECHNIC motors can be hacked.**

13 **Hmm, there's a bit of a family resemblance here, isn't there?**

10

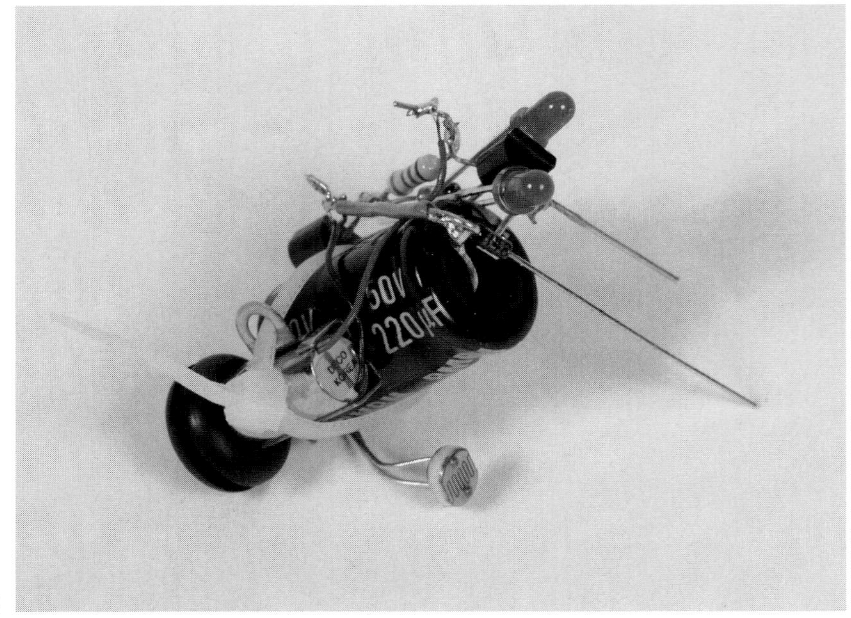

11

12

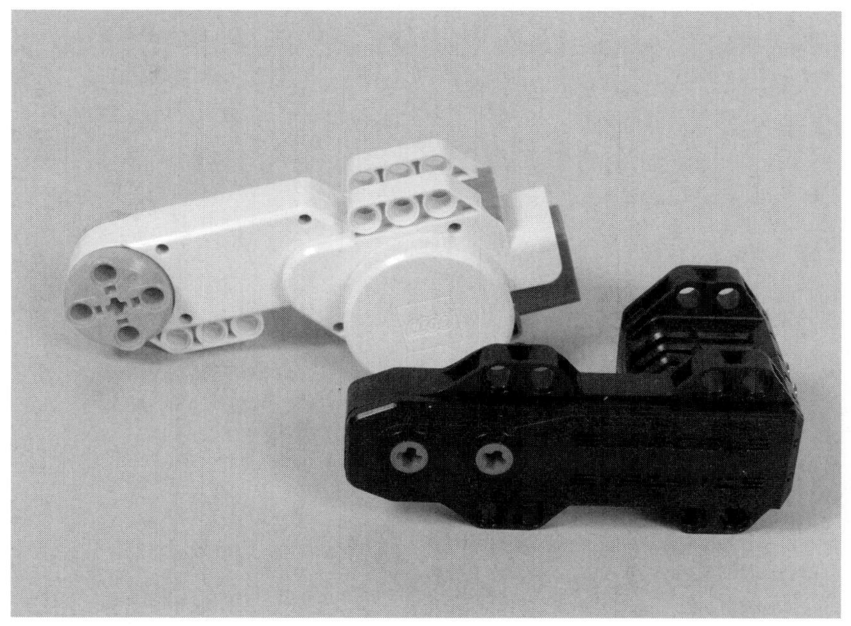

13

14 These newer TECHNIC motors even use the older RIS connectors.

14

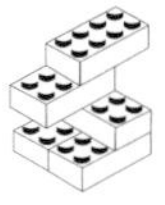

6 HMM, I SENSE SOMETHING

HOW DO YOU SEE the world around you? Aside from some fairly obvious social and political commentary, your most predictable answer would be "with my eyes." In answering this question, you've indicated that sight is a faculty for perceiving the world around you. This faculty for perception is commonly called a sense.

Most of us interact with the world through five senses—sight, smell, hearing, taste, and touch. OK, there might be some small percentage of you that have a "sixth sense." But most of us can't see dead people.

Likewise, a strong argument could be made for hoping that a "seventh sense" could be instilled in everyone—common sense. But then there would be no audience for reality TV, so let's just hold onto those five baseline senses, shall we?

A TOUCHY FEELY KIND OF ROBOT

Your LEGO MINDSTORMS NXT robot design kit comes with four senses of its own (see Figure 6-1). Unlike your bodily senses, however, the robot's senses are called sensors (see Figure 6-2). The four NXT sensors are: light, touch, sound, and ultrasonic.

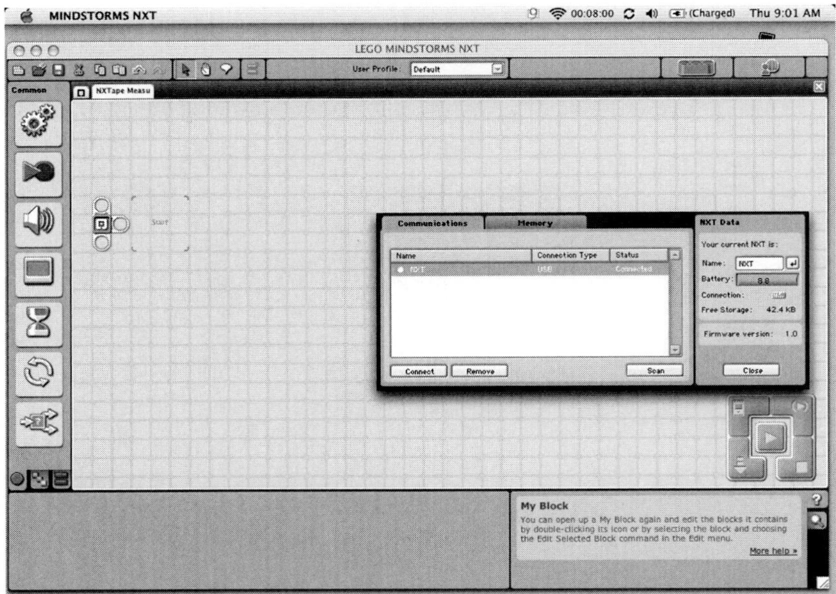

6-1 Keep your finger on your NXT Brick's pulse with the MINDSTORMS NXT software's NXT window button.

Light

The NXT light sensor is designed to respond to two types of light: reflective and ambient. In the reflective light mode of operation, the light sensor takes readings of light intensity based on the amount of light that is reflected back to the sensor from its onboard illumination source.

In ambient light sensing mode, the sensor turns off the onboard illumination source, called a floodlight, and reads the light intensity of the surrounding room. Switching between these two modes can be accomplished via programming. There is a 12-millisecond delay between switching modes and reading valid light intensity values.

- **SENSOR TYPE:** Analog; values converted to a digital response by the Atmel AVR microcontroller.
- **SENSOR SENSITIVITY:** Active readings (greater than 10 percent relative intensity) can be made between –35° to +35° from the center-front of the sensor with peak sensitivity at a wavelength of approximately 880 nanometers.
- **LIGHT INTENSITY VALUE RANGE:** Using a 2856-Kelvin tungsten lamp as a source, the light sensor is capable of returning values between .05 to 500 lumens per square meter.

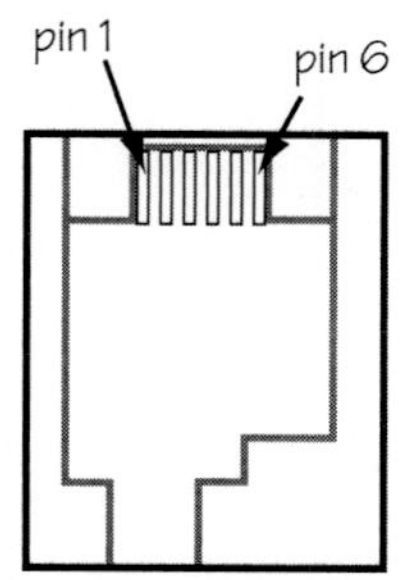

Light Sensor

pin	description
1*	Light Out
2*	GND
3*	GND
4	VDD
5*	3.3V
6	nc

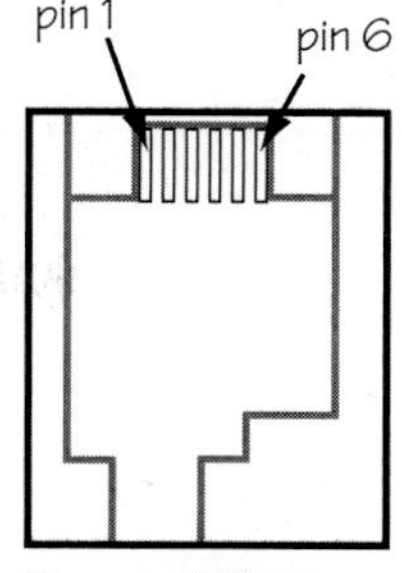

Sound Sensor

pin	description
1	Sound Level
2	GND
3	GND
4	VDD
5	Mode
6	Direct Out

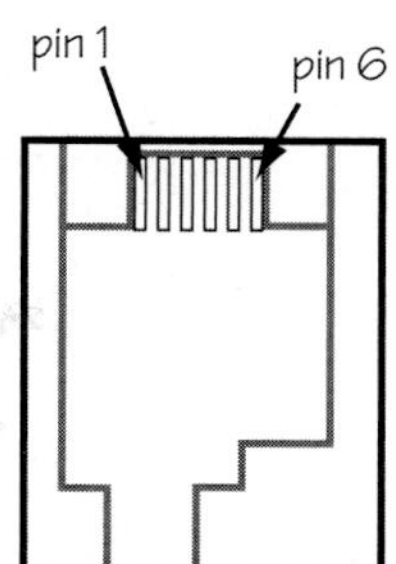

Touch Sensor

pin	description
1*	Trigger
2	nc
3*	GND
4	nc
5	nc
6	nc

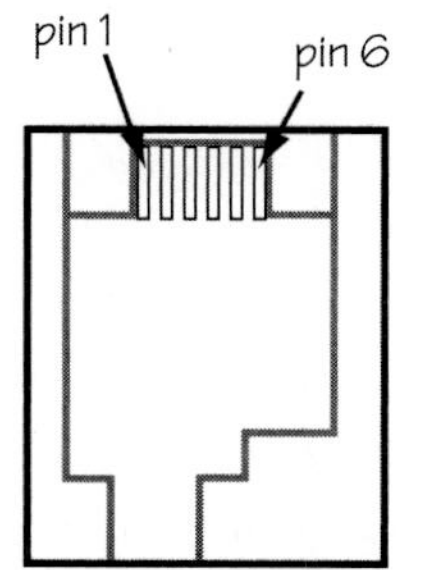

Ultrasonic Sensor

pin	description
1	9V
2	GND
3	GND
4	VDD
5	I^2C SCL
6	I^2C SDA

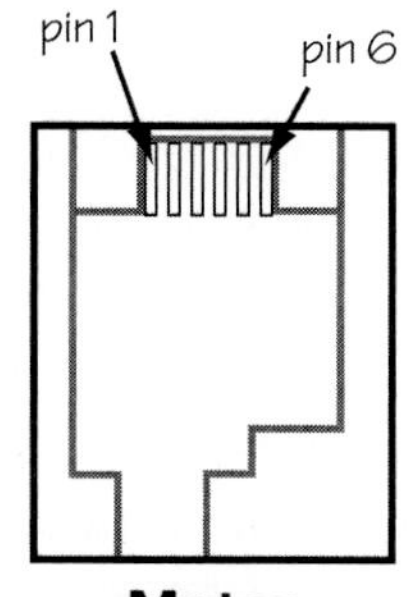

Motor

pin	description
1*	PWM Out
2*	PWM Out
3*	GND
4*	4.3V
5	Tach In
6	Tach In

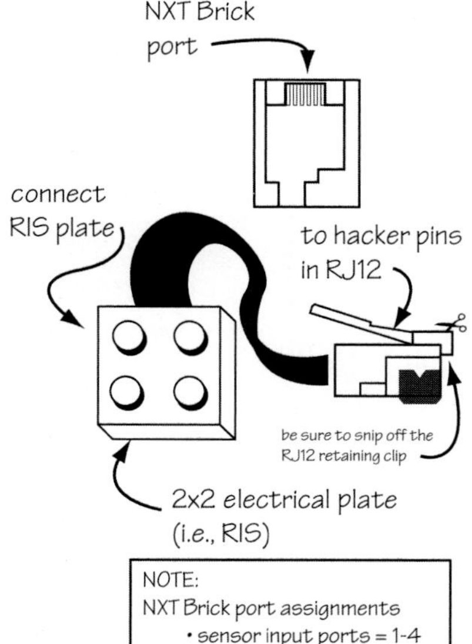

NOTE:
NXT Brick port assignments
- sensor input ports = 1-4
- motor output ports = A-C

* secret hacker pins: use them for your own projects

6-2 The hacker's guide to NXT sensors and motors.

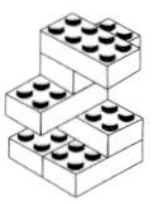

- **NXT BRICK INPUT PORT:** Each input port of the NXT Brick is predefined for a specific sensor. In this case, the light sensor should be used with Port 3 (see Figure 6-3).

6-3 **Use this simple, easy-to-build relay for hacking into the NXT Brick's light sensor port.**

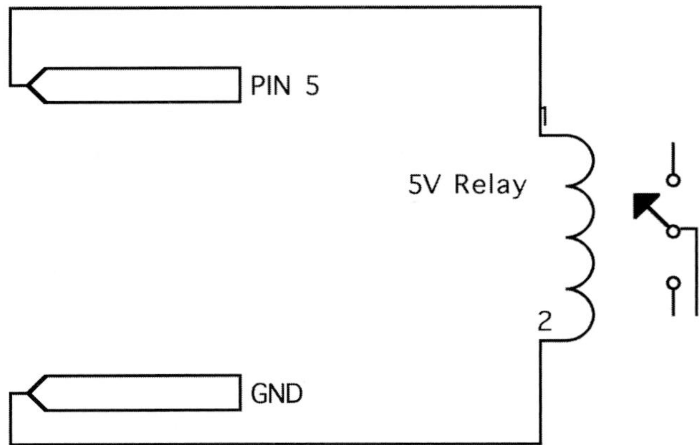

Sound

Based on an omni-directional electret condenser microphone, the sound sensor is able to measure sound pressure in either decibels (dB) or "weighted" decibels (dBA). The decibel is a logarithmic ratio used for measuring sound levels. Due to poor measurement at very low and very high frequencies, filters can be added to sound measurement equipment. The most common of these "weighting" filters is called an "A" filter and its usage is represented by the decibel measurement, dBA.

While the dBA scale is less sensitive to very low and very high frequencies, it responds very well to frequencies in the range of 3 to 6 kilohertz. This selective weighting is used to ensure that your subjective humanly perceived loudness of a sound level is roughly equivalent to the measured loudness.

Oh, and on a side note, there is one other variation in decibel measurement that you might see in a discussion of sound levels. Acoustic measurements are calibrated in decibel sound pressure level or dB (SPL). The dB (SPL) is a logarithmic ratio between a measured sound pressure level and a defined

MAKING SENSE OF NATURE

- **TASTES GREAT.** Butterflies determine whether a plant is suitable for rearing larval caterpillars by stroking the surface of a leaf with tarsal sensilla. These sensilla are sensory receptors that are located in the butterfly's feet.

- **TOMMY CAN YOU HEAR ME?** Love 'em or hate 'em, bats are an intriguing animal. As the only mammal that can fly, bats are able to navigate and hunt by using an echolocation system. Based on a series of high-frequency "squeaks" that can be uttered in bursts of 50 squeaks per second, bats are able to avoid obstacles and hunt winged insects—which is particularly handy since they are a crepuscular and nocturnal animal.

- **OHHH, THAT SMELL.** Have you ever noticed that U.S. Custom's agents (as well as other law enforcement agencies) use dogs for "sniffing" out crime? That's because dogs have over 200 million olfactory receptors and that's nothing to sneeze at! (By comparison, we humans sniff along with a paltry 5 million of these receptors.)

- **TOUCHED BY A KITTEN.** Tactile receptors located on the face of cats can serve as a navigation aid, food sensor, mood indicator, and, most importantly, keep them out of tight places. Whiskers located next to these receptors help the cat measure the diameter of an opening prior to entering... even in the dark.

- **C U SEE ME.** Scallops are the most well-endowed creatures on the planet—that is, endowed with eyes. Armed with over 100 eyes, a scallop is able to detect light and motion with the simple retina that is housed inside each eye.

reference point for the threshold of human hearing. Therefore, a dB (SPL) reading of 0 is the beginning of human hearing and heavy city traffic noise, for example, generates a reading of 90 dB (SPL).

You can switch between the dB scale and the weighted filter dBA scale through programming in the NXT Brick. Switching between these two modes will result in a 300-millisecond delay before you can obtain valid measurements.

- **SENSOR TYPE:** Analog; values converted to a digital response by the Atmel AVR microcontroller.
- **SENSOR SENSITIVITY:** Reliable, quantifiable measurements can be made between 55 dB to 90 dB. Unlike the expected 3-to-6-kilohertz frequency range for a "typical" dBA weighted filter scale, this NXT sound sensor responds to measurements between 1 and 3 kilohertz dBA.

- **NXT BRICK INPUT PORT:** Port 2.

Touch

A passive pushbutton trigger that generates a simple binary response—on or off.

- **SENSOR TYPE:** Analog; values converted to a digital response by the Atmel AVR microcontroller.
- **SENSOR SENSITIVITY:** There is a force threshold for triggering the NXT touch sensor. A force between XXX N and YYY N will activate the sensor.

- **NXT BRICK INPUT PORT:** Port 1.

Ultrasonic

Equipped with its own microcontroller, the NXT ultrasonic sensor is able to accurately measure the distance between itself and an object that is in its path. In order to accomplish this "vision" feat, the ultrasonic sensors consist of two parts: an emitter and a receiver.

The emitter generates an ultrasonic sound that is "chirped" from a piezo crystal microphone-like element. This chirp has a frequency of 40 kilohertz and a sound level of 110 dB when measured at 30 centimeters. Once this sound wave reaches an object, it is reflected back to the sensor. This reflection is called an echo.

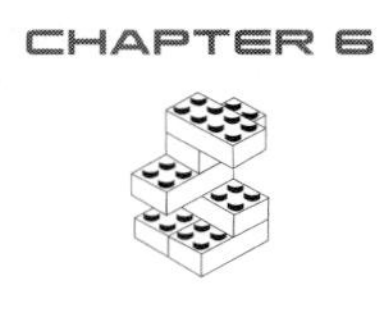

The receiver portion of the sensor detects this echo. Once this echo has been detected, the onboard microcontroller calculates the amount of time that has elapsed between the initial chirp and the reception of the return echo. A distance measurement is then calculated by the NXT Brick.

Spurious results can be received by the sensor under a number of different circumstances. For example, false echoes from other ultrasonic sensors can confuse your sensor. These bad results can be reduced, but not eliminated, by changing the number of chirps that are emitted by your sensor. Emissions ranging from single chirps to time-delayed "bursts" of chirps can be programmed into the NXT Brick. The spacing of these bursts can be controlled between 80 milliseconds and 2.5 seconds.

Please note that the default burst rate for the ultrasonic sensor is 10 to 15 bursts per second.

Other factors that can contribute to false or degraded distance readings are small objects; soft, movable surfaces; close distances; long distances; and oblique angles. Maintaining a solid, controlled mounting surface for the sensor, as well as operating the sensor under tightly controlled and predictable periods of activity, can help to lessen the effects of these unwanted variables.

- **SENSOR TYPE**: Digital; built-in microcontroller.
- **SENSOR SENSITIVITY**: The NXT ultrasonic sensor can reliably measure distances between 1 and 255 centimeters. Outside factors can adversely contribute to measurement degradation at distances greater than 78 centimeters. A deflection of approximately −10 to +10 centimeters either side of the sensor's centerline will produce valid measurements.

- **NXT BRICK INPUT PORT**: Port 4.

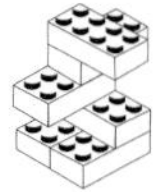

7 YES, BUT I DON'T KNOW HOW TO PROGRAM

IN A REPORT prepared by the Howard Hughes Medical Institute (HHMI) for studying the human senses, the brain is fingered as having complete creative control over the interpretation of all sensory information. Vilayanur Ramachandran, a professor of neuroscience at the University of California, San Diego, states in this report that the brain makes assumptions about the information that is gleaned from our senses. In fact, the brain will sometimes create "shapes from incomplete data," according to Ramachandran.

As a result, the HHMI report succinctly concludes that "it's all in the brain."

Not so with the NXT microcontroller-based programmable brick (see Figure 7-1). The NXT Brick is as dumb as a well, err, brick. In order to enable the NXT Brick to properly analyze your robot's sensory data and control your motors, you must write a program.

Sure, remember how you had to orchestrate code "bricks" in the older RIS RCX coding environment? Well, writing an NXT program is the same old song, just a different verse.

Back during the development of the RIS, the LEGO Group formed an evil alliance with the dreaded Vahki in conquest of Metru Nui. Oops, sorry, that's another story. Rather

7-1 The NXT Brick is a self-contained microcomputer and a programmer's dream come true.

in 1998, the LEGO Group, along with National Instruments and Tufts University, developed a programming environment called ROBOLAB.

Derived from a National Instruments (NI) product known as LabVIEW, ROBOLAB helped introduce LEGO robotics to school districts throughout the United States. Unlike the RCX coding environment packaged with the retail versions of the RIS, ROBOLAB was available in both Mac and PC versions.

So imagine the delight of school districts like Manatee School District in Bradenton, Florida where wireless-equipped Mac notebook computers had been distributed in 22 elementary and secondary classrooms. Now Dr. Tina Barrios, Supervisor of Instructional Technology for the Manatee School District, could, if she wanted to, implement a similar one-to-one program for robotics. That is, if Dr. Barrios elected to purchase a site license for ROBOLAB. Otherwise, the retail RIS was not an option. Why?

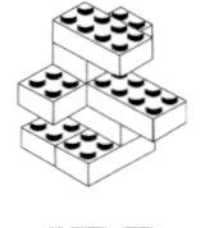

Without a doubt, LEGO RIS was a big player in the robot education marketplace and RIS had become an easy-to-use construction set for learning how to build and program a robot. Unfortunately, for Mac-equipped schools like the Manatee School District, LEGO packaged a PC-only programming environment with the RIS. Likewise, inquiries with the LEGO Group customer support were returned with a recommendation to use Virtual PC for Mac with the RIS retail software. That type of reply was neither an option nor was it acceptable to Mac users.

Let me set the record straight—Virtual PC stinks. Originally developed by Connectix Corporation, Virtual PC was acquired by Microsoft Corporation in February 2003. From its initial release, Virtual PC has been a kludgy, buggy, temperamental emulator that could take a sizzling PowerPC and turn it into a pathetically slow anemic Windows OS–emulated system.

There's got to be a better way. Leave it to the LEGO Group and National Instruments to find that better way.

In 2005, the LEGO Group and National Instruments again joined forces to thwart the attack of the deadly Piraka. Jeez, wrong story again; sorry. Revamping the MINDSTORMS robotics line in 2005, the LEGO Group again enlisted the assistance of NI for developing a "radically updated programming environment for a new MINDSTORMS product."

Today, we all know that product is the NXT robot design kit. Things were going to be different with the MINDSTORMS NXT kit, however.

First of all, the new NXT programming environment would be both Mac and PC compatible. And this compatibility would be extended to both the education version and the retail product line. More importantly, a new LabVIEW software "engine" would have to be created for dealing with advanced NXT features like 32-bit processing, Bluetooth wireless communication, an improved tachometer-based motor, and new analog and digital sensors.

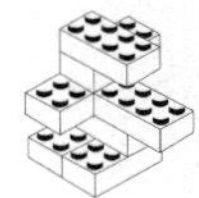

A ROOM WITH A LABVIEW

What the heck is a LabVIEW? In a nutshell, LabVIEW is, according to National Instruments, a "software tool for designing test, measurement, and control systems." From simple oscilloscopes to the NXT Brick, LabVIEW has become a powerful means for digitally analyzing our analog world. National Instruments calls this capability *virtual instrumentation*. And there's no better way to control a robot than through the measuring of sensors coupled to the controlling of responses all wrapped up inside a virtual instrument.

In the words of those Billund boys (and girls), the new NXT LabVIEW-based programming environment would have to inspire "creativity and innovation in children." Now those are some pretty big shoes to step into. Or, if you'd like to read the more wordy and more boring "official" statement from the LEGO Group, here's what Søren Lund, director of LEGO MINDSTORMS, had to say at the 2006 Consumer Electronics Show (CES) in Las Vegas, in one of the longest, single sentences ever uttered at CES:

> Using the sophisticated NI LabVIEW engine allows us to maintain everything users appreciate about the current MINDSTORMS experience, but then go the extra mile to provide a tool that is easy enough for a 10-year-old to master on a surface level and technical enough for an adult user to be challenged and inspired to create.

Z-z-z-z; whew, you caught me. Please excuse me, but I'm recovering from a "Vegas blue plate special."

You could get another taste (albeit, a low-calorie version) of this same sound byte from Ray Almgren, Vice President of product marketing and academic relations at NI. After gushing over LabVIEW, Almgren threw this great little chestnut onto the NXT editorial bonfire:

> We are fortunate to work with a company whose products are inspiring children to be innovative and creative and possibly pursue careers in science and engineering.

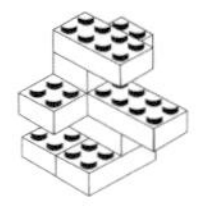

Now that's more like it—"inspiring children" to "pursue careers." National Instruments took the words right out of the LEGO Group's mouth.

CRACK THIS CODE DA VINCI

There are two methods that you can use for programming your NXT Brick: NXT programming and software programming. For your coding pleasure, you can review the complete NXT programming code set in Appendix A.

Programming your robot with either one of these methods couldn't be easier. Here, let me show you.

In NXT programming, it is the classic case of "look Ma, no hands." NXT programming does not require a computer for programming your robot. Just press a couple of buttons on the NXT Brick (e.g., as few as five button presses can write a pretty robust program) and select Run.

Away goes your robot. "Come back, Shane."

The NXT programming language itself consists of 25 buttons or commands that can be entered directly with the NXT Brick buttons. While you are restricted to a program length of *five* commands, a fairly sophisticated program can be written anytime, anywhere, without the umbilical cord of a tethered computer dragging down your innovation and creativity.

OK, smarty pants, you could be using a wireless Bluetooth connection between your NXT robot and your computer, but the point remains the same...you don't need to lug around a computer for programming your creation when using NXT programming. There, are you satisfied, now?

Alright, you want a bigger, more powerful NXT program, then start your computer and launch your LEGO MINDSTORMS NXT software. You have more commands, more precision in the configuration of your motors and sensors, and you aren't limited to a program length of five commands. You are free to be innovative and creative.

Gone are the programming bricks in the RCX days. In NXT software programming, you have programming blocks. Humorously, these programming blocks have the distinctive look of LEGO Technic elements.

Although it sounds deceptively simple, in software programming, just drag and drop the programming blocks onto the software's work area and attach them to each other. Click, that's it. Not quite, here's the rub. Becoming inno-

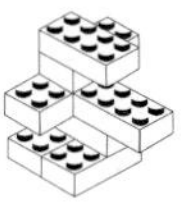

vative with your robot's programming requires that you be creative in your usage of the programming blocks.

Becoming a creative programming artist requires the use of a palette. All of the available programming blocks are held on the programming palette. There are three palettes: the common palette, the complete palette, and the custom palette.

While the complete palette is self-explanatory (i.e., it contains *all* of the programming blocks), the common palette is a subset of the complete palette. On the common palette you have the following programming blocks available for writing your code: Move, Record/Play, Sound, Display, Wait, Loop, and Switch.

But wait, there's more.

Best among these three programming palettes is the custom palette. The custom palette lets you "roll your own" programming palette with a population of your favorite programming blocks. Or, if your friend has a really terrific custom palette, you can download her palette, and use it in your next software programming project.

All of this seems daunting, doesn't it? Remember, no matter how you build your bot or sling your code, familiarity breeds creativity. Now go code something.

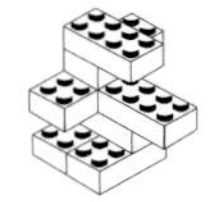

HOW TO USE THE MINDSTORMS NXT SOFTWARE

1

2

1 Today's MINDSTORMS NXT software is a vast improvement over the older RIS software.

2 Based on LabVIEW technology, you don't even need a manual to write your first program with this software.

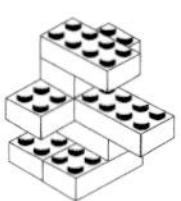

3 Enter a name for your project and away you go.

4 Select a programming block and drag it onto the work area.

5 Adjust the block's properties with the configuration panel along the bottom edge of the screen.

6 Add another programming block and they automatically link together with a TECHNIC-like beam.

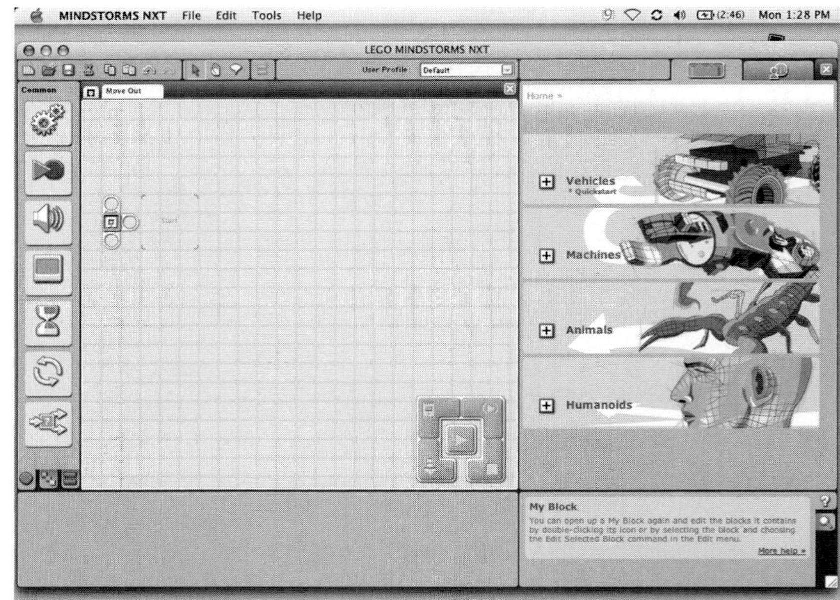

3

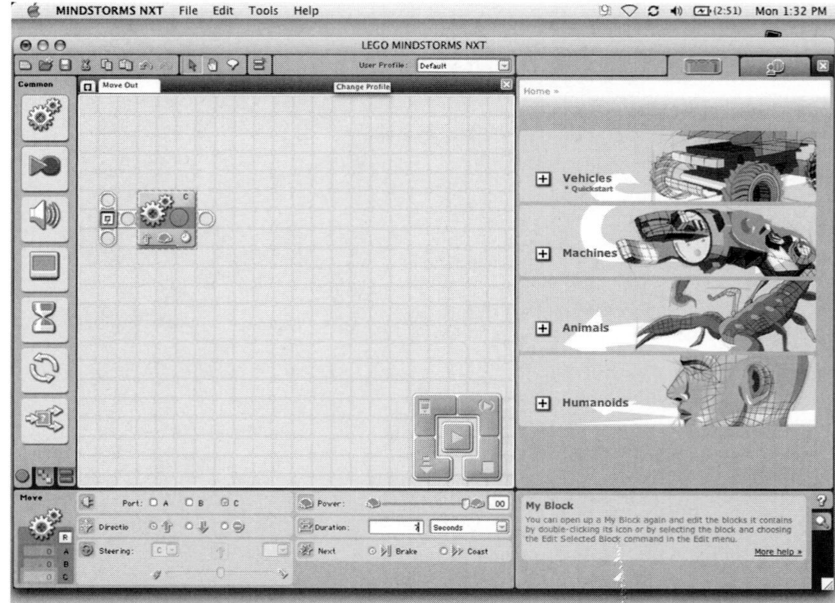

4

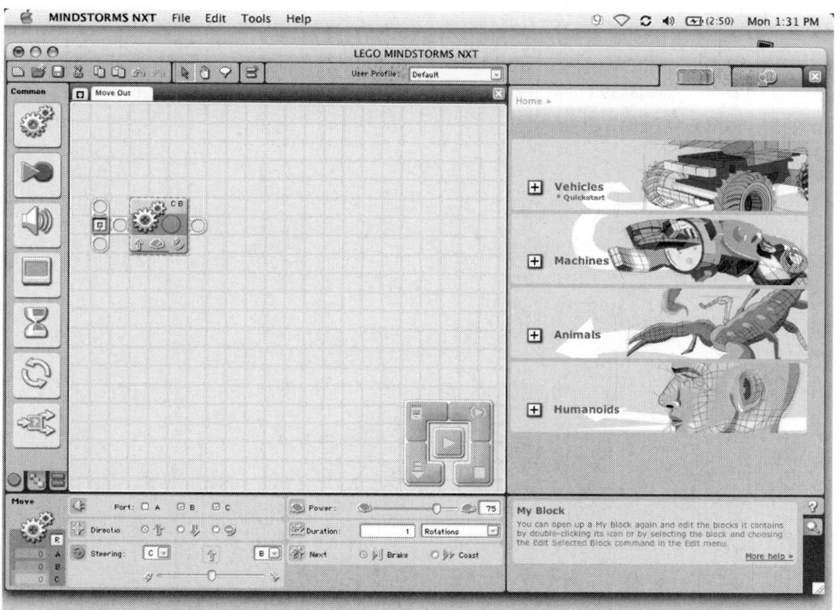
MINDSTORMS NXT File Edit Tools Help
(2:50) Mon 1:31 PM
LEGO MINDSTORMS NXT
User Profile: Default
Common
Move Out
Home »
Vehicles
* Quickstart
Machines
Animals
Humanoids
Move
Port: A B C
Power: 75
Duration: 1 Rotations
Steering: C B
Next Brake Coast
My Block
You can open up a My Block again and edit the blocks it contains by double-clicking its icon or by selecting the block and choosing the Edit Selected Block command in the Edit menu.
More help »

5

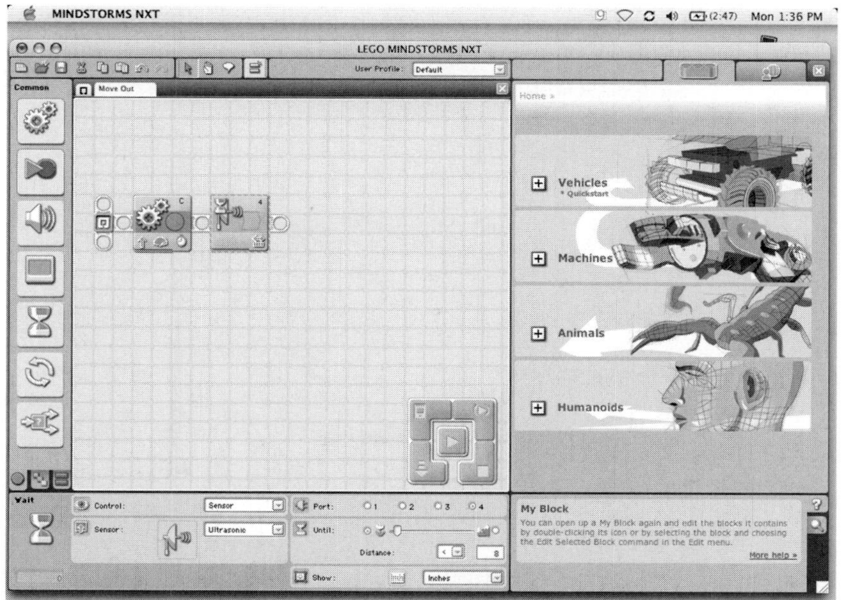
MINDSTORMS NXT
(2:47) Mon 1:36 PM
LEGO MINDSTORMS NXT
User Profile: Default
Common
Move Out
Home »
Vehicles
* Quickstart
Machines
Animals
Humanoids
Wait
Control: Sensor
Sensor: Ultrasonic
Port: 1 2 3 4
Until:
Distance: < 8
Show: Inches
My Block
You can open up a My Block again and edit the blocks it contains by double-clicking its icon or by selecting the block and choosing the Edit Selected Block command in the Edit menu.
More help »

6

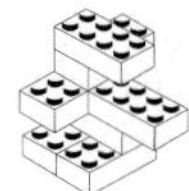

7 Add variables with the Edit Variable menu option.

8 Enter a variable name and parameter type.

9 Your programs can become lengthy. If so, just hide the Robo Center.

10 There, now you can see all of your programming glory.

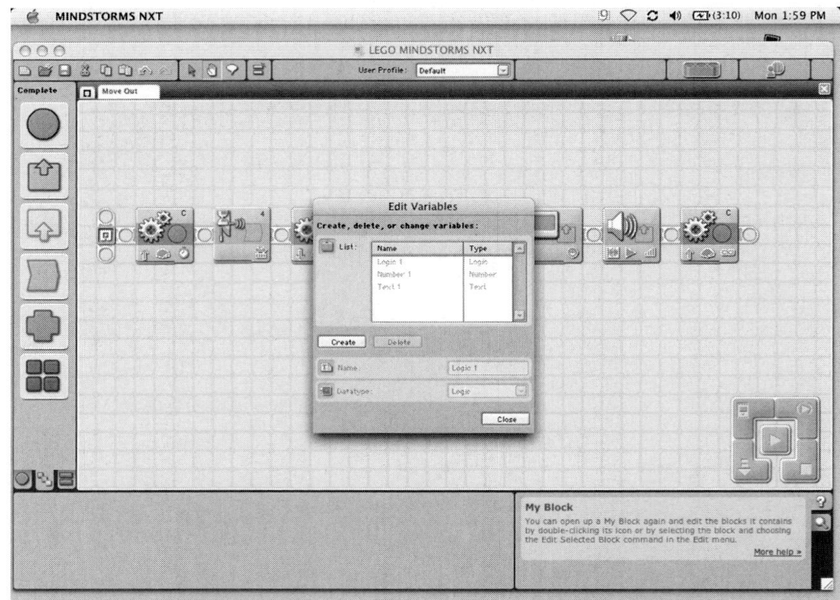

7

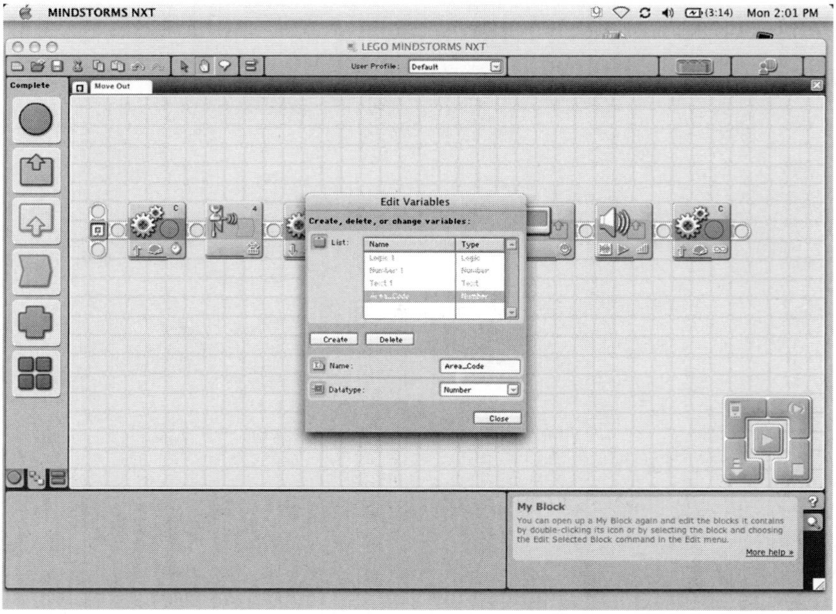

8

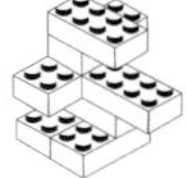

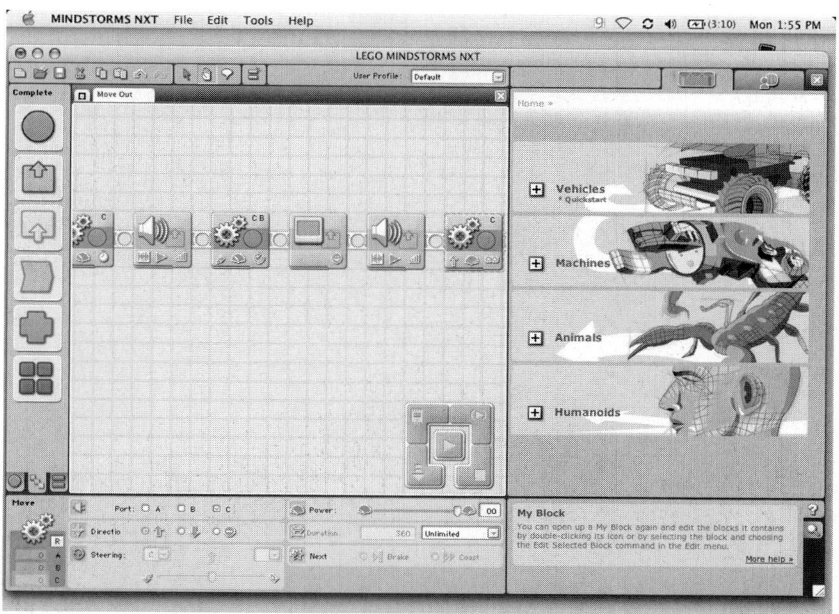

9

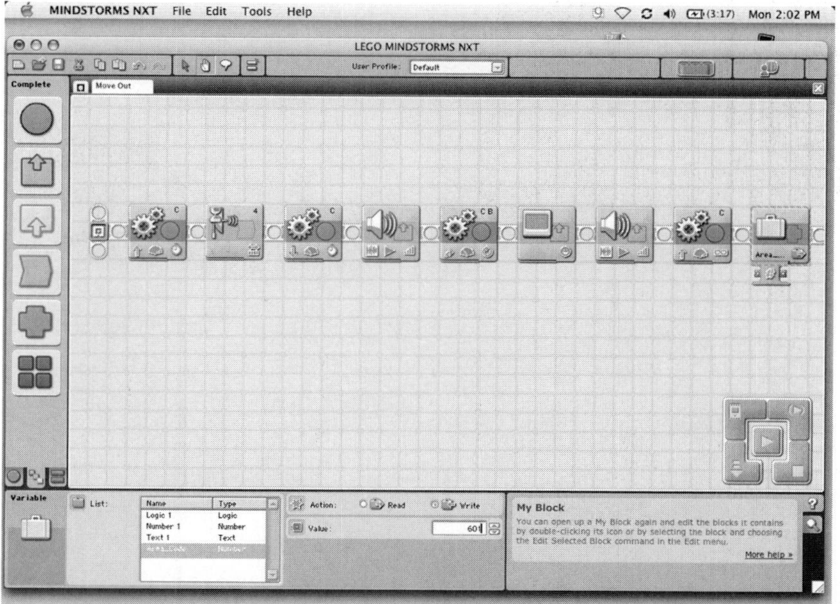

10

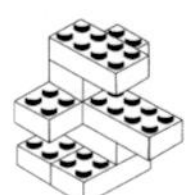

11 If you need more programming options, just select the Complete Palette from Programming Palette along the left side of the screen.

12 Now download your program to the NXT Brick and test it.

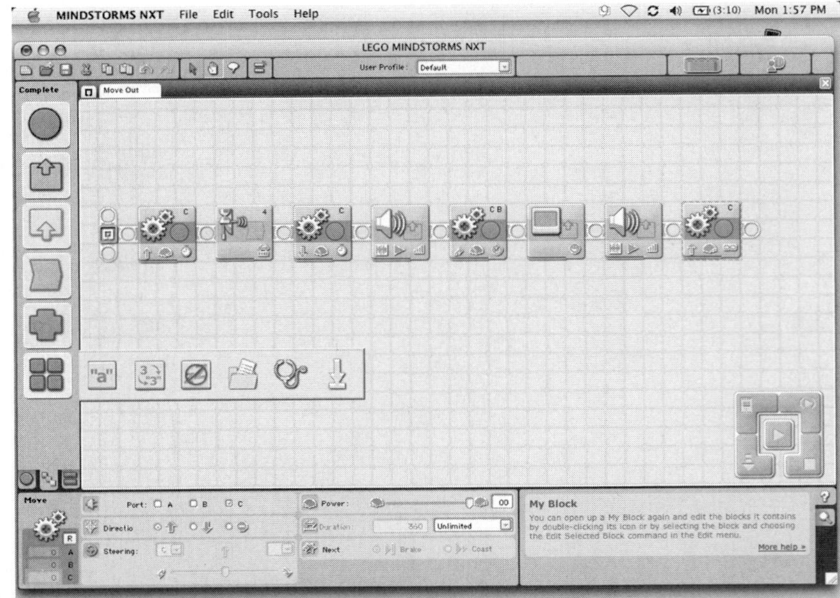

11

12

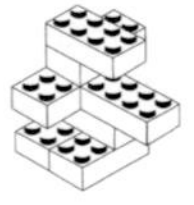

8 TESTING, TESTING: OH, TROUBLE SHOOT

THERE IS NOTHING worse than something that doesn't work correctly. Whether it's your automobile or your NXT TriBot robot, if it ain't working properly, then you'll get frustrated with this lack of performance and possibly resort to some rather rash action.

By rash action, I'm thinking about a fellow who I once saw in an office who pounded a hole in his keyboard and threw the monitor out of his cubicle. Why this extreme response? Well, aside from the guy having a permanently "lit fuse" personality, he was frustrated with some code errors in a program he was writing. Needless to say his tenure at that firm was abruptly terminated after that outburst. Ouch.

But you know what? Secretly, many of us could "feel his pain." Granted, we haven't KO'd our computer equipment, but, at one time or another, we have all been frustrated with code errors.

How you deal with this frustration is the key to success in life, as well as a good, strong mental health. Yuck, I sound like some preachy parent. That's the first and the last bit of life counseling that you'll hear from me. Deal?

Resisting the obvious Howie Mandel game show setup line, tracking down errors in your NXT robot designs can, generally speaking, be focused on either a software or a hardware problem. Sometimes, your access to the NXT Brick will be blocked by a nonresponsive system (i.e., the running icon stops spinning, a continuous tone is emitted from the NXT Brick's speaker, your robot runs into a wall, etc.). At this time, doctor, you must arm yourself with a trusty straightened paperclip and do the following to your Brick patient:

1. Locate the NXT Brick.
2. Make sure the NXT Brick is powered ON.
3. Flip the NXT Brick over with the USB port pointing up.
4. Find the farthest left TECHNIC hole along the upper edge of the battery box.
5. Insert your straightened paperclip into this hole and push the Reset button.
6. Pass GO, but do not collect $200. You must now reload your NXT Brick's firmware, as well as the program that put you in this pickle in the first place.

Now that you've reset the NXT Brick, it's time to find out what went wrong. Following is a simple flowchart that will help you with your troubleshooting chore.

IS IT A ROBOT PROBLEM?

Yes; is the low battery icon flashing?

  Yes; replace the batteries.

Yes; did the robot turn off unexpectedly?

  Yes; deactivate Sleep mode in the NXT Brick Setting menu.

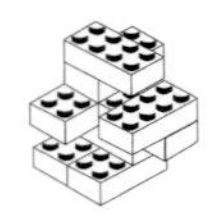

BREAKING AND ENTERING

Following the release of the LEGO MINDSTORMS NXT robot design kit in August 2006, the LEGO Group made two remarkable announcements. First, users would be able to download advanced software, hardware, and Bluetooth documentation for the NXT Brick. Second, and more exciting, was the simultaneous announcement that the NXT Brick firmware would be released as open source. Now third-party developers had all of the tools necessary for shaping the "NXT" thing to come in robotics. Or, did they?

Buried deep inside a confidential document published by the LEGO Group was this ominous statement: "Remember that when the NXT (Brick) is disassembled or 3rd part (sic) firmware is used together with the NXT all warranty (sic) is no longer valid." Gadzooks.

There were two other humorous footnotes from this confidential document related to this same issue.

First, there was this precaution regarding the destruction of the NXT Brick: "When developing 3rd part (sic) firmware for the NXT, developers needs (sic) to be careful when addressing the hardware as wrong initialization can destroy parts of the hardware..." Gosh, I hate it when that happens.

Better yet was this demonstration of embedded systems developer humor: " [ARM7 power will be shut down]...if the ARM7 processor doesn't send back the following messages to the AVR microcontroller over the I^{2}C communication between the two processors within 5 minutes after startup:"

\xCC" "Let's samba nxt arm in arm, (c)LEGO System A/S

Evidently, those Billund folks must watch way too much ballroom dancing. Or, are they referring to a network file system boot with the Atmel ARM7 microcontroller? Nah, they're hooked on ballroom dancing.

IT'S YOUR BAG

Have you ever been looking for a specific tool in your workshop, like a screwdriver, only to find a hammer, instead? Then you did the unthinkable—you hammered a screw into something. Ugh. What you need is a better tool storage system. Then what you really should get is an 8-inch Electrician's/Maintenance Tote (Model #22128) from McGuire-Nicholas.® It's a great transportable tool storage solution (see Figure 8-1).

McGuire-Nicholas is a product label of the Rooster® Group. A well-known supplier of plastic storage systems, suspenders, and back support belts for carpenters, electricians, plumbers, and other Do-It-Yourself (DIY) workers, you can purchase McGuire-Nicholas products at your favorite home improvement supplier.

Built from a rugged-wear material that McGuire-Nicholas calls Toughwear,™ this tote sports over 30 pockets, loops, hooks, and latches for holding just about every robot-building tool you own. Additionally, an included plastic parts organizer can be used for holding those small parts and fasteners that typically slide around and hide inside other tool storage bags.

One feature that we really like about this tote is that it stands upright when placed flat on the ground. Then just push aside the oversized carrying handle and you have easy access to the four big pockets that are ensconced inside the tote's main body.

If you're looking for a storage solution that is a little less vertical and more horizontal, the 14-inch Tool Bag with Plastic Tray (Model #22314) offers fewer loops, hooks, and latches, but more "super-sized" pockets for holding your batteries, wheels, and servo motors. All wrapped up in an easy-to-transport bag.

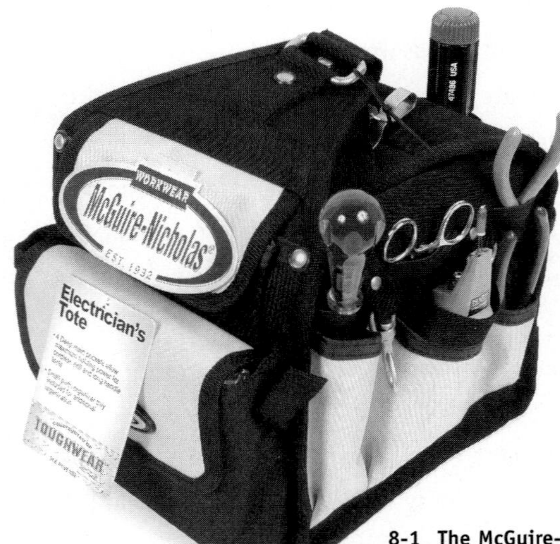

8-1 The McGuire-Nicholas Electrician's/Maintenance Tote is great for holding your hacking tool arsenal.

IS IT A MOTOR PROBLEM?

Yes; are the motors running?

No; make sure that all motors are connected to output ports A, B, and C.

Yes; are the motors not synchronized properly?

Yes; make sure that the two motors are plugged into output ports B and C.

IS IT A SENSOR PROBLEM?

Yes; is the sensor working?

No; plug the touch sensor into input port 1.

No; plug the sound sensor into input port 2.

No; plug the light sensor into input port 3.

No; plug the ultrasonic sensor into input port 4.

Yes; are you getting an incorrect reading?

Yes; use the View menu to select the sensor and its corresponding input port, then read the data generated by the sensor.

IS IT A COMPUTER PROBLEM?

Yes; is your connection USB?

Yes; is the USB icon properly displayed on the NXT Brick's screen?

No; recheck your USB connection.

No; is your connection Bluetooth?

Yes; is the Bluetooth icon properly displayed on the NXT Brick's screen?

No; turn on Bluetooth.

Good luck.

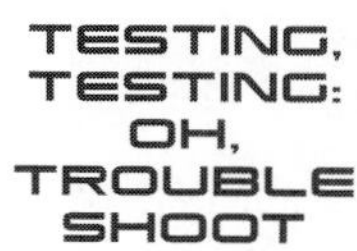

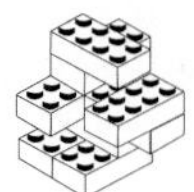

COLOR ME PORSCHE (RED)

No matter what your level of involvement is with LEGO MINDSTORMS NXT robotics, one thing is for sure, after about one year of building bots you will have acquired a substantial warehouse of parts, components, and elements. Now your real problem arises...where to store all of this stuff.

It's not just a matter of ferreting your parts collection away in a set of plastic cubbyholes, you have to be able to find the right part at the right time. What you need is a parts organizer storage system.

As the leader in the manufacture of tackle boxes, Plano Molding Company has over 50 years worth of experience in designing practical and useful plastic storage systems. This experience is readily apparent in their incredible Stow 'N Go™ product line.

Starting with the single-sided Stow 'N Go Organizer (5230) you will be able to segregate up to 27 different parts. Protected with a smooth-operating snap closure system that Plano Molding calls Lock-Jaw,™ this organizer has a clear hinged lid for quick "see-through" parts identification. Painted in a fancy color

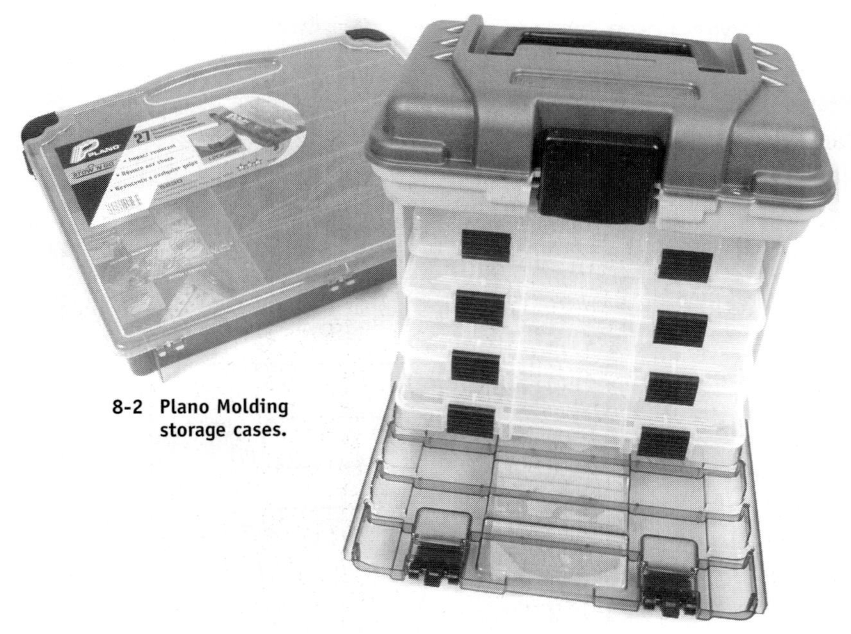

8-2 Plano Molding storage cases.

called Porsche Red this impact-resistant organizer will make you feel like your components collection is being transported in a Hummer (see Figure 8-2).

Best of all, the Stow 'N Go Organizer will hold the entire NXT kit (8527) in one neat, orderly, portable package (see Figure 8-3).

If you were raised on a tackle box mentality, however, the Stow 'N Go Pro Rack Organizer (1354-20) will give you an instant familiarity check. Looking a lot like a tall tackle box, the Pro Rack Organizer is actually four Plano Molding 3500 series StowAway® utility boxes housed inside a brilliantly designed, smoked plastic drop-front lid. Just flip two safety latches, drop the front lid, and slide out one of the four utility boxes. Access to your beloved LEGO Technic elements has never been so easy.

Additionally, there is an open bay storage bin located in the Pro Rack Organizer's lid. An oversized snap closure latch keeps this lid closed during transport.

If you're one of the original MINDSTORMS RIS owners, then you have amassed a much larger parts collection. Luckily, Plano Molding makes a couple of larger storage solutions (e.g., Stow 'N Go Double Organizer; 5232 and Stow 'N Go Pro Rack Organizer w/3600 boxes; 1364-20)—just right for all of your big boy LEGO toys.

8-3 This Stow 'N Go Organizer is great for organizing your NXT kit.

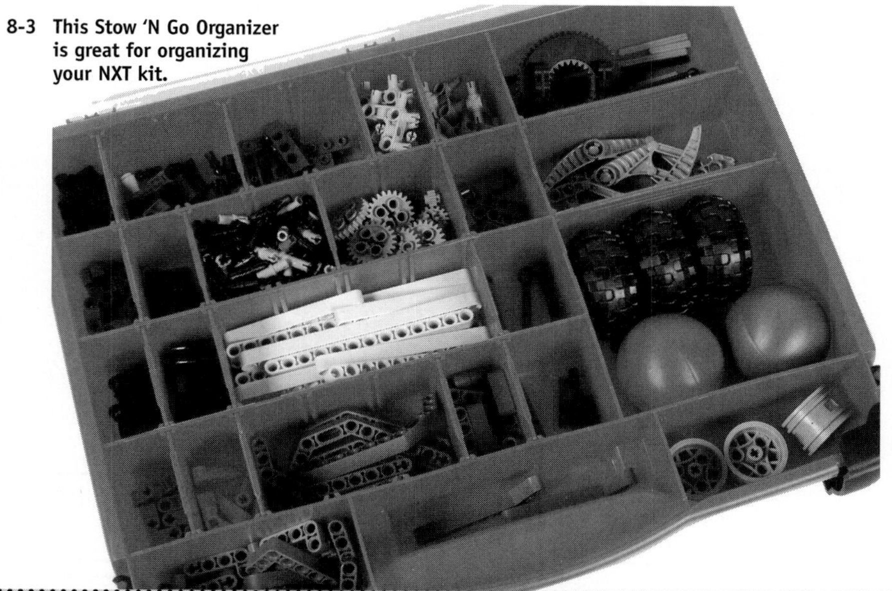

HOW TO BUILD A DIGITAL TAPE MEASURE WITH YOUR NXT KIT

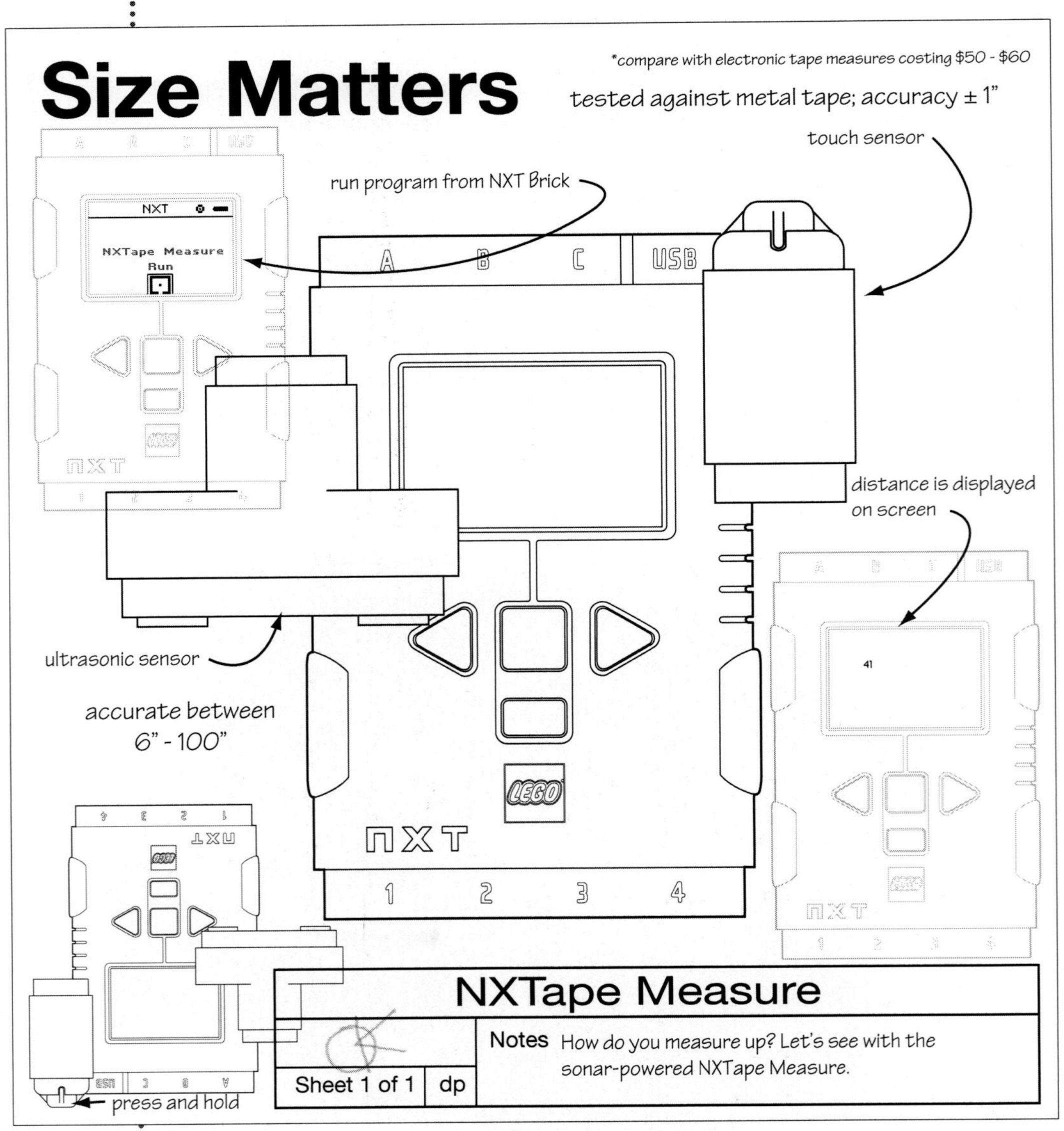

1

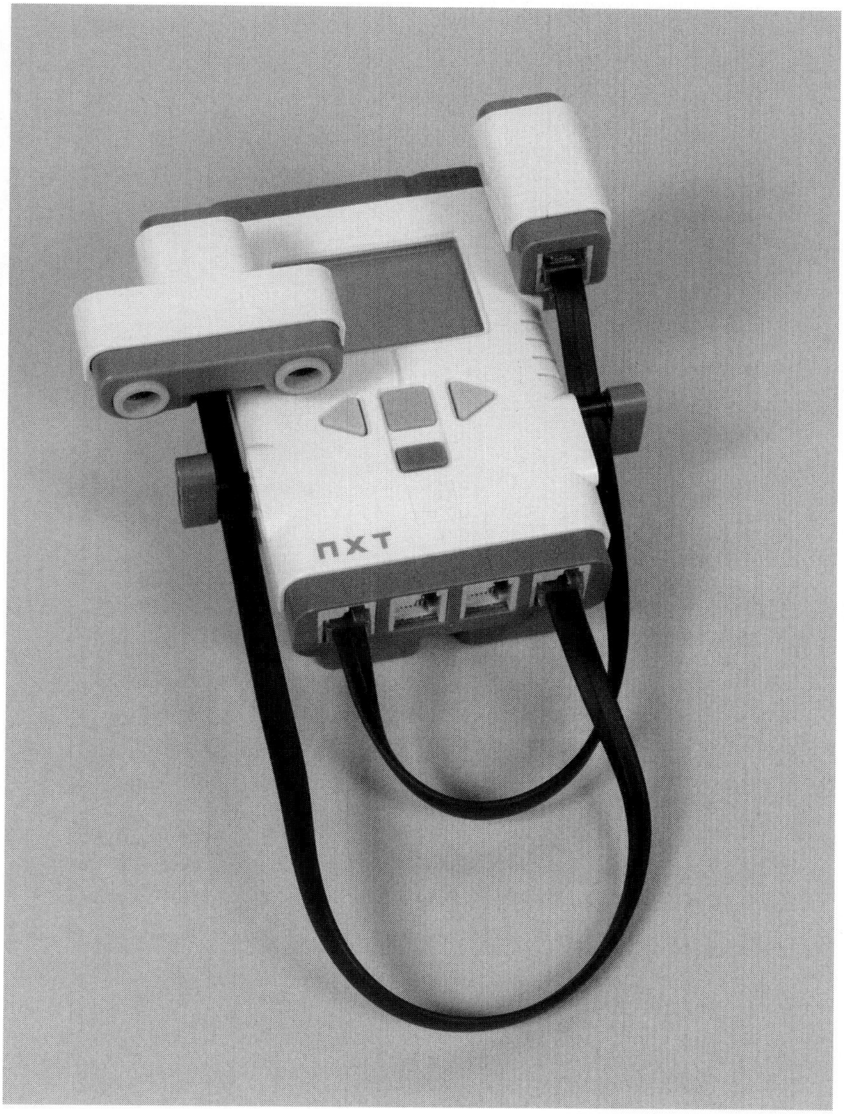

1 **Plan for the NXTape Measure.**

2 **Wired up and ready to go.**

2

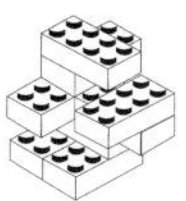

3 If you have trouble keeping your NXT cabling where you want it, try these TECHNIC cable holders.

4 These TECHNIC connectors are ideal for holding cables in place.

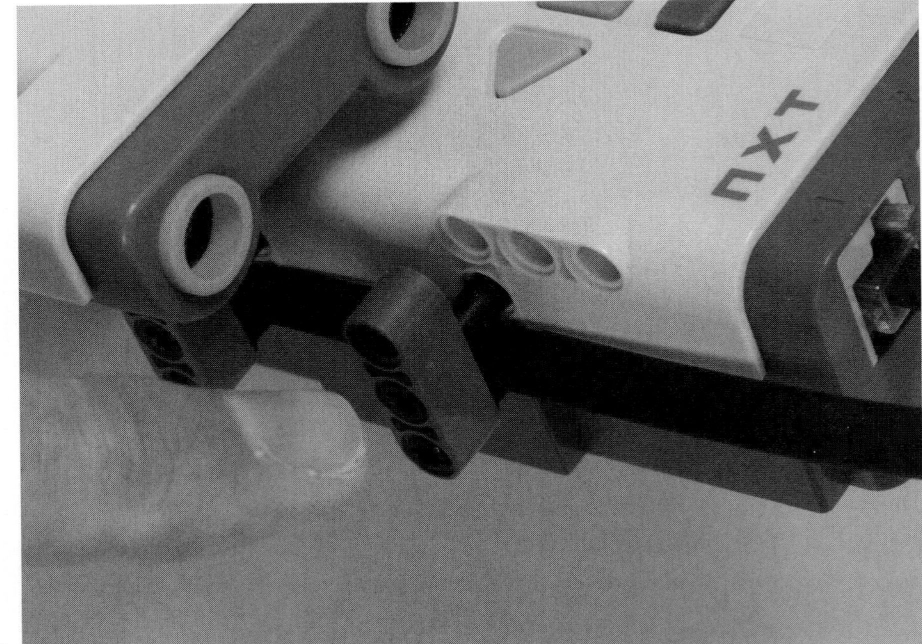

3

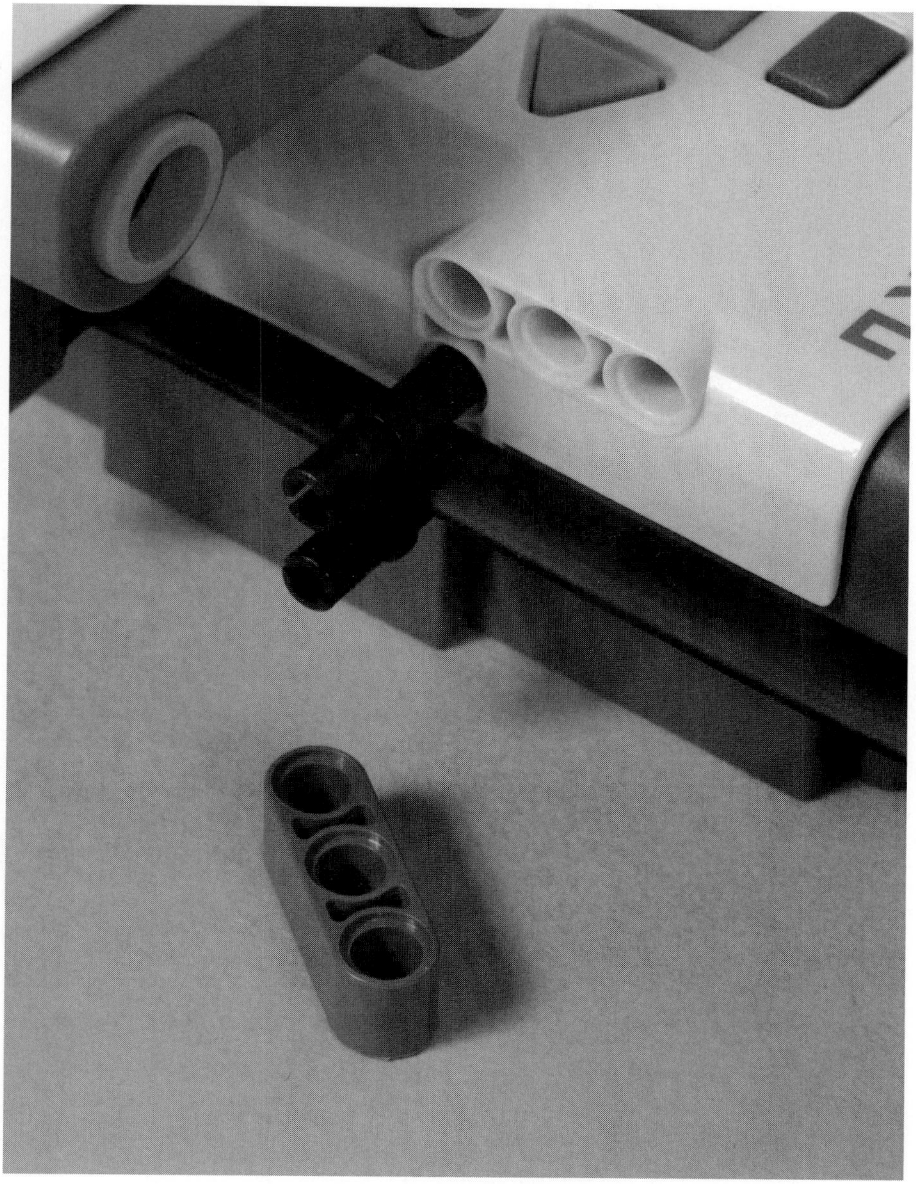

4

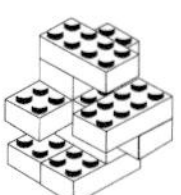

5 Parts for building the NXTape Measure.

6 Attach the sensors. Remember, they go in opposite directions.

7 When operating the NXTape Measure make sure the touch sensor comes into contact with the surface you are measuring.

8 Press the touch sensor down and the NXTape Measure will give you a readout for the distance between two surfaces.

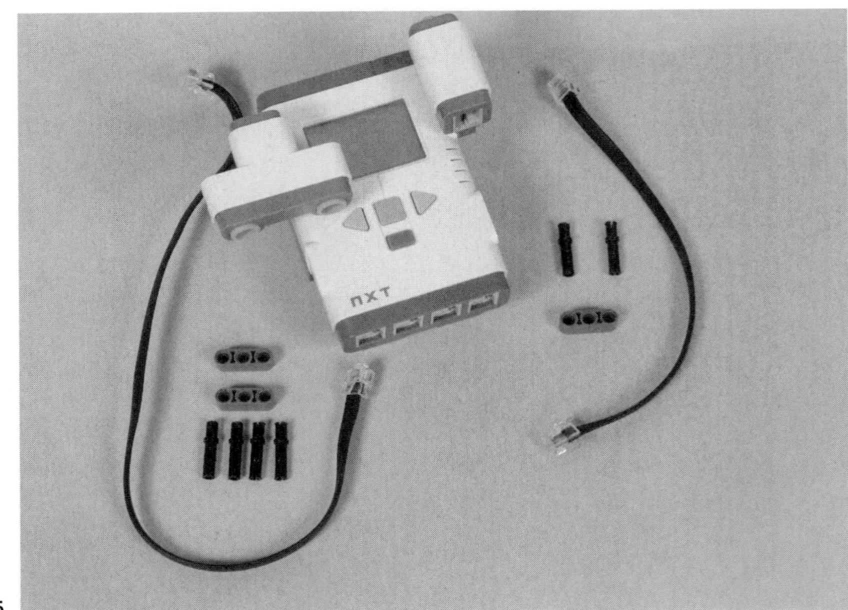

5

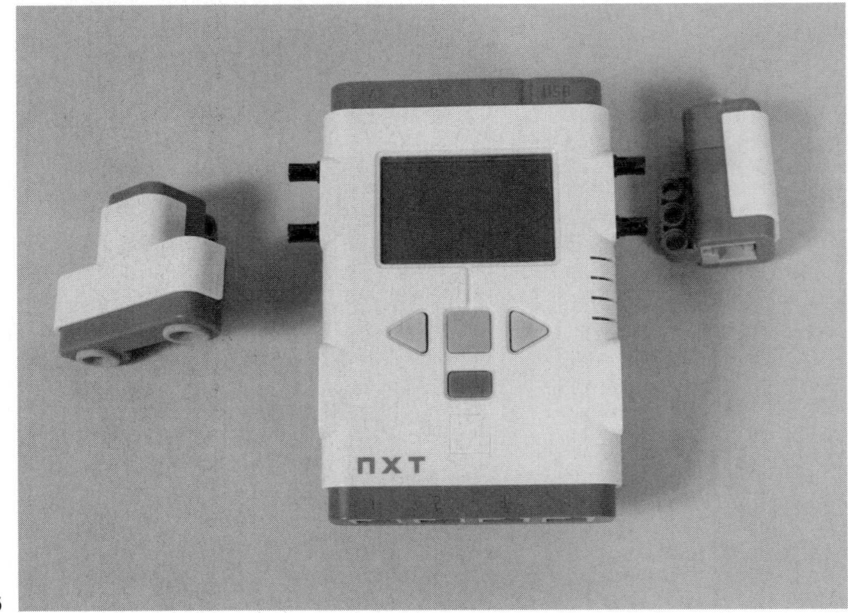

6

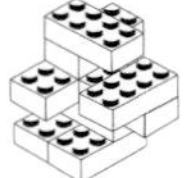

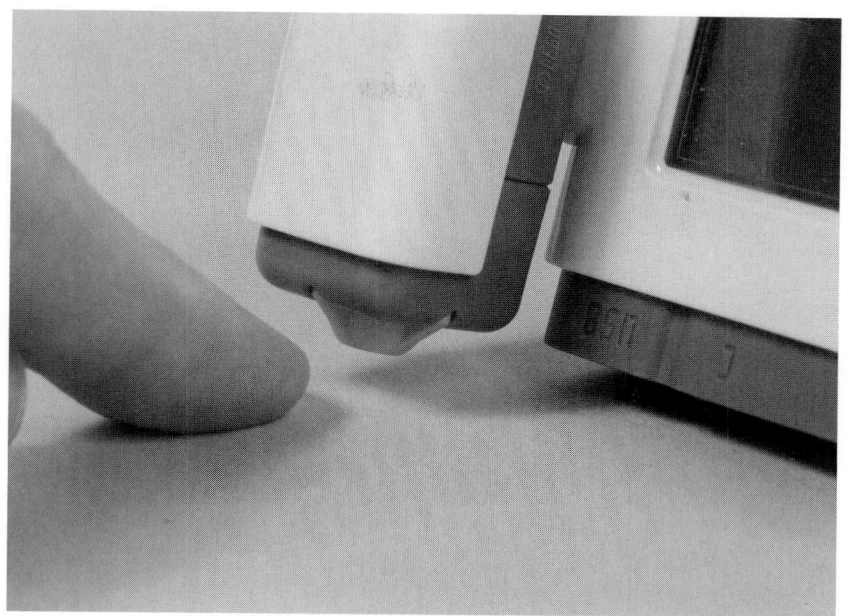

7

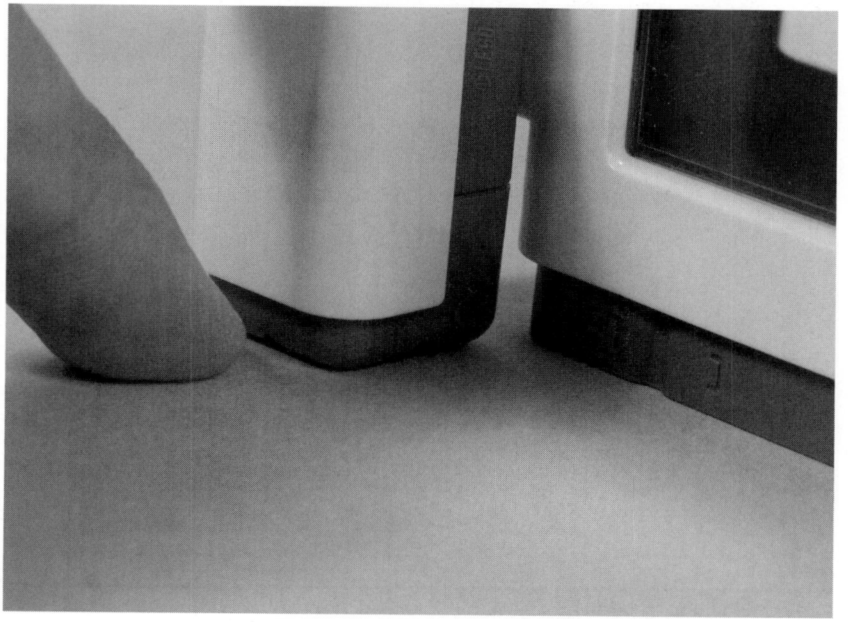

8

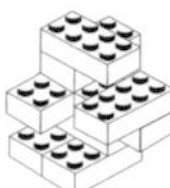

9 Hold the NXTape Measure in contact against the surface until you are done recording the measurement. Releasing the touch sensor will reset this digital tape measure.

10 Now program the NXTape Measure. Begin by connecting the NXT Brick to your computer.

11 If you require more sophisticated measurements, you can add variables to your program.

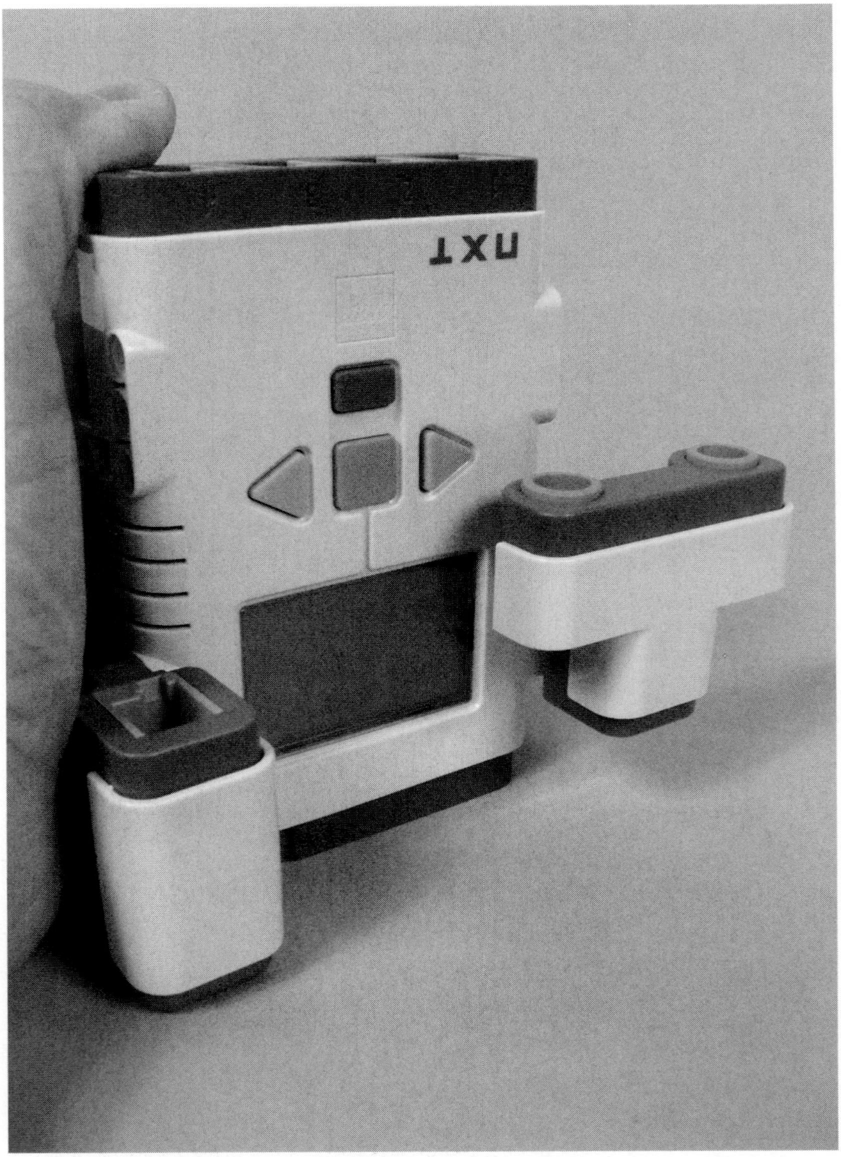

9

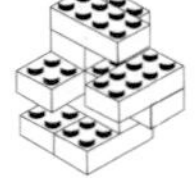

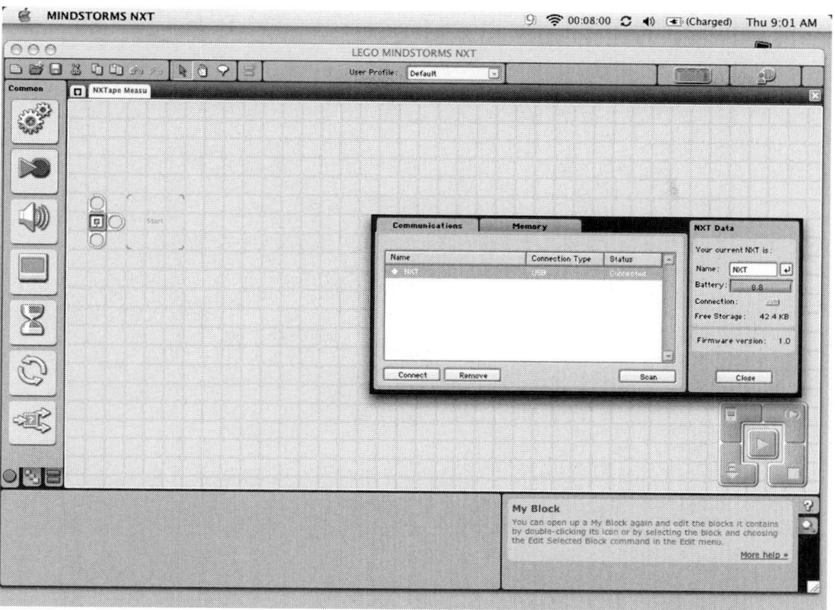
MINDSTORMS NXT
00:08:00
(Charged)
Thu 9:01 AM
LEGO MINDSTORMS NXT
User Profile:
Default
Common
NXTape Measu
Communications
Memory
Name
Connection Type
Status
NXT Data
Your current NXT is:
Name:
NXT
Battery:
Connection:
Free Storage:
42.4 KB
Firmware version:
1.0
Connect
Remove
Scan
Close
My Block
You can open up a My Block again and edit the blocks it contains by double-clicking its icon or by selecting the block and choosing the Edit Selected Block command in the Edit menu.
More help »

10

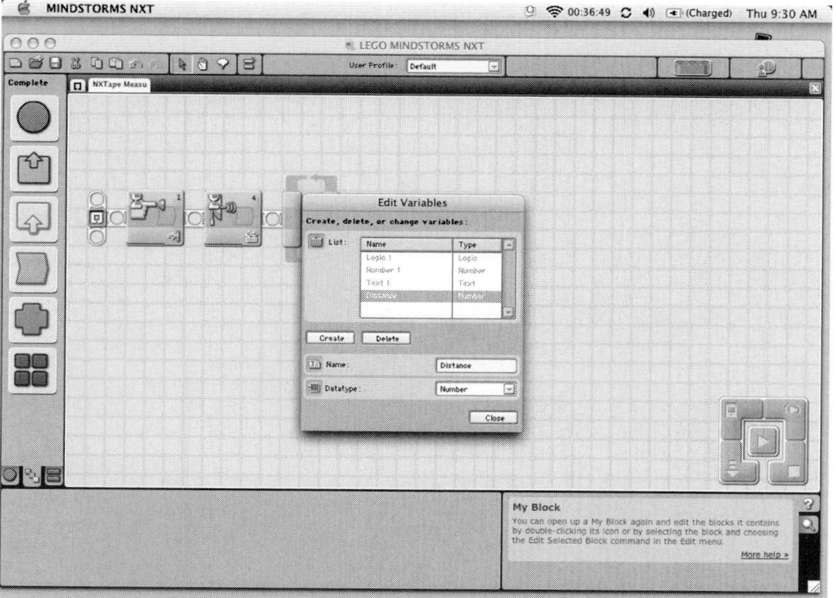
MINDSTORMS NXT
00:36:49
(Charged)
Thu 9:30 AM
LEGO MINDSTORMS NXT
User Profile:
Default
Complete
NXTape Measu
Edit Variables
Create, delete, or change variables:
List:
Name
Type
Logic 1
Logic
Number 1
Number
Text 1
Text
Create
Delete
Name:
Distance
Datatype:
Number
Close
My Block
You can open up a My Block again and edit the blocks it contains by double-clicking its icon or by selecting the block and choosing the Edit Selected Block command in the Edit menu.
More help »

11

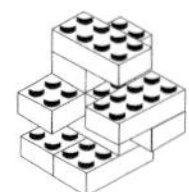

12 Test your program with frequent downloads to the NXT Brick.

13 Use data wires for creating relationships between the sensors, display, and variables.

14 You can "expose" properties and variables for each program block by clicking on the lower edge of the block.

15 The completed and working program for NXTape Measure.

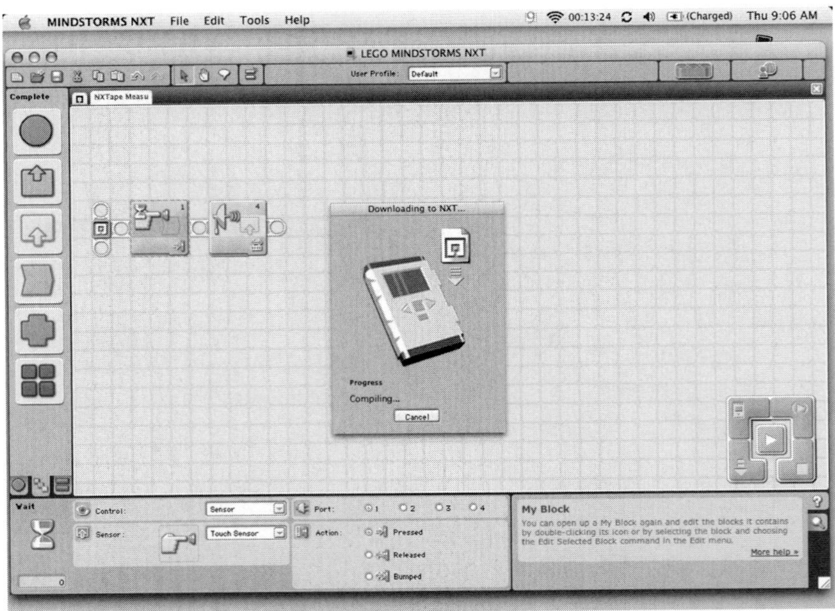

12

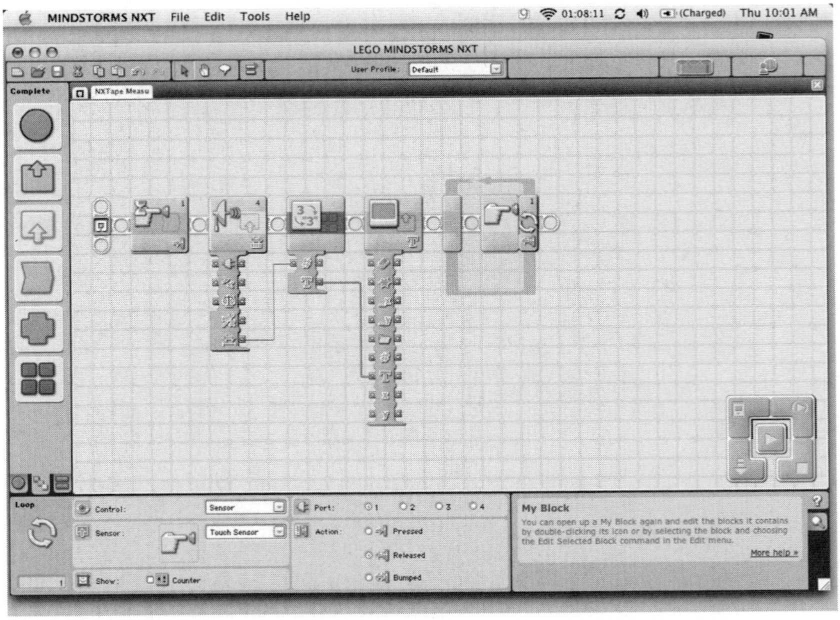

13

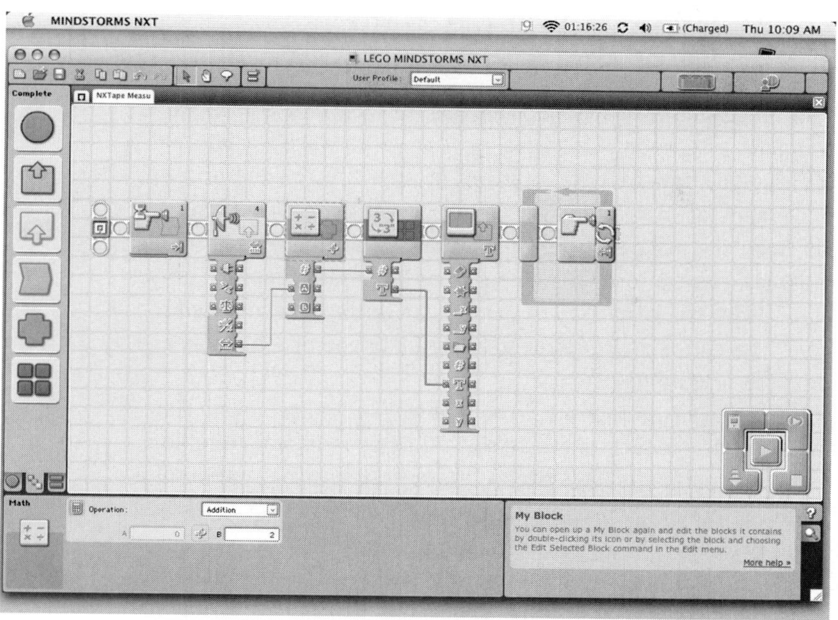
MINDSTORMS NXT
01:16:26
(Charged)
Thu 10:09 AM
LEGO MINDSTORMS NXT
User Profile:
Default
Complete
NXTape Measu
Math
Operation:
Addition
A
0
B
2
My Block
You can open up a My Block again and edit the blocks it contains by double-clicking its icon or by selecting the block and choosing the Edit Selected Block command in the Edit menu.
More help »

14

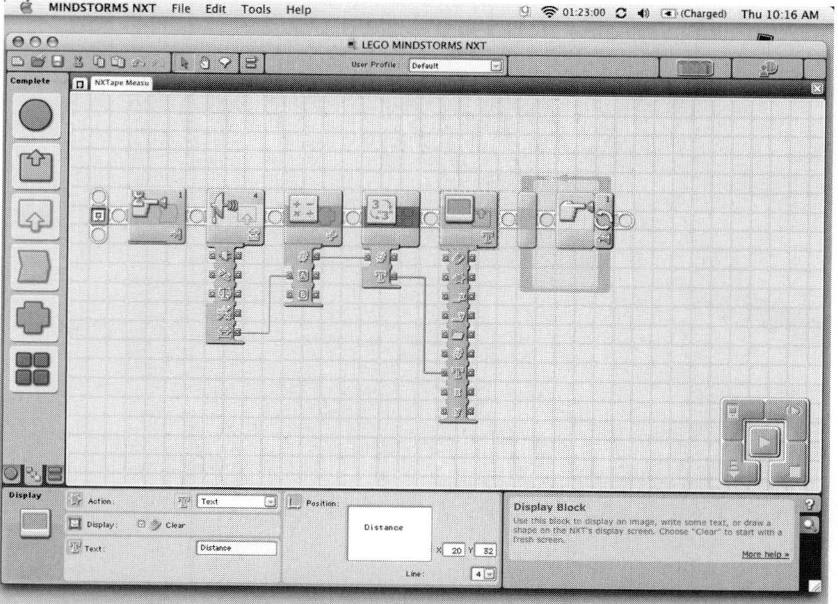
MINDSTORMS NXT
File
Edit
Tools
Help
01:23:00
(Charged)
Thu 10:16 AM
LEGO MINDSTORMS NXT
User Profile:
Default
Complete
NXTape Measu
Display
Action:
Text
Display:
Clear
Text:
Distance
Position:
Distance
X 20
Y 32
Line:
4
Display Block
Use this block to display an image, write some text, or draw a shape on the NXT's display screen. Choose "Clear" to start with a fresh screen.
More help »

15

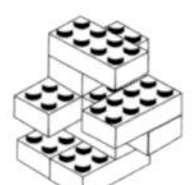

SOME YO-YO'S IDEA OF A TOOL TOTE

Who didn't grow up playing with a yo-yo? This was the ideal diversion to while away a lazy summer afternoon. Of course, if your yo-yo wasn't a Duncan,® then it wasn't a yo-yo. Now that we've grown up, our toys have grown up, too. Luckily, the manufacturer of Duncan yo-yos is still there by your side.

Flambeau,® Inc., the makers of Duncan toys and a member of the Nordic Group of Companies, has captured the elegance of the classic Imperial yo-yo design, coupled it with the versatility of the Duncan Butterfly, and added the scientific marvel of a corrosion inhibitor to a terrific lineup of tool totes and parts organizers (see Figure 8-4).

Populated with drawer liners, totes, tool chests, and parts boxes, Flambeau's contractor-grade products offer a solution to every storage problem. Unlike other tool and parts storage systems, however, the incorporation of a corrosion inhibitor into their storage products makes Flambeau bags and boxes ideal for keeping rust away from your tools.

Called Zerust,® this corrosion inhibitor is infused directly into the fabric and plastic of all Flambeau "Z" products. Flambeau Zerust is a patented prod-

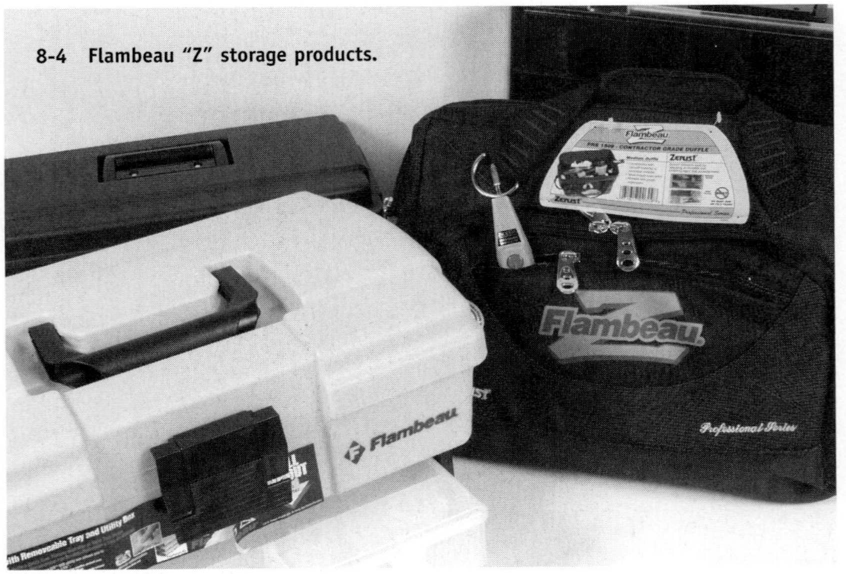

8-4 Flambeau "Z" storage products.

uct based on an advanced polymer technology. Zerust works by generating an invisible, pervasive "vapor shield" that cloaks your tools in a protective layer that screens out harmful oxidation processes.

According to Flambeau, Zerust will inhibit rust for up to five years. Furthermore, the company claims that this anti-rust technology is nontoxic, odorless, FDA approved, and, best of all, maintenance free. Just open the Zerust bag or box, drop your tools inside, and forget about pulling out a rusty driver bit, hammer, or wire cutter ever again.

If you're looking for a spacious tote for lugging around your assorted battery chargers, test equipment, and sundry robotics parts, the Flambeau Z Medium Duffle (PRS 1509) is a great choice (see Figure 8-5). When coupled with a suitable parts box (Model 5007; which fits snuggly inside the duffle bag's bottom), this tool tote is able to transport a sizeable inventory of tools, batteries, and parts—becoming a veritable mobile robotics development workshop. Thankfully, there's even an inside pocket that can accommodate a Duncan Imperial or Butterfly yo-yo. Just the ticket as you while away the hours recalling the idyllic days of your youth, err, reflecting on an answer to a nagging robotics problem.

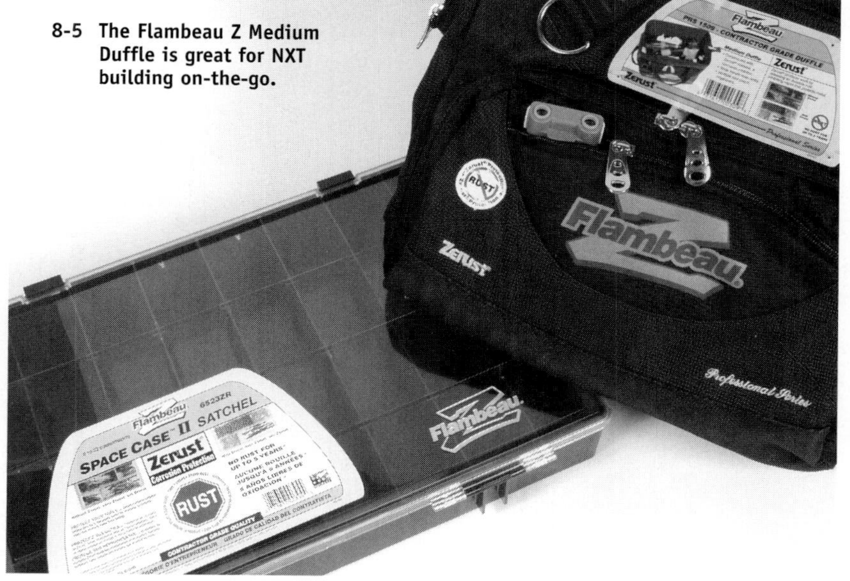

8-5 The Flambeau Z Medium Duffle is great for NXT building on-the-go.

PART III: KATHERINE'S BEST HACKING PROJECTS

OK, 'NUFF SAID, here's a collection of my 10 best NXT hacking projects. No matter what your level of hacking expertise, there's an NXT hack here for you. Whether you're a solder slinger or a code banger, you will find a project that piques your inner hacking self.

First, I am obligated by common sense to restate the obvious:

Although I have personally tested each and every one of the projects in this book and found them *all* to be fun, educational, and reasonably easy to reproduce, *any* disassembly, modification, or hacking of *any* LEGO MINDSTORMS NXT *electronic* part or component **WILL RENDER YOUR WARRANTY NULL AND VOID**.

I have *never* found this to be a bad thing, but your opinion may differ from mine.

Likewise, any *mistake* in the execution, soldering, or programming of any of the projects in this book *could* destroy your NXT Brick and its associated sensors. Please, please don't attempt any project in this book without **FIRST THOROUGHLY READING AND UNDERSTANDING THE INSTRUCTIONS** for each project.

Finally, the McGraw-Hill Companies and Dave Prochnow will neither assume nor be held liable for any damage caused to anyone or anything that is associated with the disassembly, modification, or hacking of any gadget, gizmo, or electronics equipment.

PROJECT LIST

Phew, now that I have those precautions off of my chest, here is an annotated summary of the projects that are contained in this section:

"PHAT" PHOTOVOLTAIC POWER PLANT.

Never buy another AA battery, again. Operate your NXT Brick for FREE with electricity from the sun.

I C U.

Give your next robot a pair of eyes. OK, at least give it one good eye, a video eye.

SPEAKING IN TONGUES.

Are you sick and tired of listening to those "canned" phrases that are preprogrammed into the NXT software? Add some voice talent. Add your own voice, messages, and phrases to the NXT Brick.

THE THING WITH TWO HEADS.

Gosh, wouldn't it be great if you had another microcontroller that you could add to the NXT Brick? You know, like a brain transplant or brain implant.

WELCOME TO THE ZIGBEE FOLLIES.

That's just what we need—another wireless standard. Well, if you could send data to your computer for controlling your NXT Brick, then maybe another wireless link couldn't hurt.

EL NXT.

Can you believe a robot without blinky eyes? Of course, not. So add your own electroluminescent wiring and pimp your bot.

FROM FEET MADE OF TREAD.

Wheels or walk; take your pick. Those are your only mobility options with your NXT robot unless you hack some better shoes onto your beast.

PUMP ME UP.

Learn how to shape up with a piston pump to change your, err, shape.

PICTURE THIS.

Attention: you now are entering a "no solder zone." Put a face on this bot with custom graphics designed in the Ricoh FAX document format. Yeah, you've got to read this one to believe it.

GETTING A BIT LONG IN THE BLUETOOTH.

No phone, no computer, no Bluetooth...wrong. Build your own Bluetooth controller for your NXT Brick.

A POTPOURRI OF HACKER HINTS

1 ERECTOR® sets by MECCANO® are an excellent source of hacker's parts.

2 Parts from ERECTOR sets are a good fit to the studs of LEGO bricks.

3 A lot of the ERECTOR set hardware can be used for fastening LEGO bricks together.

4 You can attach an ERECTOR beam to a LEGO brick by drilling a hole into a plastic stud.

1

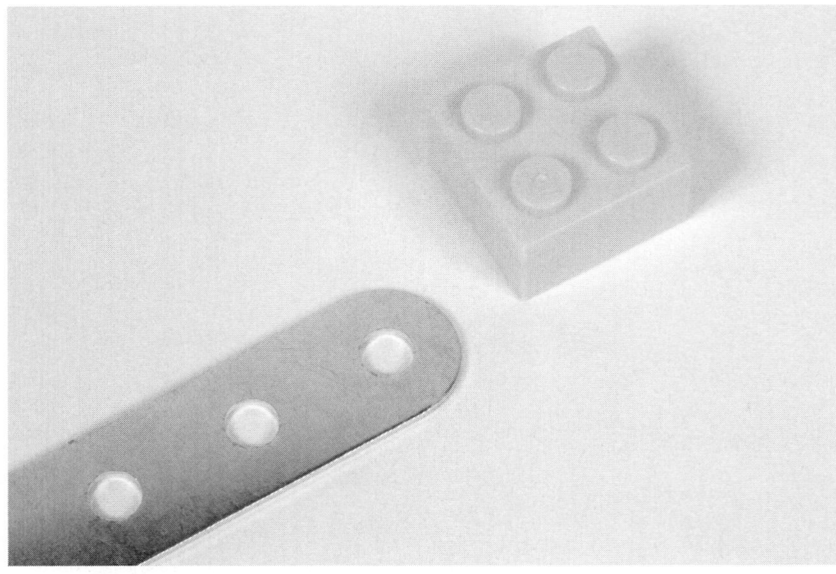

2

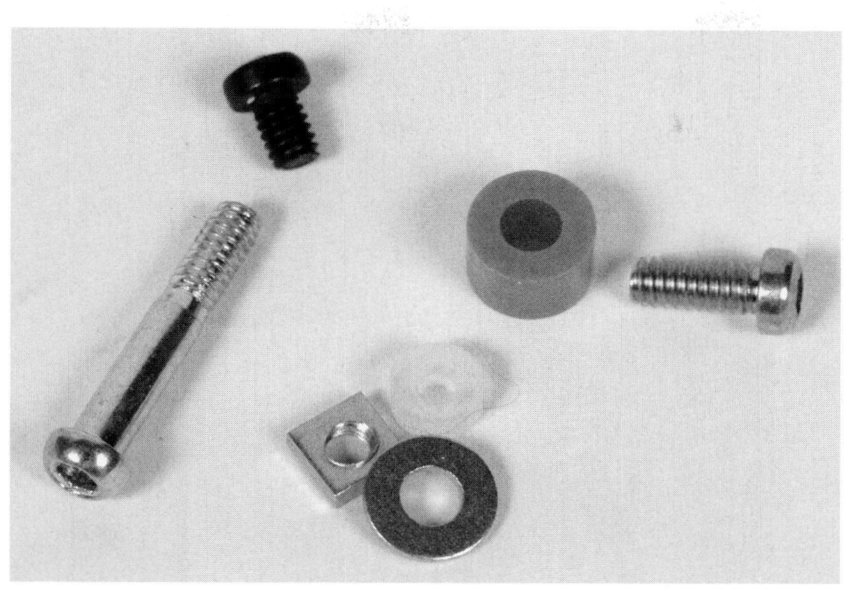

3

4

5 Clean up the opening with a hacker's best friend—the reamer.

6 A perfect hole.

7 Use ERECTOR fasteners for joining your parts together.

8 The tools in ERECTOR sets will help with connecting these parts together.

9 That other plastic brick manufacturer—MEGA BLOKS.® This set was a steal at less than $5.00.

5

6

7

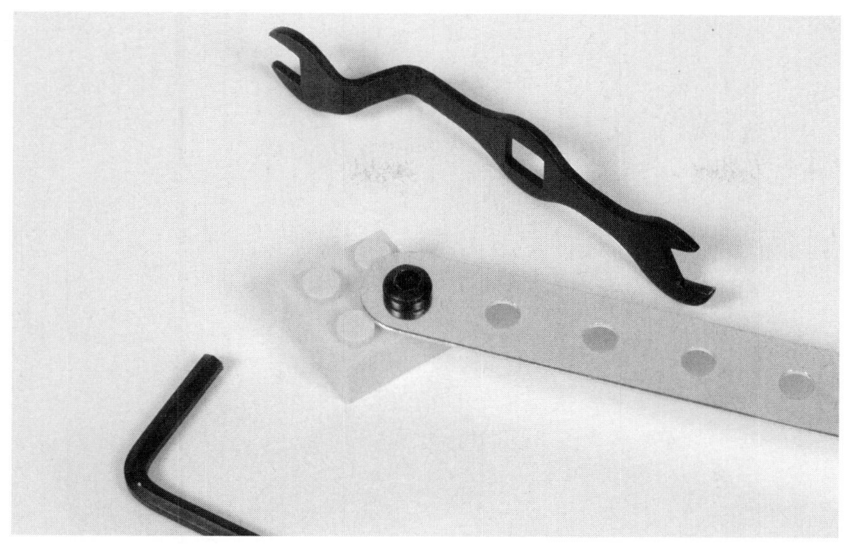

8

MEGA BLOKS
JUST BUILD
1760
Building Imagination
Ideas Book
600
WARNING:

9

10 Which brick is which—LEGO (left) and MEGA BLOKS (right). Or, was that, MEGA BLOKS (left) and LEGO (right). Gee, I forget.

11 And they said it wouldn't work—a perfect match.

12 Now where did that LEGO brick go? Oh, there it is...hiding in all of those MEGA BLOKS bricks.

13 Mix and match; it's all up to you.

10

11

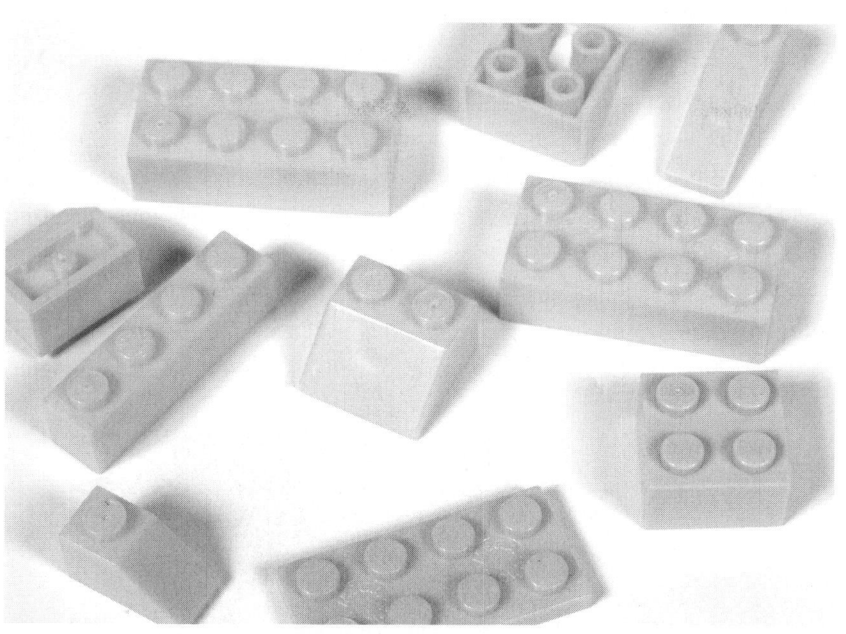
12

13

PHAT PHOTOVOLTAIC POWER PLANT

Behold, I give you power from the sun. Solar power, that is. Or, more specifically, electrical energy from a photovoltaic (PV) cell (see Fig. III-1).

PV cells convert sunlight into electricity with nary a moving part, consumable fuel source, or battery connection. Typically, PV cells are made of chemically refined silicon—a semiconductor. Either boron or phosphorous are mixed with the silicon to enhance the PV cell's ability to release electrons when tickled by photons. Selenium, gallium-arsenide, and cadmium telluride have been substituted for silicon in the manufacturing of PV cells with varying results.

Following the release of these electrons inside the PV cell, the building

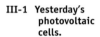

III-1 Yesterday's photovoltaic cells.

electrical current is funneled into a set of attached wires. Additional cells can then be wired together for generating more current and higher voltage. Groups of PV cells can then be arranged side-by-side into a rectangular shape called a module. Likewise, several modules then can be tied together forming an array.

When PV arrays get big enough for some serious power generation, they are generally mounted on the roof of a building angled south for an optimal dose of daily sunshine. Some more sophisticated PV arrays are mounted on moveable tracking systems that follow the sun's path for prolonged maximum solar exposure.

One of the most amazing advancements in solar-powered arrays has been the recent development of low-cost, high-yield flexible thin-film PV cells. Fabricated on a tough, resilient, flexible plastic polymer substrate (some even have a fabric backing for added strength), these thin-film PV cells are able to generate voltages roughly equivalent to their rigid counterparts.

A vast improvement over the rigid PV cells, however, these thin-film arrays are typically supplied in sheets or rolls, can be connected together, and are weather-resistant. Based on its ease of operation and fabric-like qualities, thin-film PV cells can be easily integrated into portable lightweight products.

Even the U.S. Army has begun to notice thin-film PV arrays and studies are currently underway regarding the incorporation of these flexible power supplies into tents, backpacks, and, even, uniforms.

Lacking a Department of Defense budget, however, you might want to lower your solar sights to a PV cell that will fit on a penny. This new power plant is much more financially accessible, too. It is actually hidden inside an OSRAM Opto Semiconductors' silicon photodiode (BPW 33). The BPW 33 is priced at around 25¢ each (see *The Electronic Goldmine* Web site), but this photodiode also leads a double life (see Fig. III-2).

It is able to both sense the intensity of light *and* supply current when held

III-2 Today's photovoltaic cell wonders.

in direct sunlight. While this former attribute would be a marvelous feature for building a new NXT light meter, it is the second attribute that interests us here.

This compact 2-pin silicon photodiode acts like a tiny PV cell. Encased inside a plastic case, the BPW 33 responds to light spectral ranges from 350 to 1100 nanometers with optimal results achieved at 800 nanometers. Furthermore, these 3/16-inch square cells have a low reverse current (typically 20 pA) with a short circuit current of 72 μA. Best of all, the BPW 33 is able to generate an open circuit voltage of .44 to .475 volts at 1.86 mA.

These are tremendous performance specifications that roughly translate into eight BPW 33 photodiodes wired in series being able to power many of your NXT Brick projects...for FREE. In fact, with 18 to 20 BPW 33s wired in series you could even power your NXT Brick—at the cost of a scant $4.50. That's less than the cost of one pack of six AA batteries.

HOW TO BUILD YOUR OWN SOLAR-POWERED BATTERIES

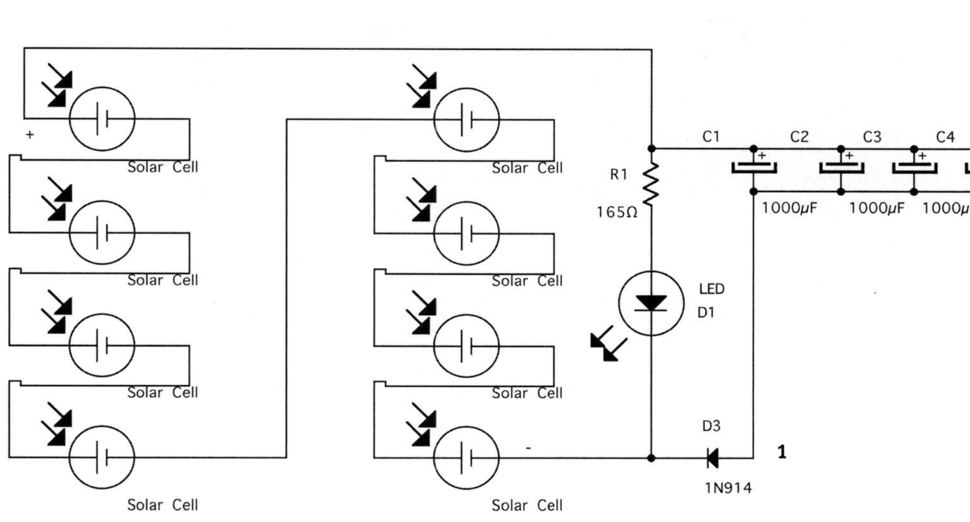

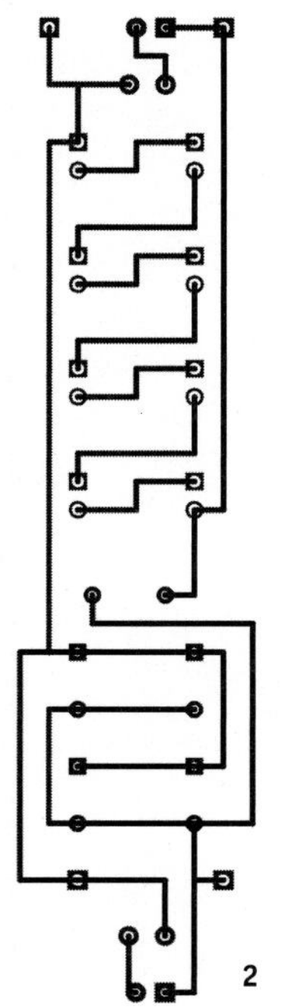

1 **Schematic diagram for a solar-powered battery.**

2 **A sample printed circuit board (PCB) for building a solar-powered battery. Refer to 1 for parts layout.**

3 Only a handful of parts are needed for building a solar-powered battery.

4 If you elect to etch your own PCB, you can use conventional single-sided copper-coated board.

5 Or, you can opt for a thinner board.

6 A thin copper-coated board sold by The Electronic Goldmine is thin enough to cut with a pair of snips.

3

4

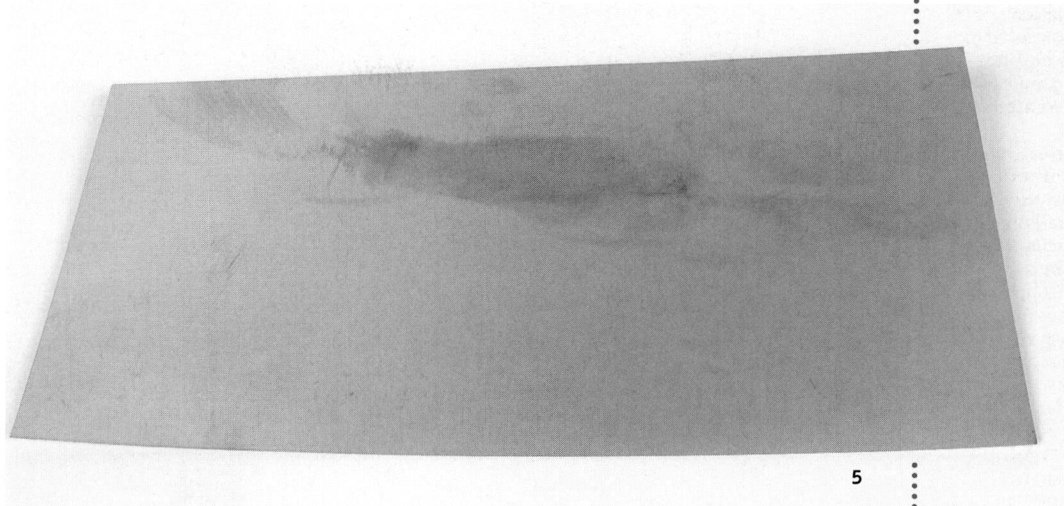

5

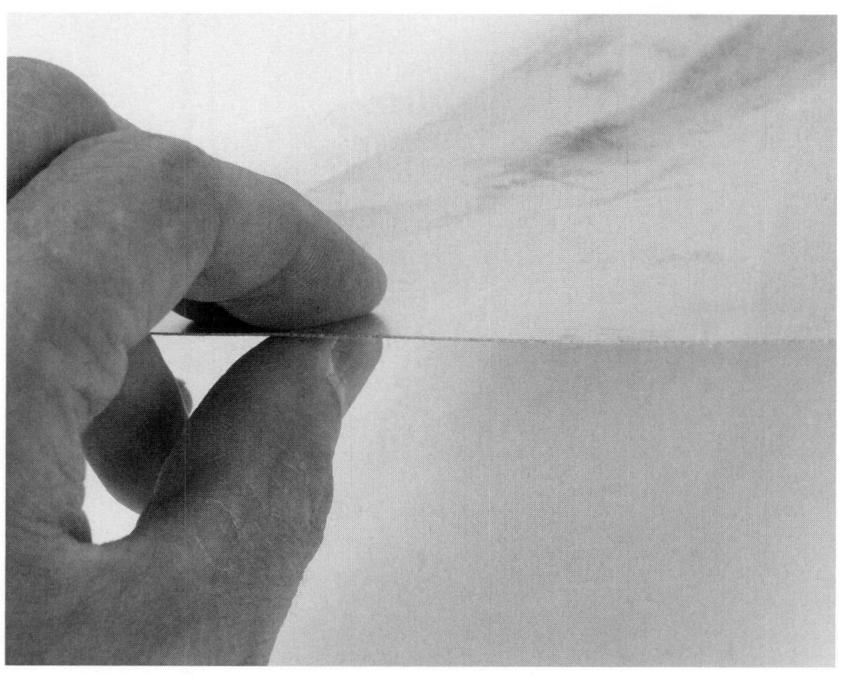

6

7 Use aviation snips for cutting a clean outline of this thin copper-coated board.

8 Press-n-Peel enables you to transfer your PCB tracings and pads directly onto a copper-clad board by using your laser printer.

9 Predrilled wiring boards are ideal for building your solar-powered battery project.

10 Cadmium-sulfide photocells can be added to the circuit in 1 for switching between stored energy and "live" solar energy during periods of peak sunlight.

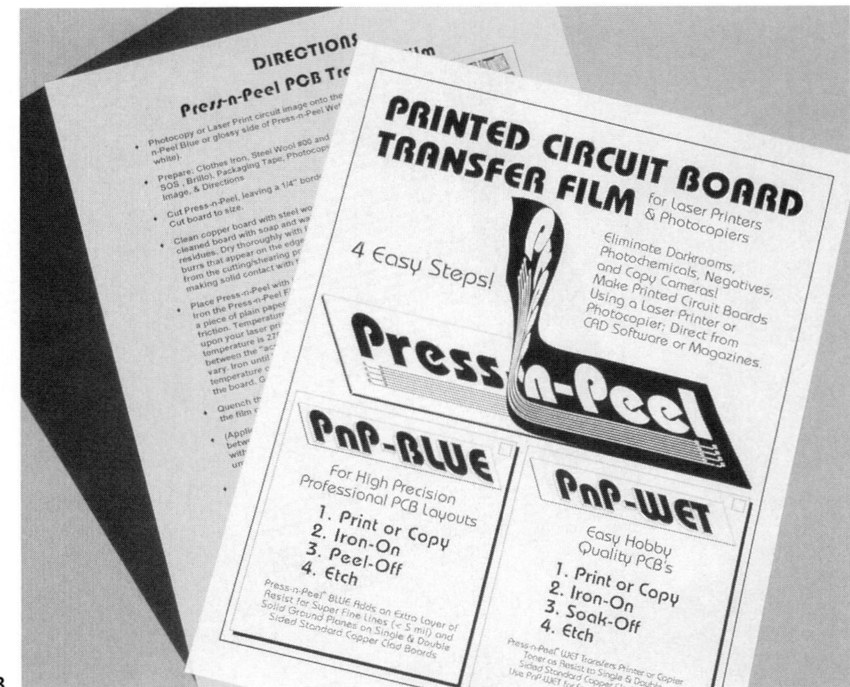

8

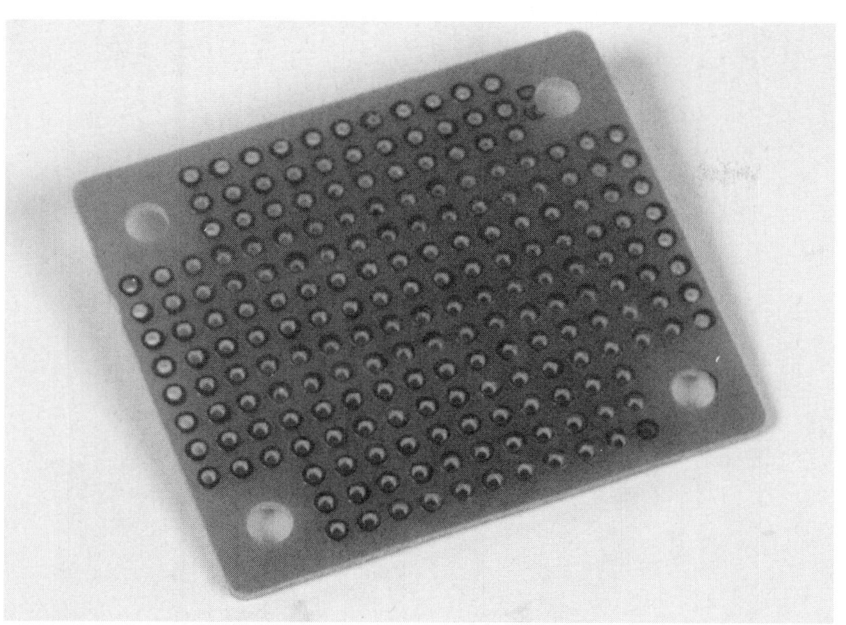

9

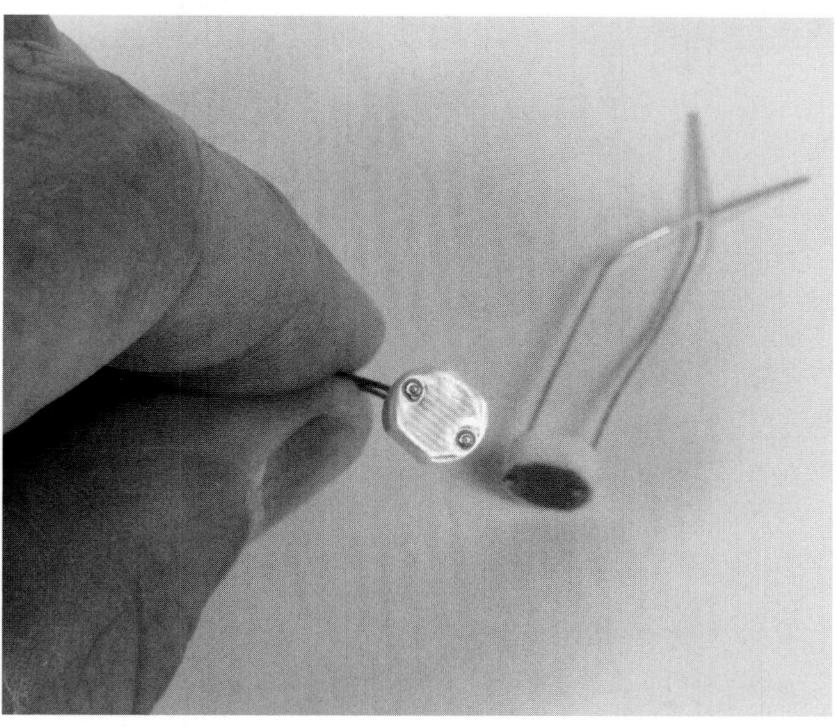

10

11 Start adding your BPW 33 photodiodes.

12 Each photodiode should be soldered together. Follow the wiring diagram in 1.

13 Add the electrolytic capacitors.

14 Add the LED indicator lamps.

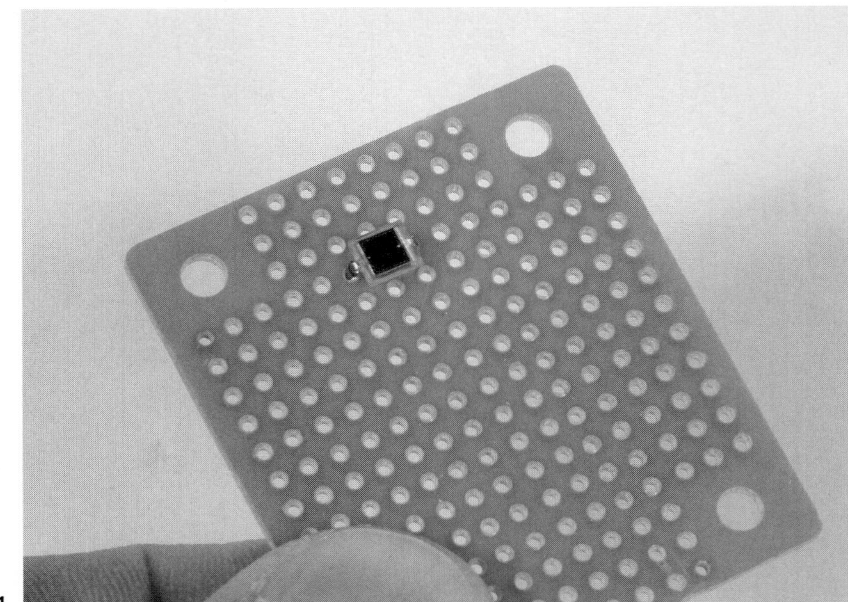

11

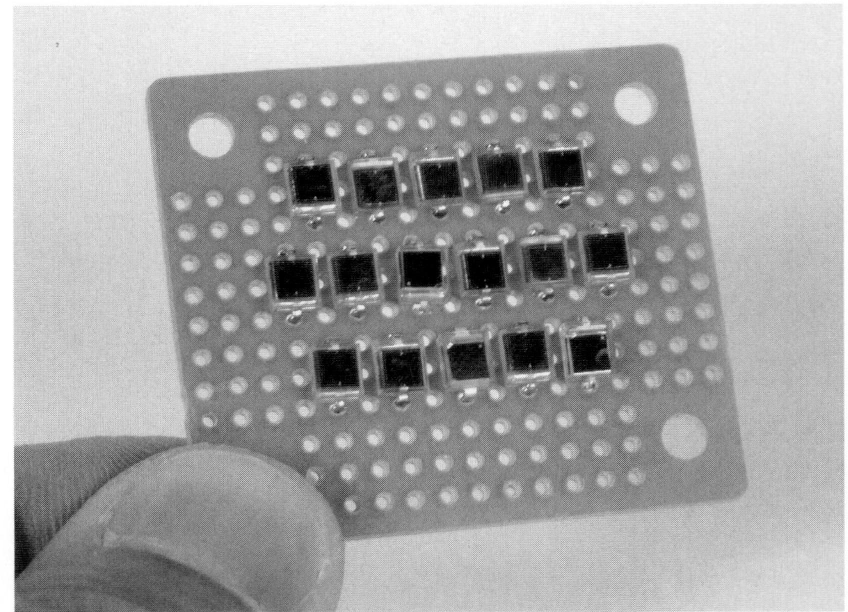

12

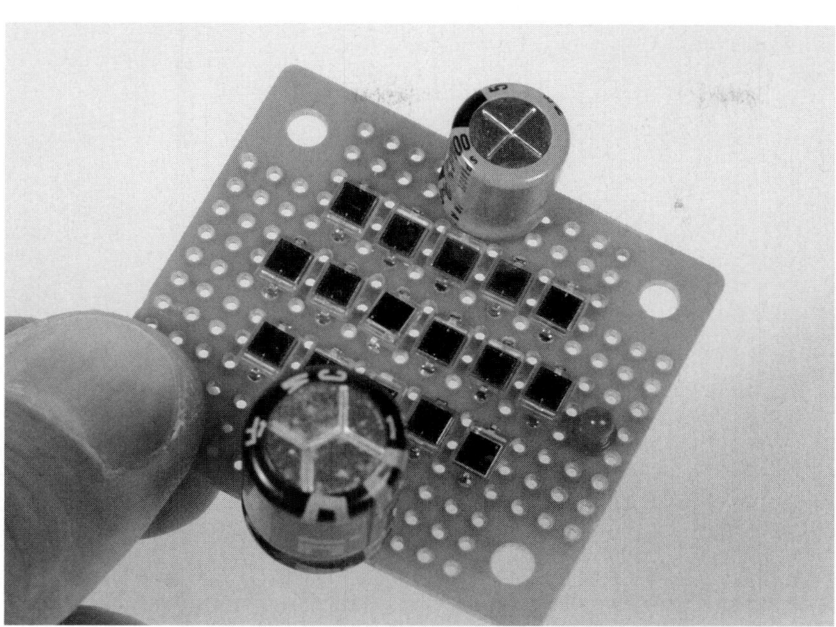

13

14

15 **Add the diode.**

16 **Optionally, you can add a photocell to this circuit. Now, test your handiwork with a multimeter prior to replacing your NXT batteries with this new solar-powered battery.**

15

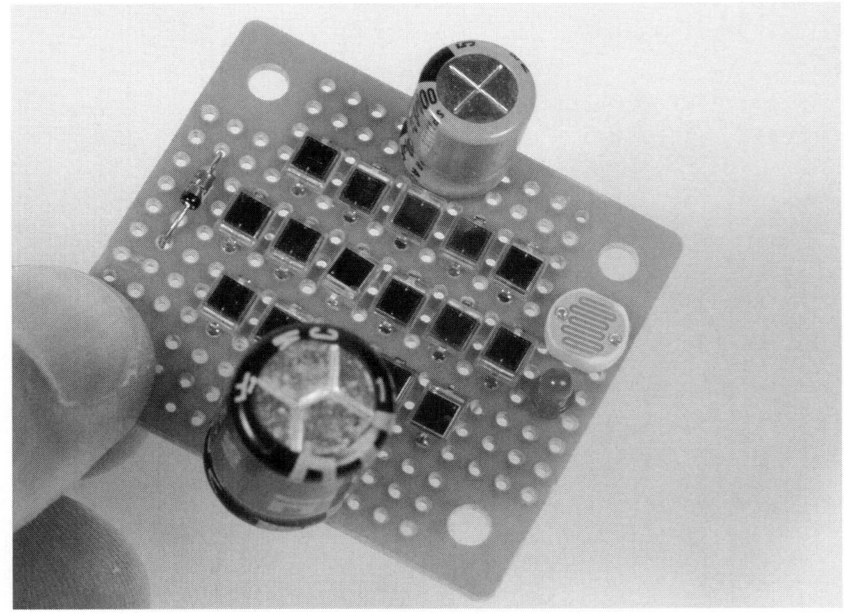

16

ICU

Occasionally you stumble across a real treasure when shopping for surplus electronics. Such is the case of the "no-name" wireless video camera kit (203CA). Yup, that's all there is for a brand name, make, and model number. Granted, there are a lot of manufacturers who package a video camera bearing this designation, but there is no clear-cut brand name for this product.

For example, if you're a reseller of electronics, you can buy a version of this camera from Shenzhen Lianyida Science Co., Ltd. This camera is packaged in quantities of 20 and is shipped to vendors for resale.

On the home front, retailers like Jameco Electronics RobotStore sell a "mini wireless color camera" that bears a strong resemblance to the 203CA wireless video camera for $89.95.

Most vendors, however, sell the 203CA wireless video camera as a "spy" or "mini spy" camera. Typically, this bundled DIY spy "kit" sells for less than $80 and features the camera, a video receiver, and two power supplies (e.g., one for the camera and one for the receiver).

Regardless of where you find a 203CA wireless video camera or no matter how the reseller is marketing this camera, this is a powerful video system that will fit in the palm of your hand or the head of an NXT bot. Likewise, if you can find this camera for less than $60, buy it. There are many different projects that are ideally suited to this camera.

By using 900 megahertz to 1.2 gigahertz transmission frequencies you will be able to receive a video signal at distances up to 150 feet. Additionally, most 203CA wireless video cameras are equipped with a 9-volt battery connector. This connector will enable you to eliminate the DC wall transformer power supply and generate true remote video signal transmissions.

In terms of performance specifications, the 203CA wireless video camera is able to generate a resolution of 380 lines with a 60-hertz scan frequency. At an approximate aperture of f1.2, the camera is able to transmit a readable video signal at a minimum light illumination of 3 Lux. Shutter speeds are electronically varied between approximately 1/60 to 1/15,000 of a second. Both the aperture and shutter speed are automatically set by the camera.

HOW TO HACK A VIDEO CAMERA SYSTEM

1 A "no-name" wireless video camera kit (203CA).

2 This video camera can transmit both video and audio.

3 Power bricks can limit the functionality of your "wireless" video camera system.

4 Luckily, some video camera systems come with a 9-volt battery harness. This harness will help ensure that you have a totally "wireless" video camera system.

1

2

CMOS POWER
INPUT : AC 120V 60Hz
OUTPUT : DC 8V 200mA
AC-DC ADAPTOR
MODEL: MWY41-0120500DC
INPUT: AC 120V 60Hz
OUTPUT: DC 12V 500mA
CAUTION: DO NOT REMOVE
COVER.ONLY FOR RECEIVER

3

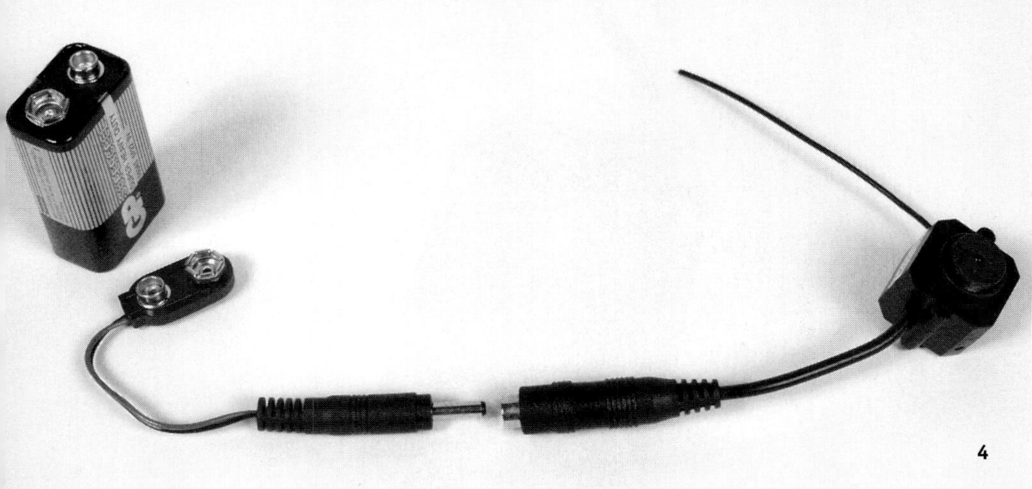

4

5 The receiver comes with a short-range antenna.

6 The antenna screws into the back of the receiver.

7 All video and audio cables were supplied with this kit.

8 This low-cost, battery-powered TV is perfect for receiving the signal from a wireless video camera system.

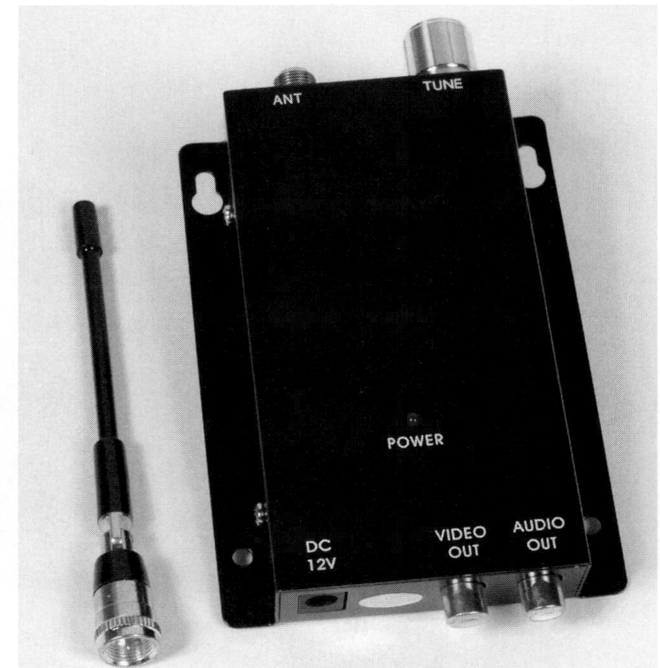

5

6

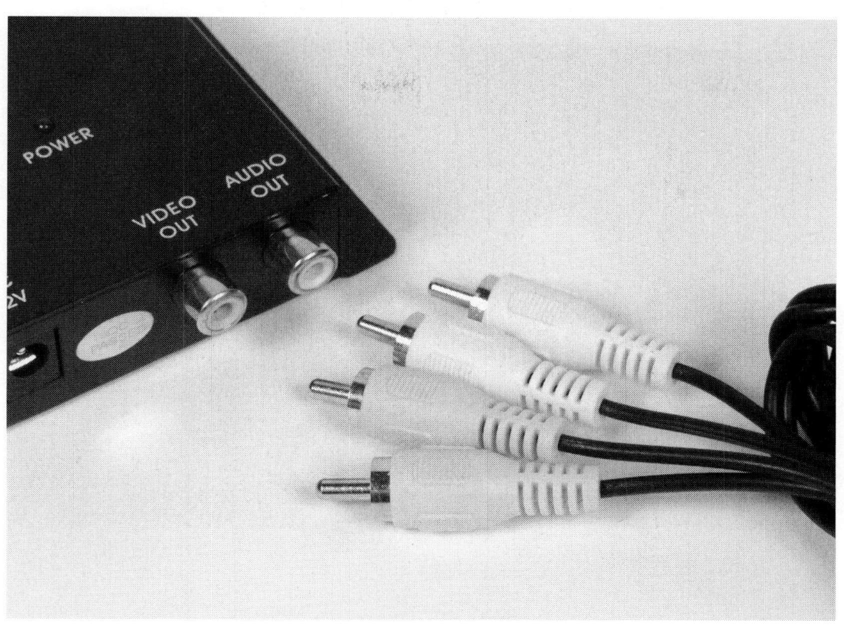

7

8

9 Make sure that your monitor (or, TV) has video and audio input phono jacks.

10 Hooked up and ready to receive some NXT video broadcasts.

11 Even with the 9-volt battery, this wireless video camera is small enough to fit inside almost any LEGO design.

12 A simple camera platform can be built from the beams and connectors inside your NXT robot design kit.

9

10

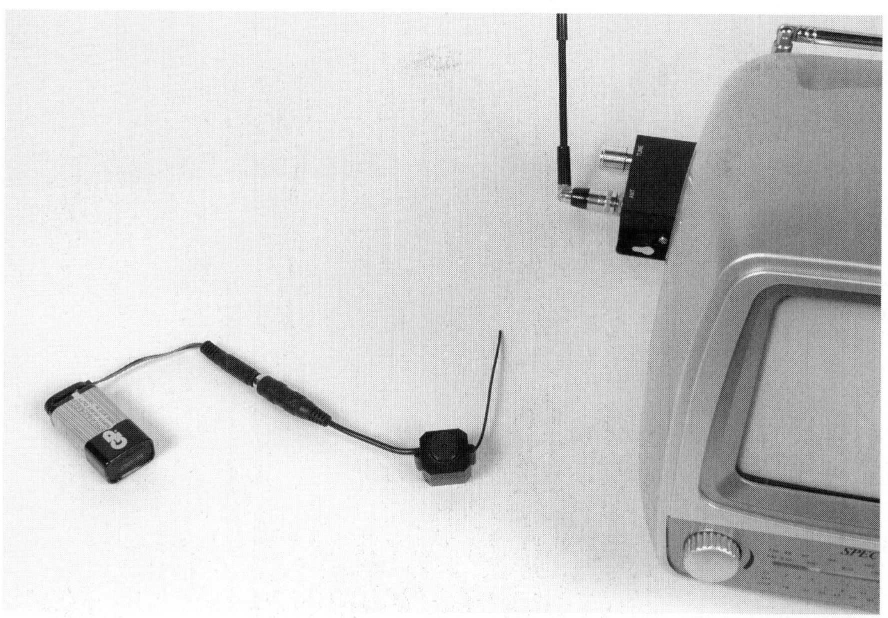

11

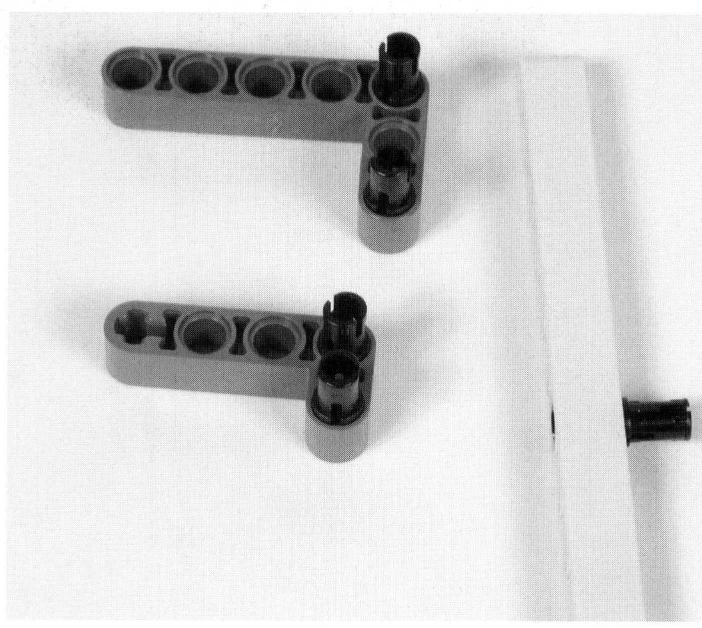

12

13 **Set up two beams with four angled beams.**

14 **Fix the video camera into one side of the holder.**

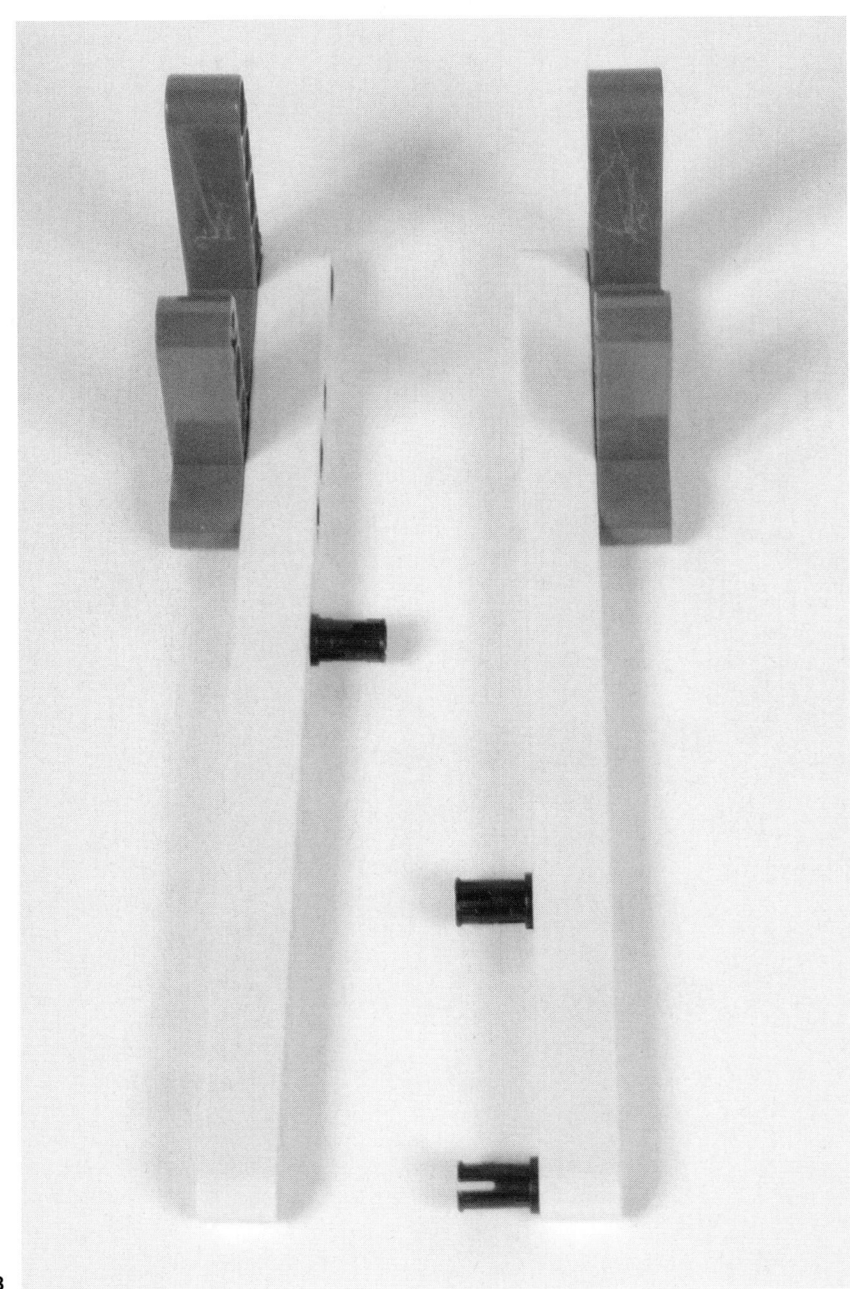

13

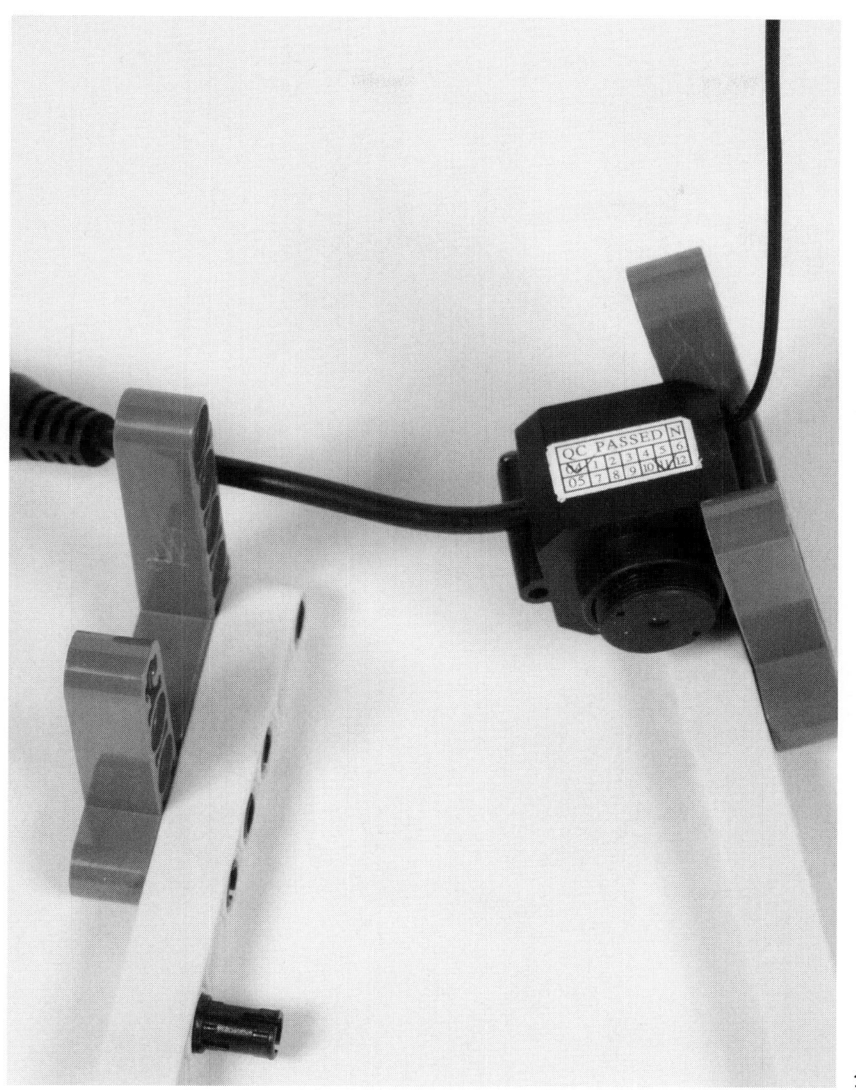

14

15 Snap the other side together.

16 Now fix the video camera assembly to an NXT motor and you can have the NXT Brick control your viewpoint for you.

15

16

SPEAKING IN TONGUES

Well, here is a circuit design that addresses the issue of the NXT speech programs with great aplomb (see Figs. III-3 and III-4). This hack combines the NXT with the nifty ChipCorder® IC from Winbond. Consider this hack to be a little bit of a hack combined with a serious circuit design and all wrapped up in an elegant form factor. In fact, this project could be combined with a lot of other personal electronics, as well. Therefore, there is tremendous latitude for modification of the hack's concepts.

First of all, what's with this ChipCorder thing? ChipCorder is a digital, single-chip tape recorder that can provide high-quality playback without the need for battery backup circuits. This incredible chip requires only a handful of support components to record messages up to 40 seconds (e.g., for the I16xx Series) in duration and then play the message back on a standard 8-ohm speaker. Additionally, Winbond claims that the message has a retention of 100 years, can be overwritten 10,000 times and can cost less than $20 (depending on the exact ChipCorder series device).

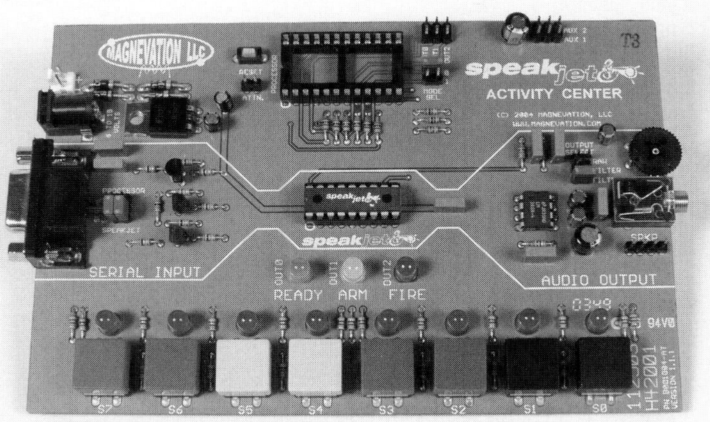

III-3 SpeakJet is a simple-to-use speech synthesizer.

Chances are that you might already have a ChipCorder in your house, right now. Many promotional key chains, picture frames, greeting cards, and stuffed toys that speak high-quality human phrases use ChipCorder devices for playback. Even some warning alarms in automobiles and industrial control centers rely on ChipCorder for signaling a clear and understandable signal. "Hey stupid, STOP" gets a person's attention a lot better than "Beep, beep."

The maker of ChipCorder, Winbond Electronics Corporation America, a wholly owned subsidiary of Winbond Electronics Corporation of Hsinchu, Taiwan, is based in San Jose, California. Started in 1990, the American branch began distributing signal conditioning devices for consumer and industrial markets. With the 1998 acquisition of Storage Devices, Winbond America quickly became a market leader in silicon voice recording and playback IC solutions—namely, ChipCorder.

Additional acquisitions of Bright Micro Electronics and the Thin Film Transistor (TFT) LCD division of Cirrus Logic,® based in Austin, Texas, further entrenched Winbond America's contribution to award-winning voice and speech chip solutions and state-of-the-art TFT LCD-driver ICs. Four new product lines are aimed at bolstering Winbond America's speech products: the WTS70X series, the industry's first single-chip IC solution that converts Text-To-Speech (TTS); the W68xx series, a family of voice CODEC chips aimed at telephony, communications, and consumer applications; the WMS72xx series, a family of 256-tap, nonvolatile, digitally programmable potentiometer ICs aimed at communications, industrial, and consumer applications; and the I16xx ChipCorder series of single-message record/playback ICs.

The I16xx is the first series of ChipCorder devices designed to operate from 2.4 to 5.5 volts. Furthermore, the I16xx series features 6.6 to 40 seconds in record/playback duration, pushbutton operation, LED indicators, nonvolatile message storage, and an integrated speaker driver, which provides both PWM and current-mode speaker outputs. By varying a user-determined external oscillator resistor, the I16xx series (1610, 1612, 1616, and 1620) can be programmed for a 4- to 12-kilohertz sampling frequency. Likewise, this variable sampling frequency also determines the length of recording time. Finally, this is a fully integrated system-on-a-chip with support functions that include: AGC, microphone preamplifier, speaker drivers, oscillator, and memory. All of this in a neat 16-pin DIP—perfect for hacking into an NXT Brick.

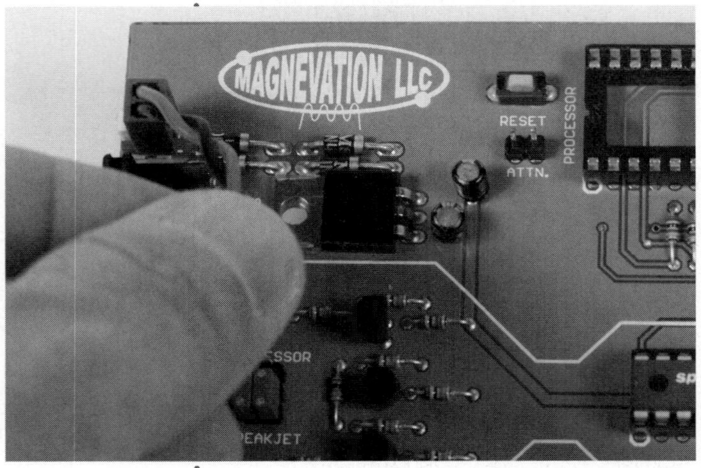

III-4 Using its own power supply, SpeakJet is worthy of hacking into your next NXT project.

In this hack, we are going to turn the NXT Brick into a digital message center. By inserting a ChipCorder I1610 device (with support components) inside a LEGO container, you can instantly record a short message that can be played back on command. If you elect to switch to a different ChipCorder series IC, you can add several of these digital recording devices in series and make a variety of messages that can be individually played back. You are only limited by the space available inside your LEGO design (or, other electronic project).

A HACKER'S GUIDE TO WINBOND CHIPCORDER

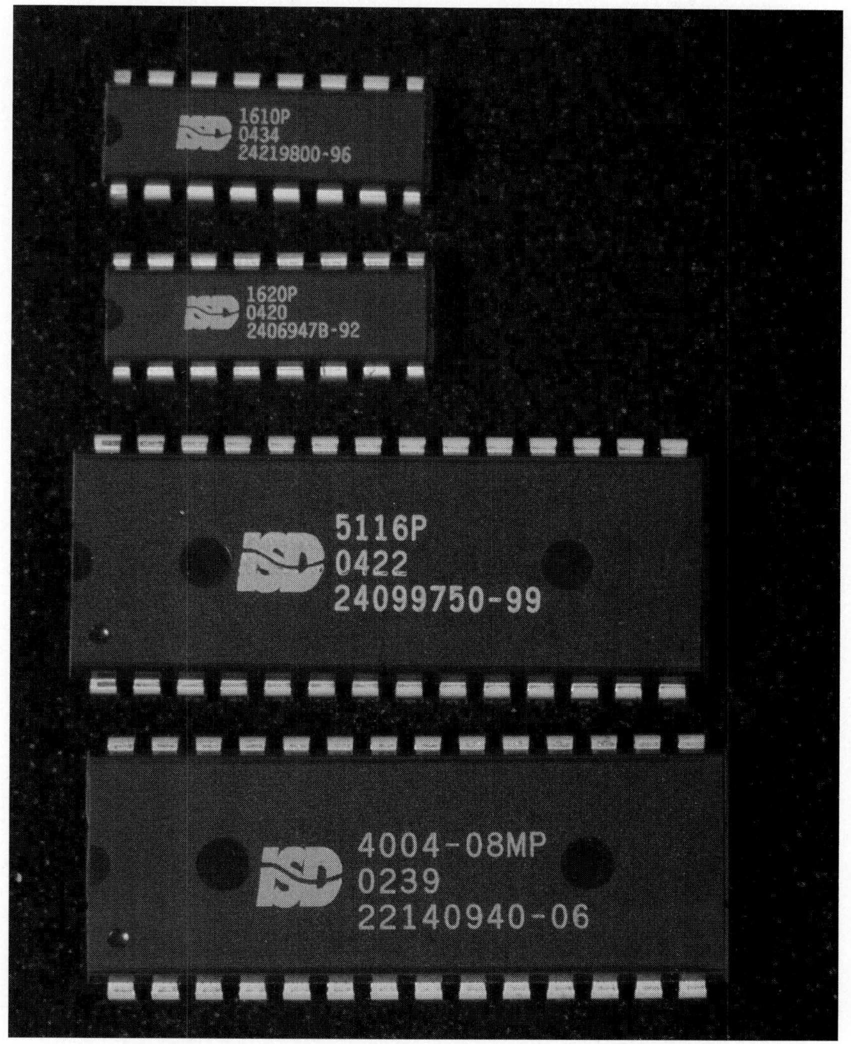

1 The ChipCorder family of digital recording circuits from Winbond.

2 A complete circuit for recording and playing back 16-seconds worth of speech... your own speech.

3 You can easily drop this ChipCorder circuit into any small enclosure.

4 Even this IC carrier could be used for housing the entire family of ChipCorder circuits.

2

3

4

HOW TO HACK A BIONICLE VOICE CHANGER

1 Bionicle Voice Changer. It was a bargain for under $6.

2 This helmet should come in very useful around Halloween time.

3 There is a microphone in the helmet and a speaker inside the sound effects unit.

1

2

3

4 There is a speaker, amplifier, and sound effects unit all tucked away inside the box which hangs from your belt.

5 The helmet's microphone.

6 Ooo, now that's scary. But she sounds great.

7 Open the speaker unit.

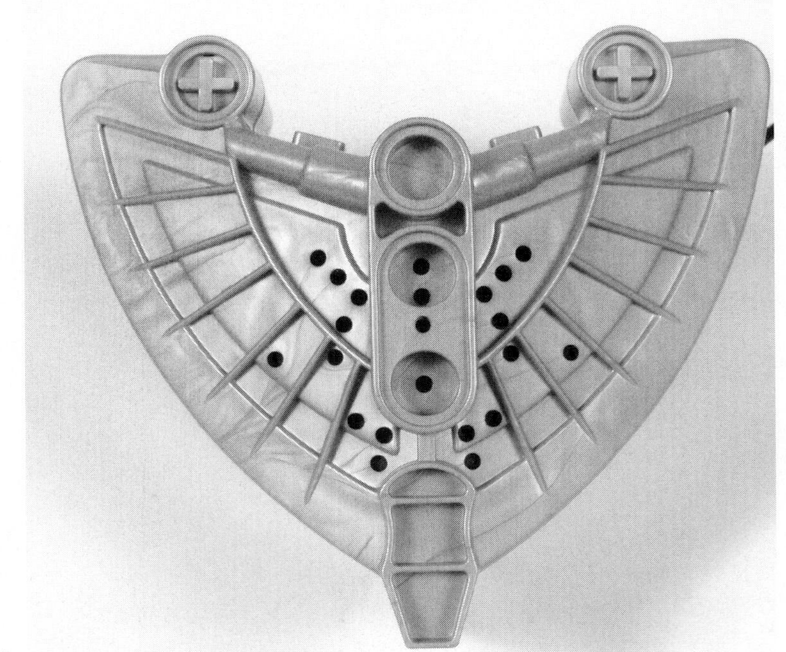

4

5

6

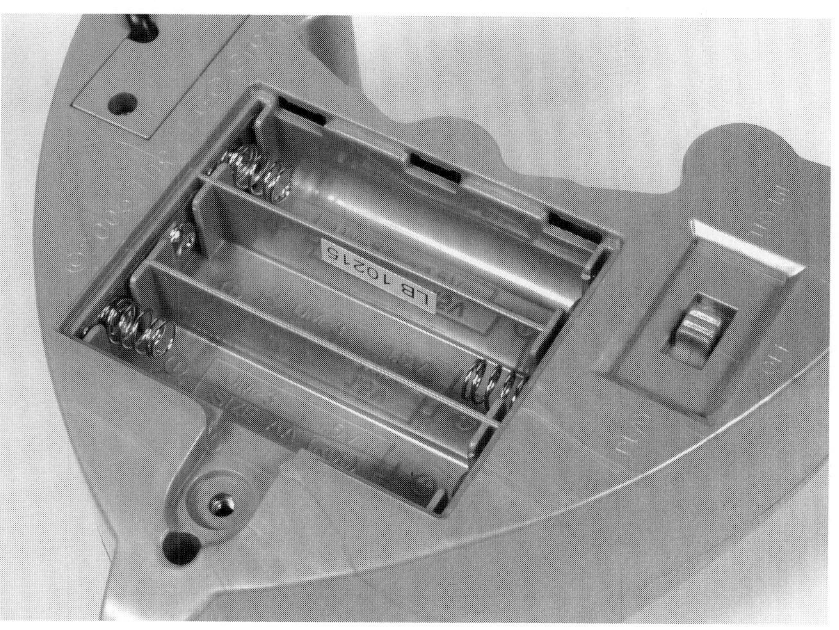

7

8 There are two circuit halves.

9 The main circuit board.

10 The microphone interface.

11 The "canned" speech is programmed into this circuit board.

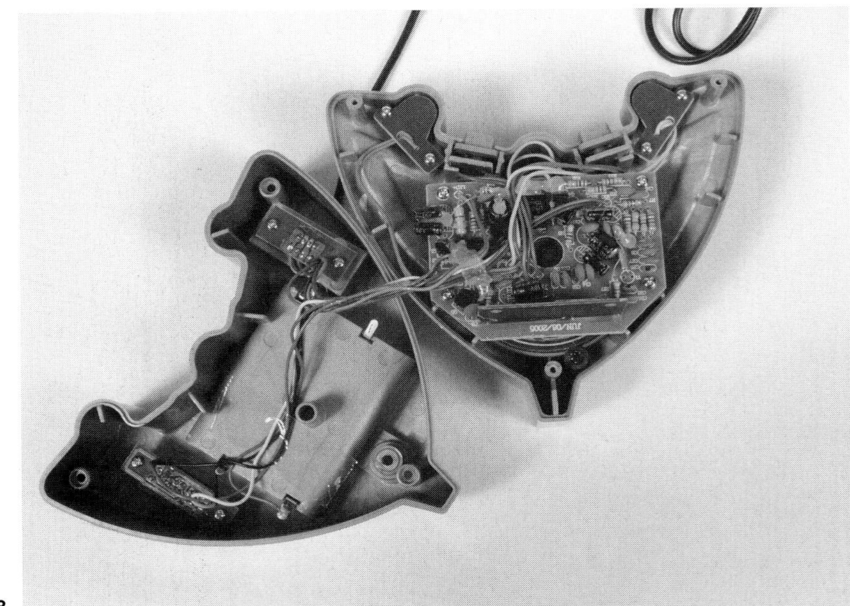

8

9

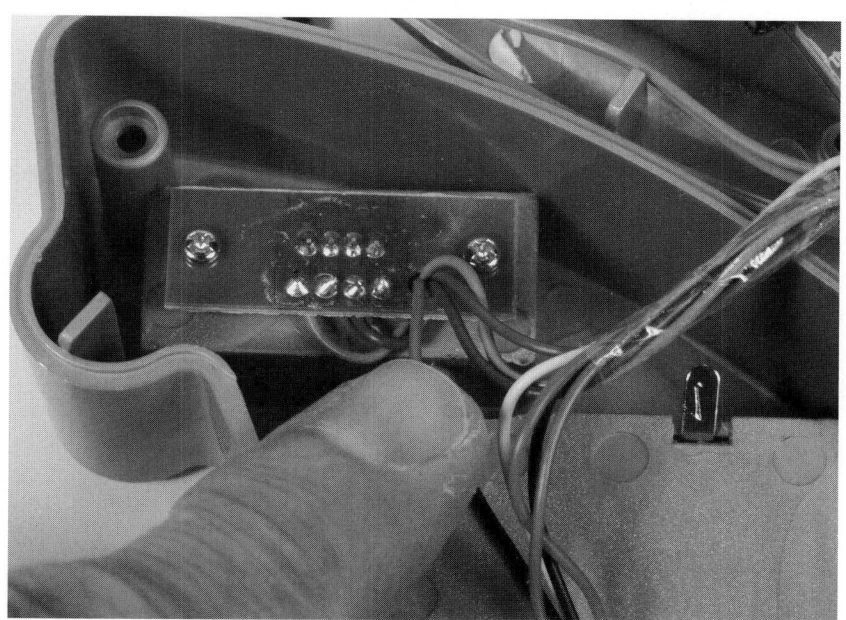

10

11

12 Use your own microphone for "changing" the voice of your NXT Brick.

13 Just attach your microphone to the speaker grill of the NXT Brick and you can route your robot's sounds through the hacked Bionicle Voice Changer.

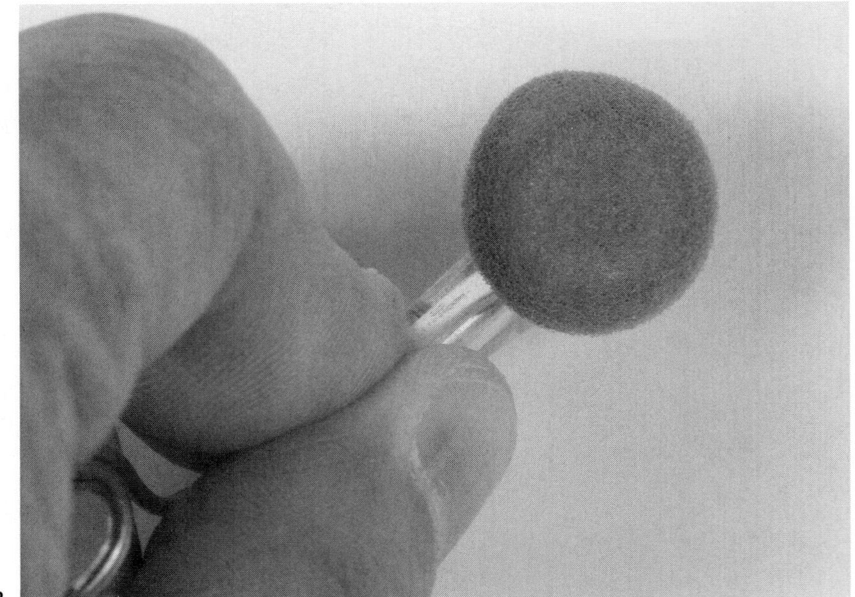

12

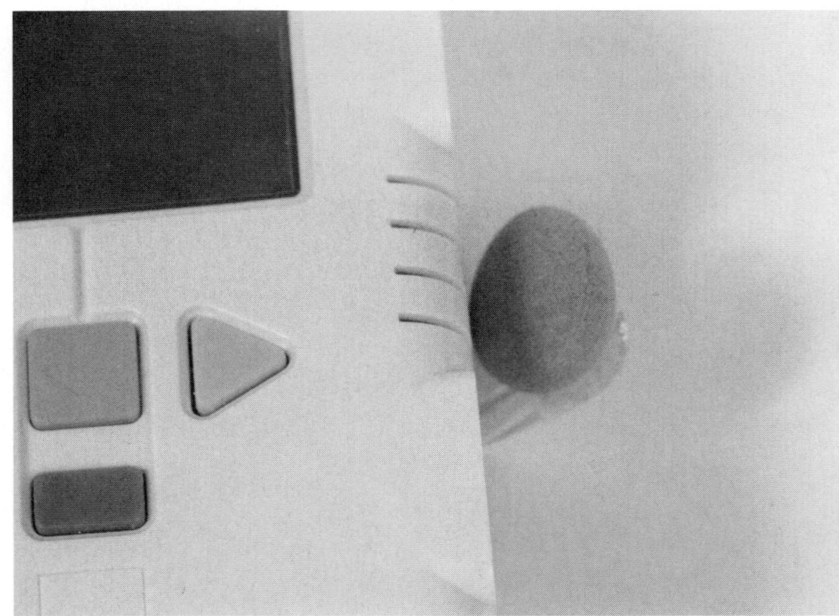

13

THE THING WITH TWO HEADS

Is your NXT creation in need of some extra control? Here, presented in alphabetical order, is my heavily edited "short list" of the best in robot brains.

BASIC STAMP® 2

In 1992, Parallax kick-started the fledging DIY robot movement with the release of the BASIC Stamp® Rev. D. By the end of 1998, Parallax had sold over 125,000 of these BASIC Stamp modules. Today the BASIC Stamp microprocessor line includes eight different types of BASIC Stamp modules with varying capabilities.

One of the most common Stamps is the BASIC Stamp 2. This 24-pin DIP module includes an onboard processor, memory, clock, and interface (via 16 I/O pins). Controlled by a derivative noncomplied BASIC programming language known as Parallax BASIC (PBASIC) which contains 42 commands, the BASIC Stamp 2 is able to monitor and control motors, timers, switches, sensors, relays, and valves via simple access to its I/O lines.

BASICX-24™

Another all-in-one BASIC programming module is from NetMedia. Packaged in a similar 24-pin DIP, the BASICX-24™ houses 21 user-programmable I/O lines, onboard status LEDs, 32 kilobytes EEPROM, 400 bytes of RAM, and a serial/parallel programming interface.

One strong plus for the BASICX-24 module is the presence of a robust BASIC programming language implementation. The BASICX language features 80+ commands, including 7 valuable string functions. This depth in programming capability enables the development of powerful applications for enabling data logging, accessing ultrasonic range finder measurements via a Polaroid® sonic sensor, and controlling more powerful RC servos. Also, its performance specs make the BASICX-24 a screamer—able to execute 65,000 BASIC instructions per second.

BRAINSTEM® GP 1.0

For about ten years there was only one name in home-brew robotics, Acroname. Since 1994, this small Boulder, Colorado, company has worked very hard at supplying quality, hard-to-find parts at affordable prices to robot builders, hobbyists, and experimenters. Their most popular microcontroller is the epitome of this working-class ethic.

The BrainStem® General Purpose module (GP 1.0) supports five 10-bit A/D inputs, five flexible digital outputs (i.e., for a combined 10 I/O lines), a Sharp GP2D02 driver port, an IIC bus, and four high-resolution servo outputs. Measuring a scant 2 1/2 inches square, the BrainStem GP 1.0 module is a good candidate for fitting comfortably inside an NXT bot.

HANDY CRICKET

The Handy Cricket Version 1.1 is a low-cost module based on the Microchip PIC® microprocessor featuring a built-in Logo interpreter. Equipped with two motor ports, two sensor ports, two bus ports, 4 kilobytes of static memory, and a piezo speaker, the Handy Cricket connects with the host PC via a serial port IR Interface Cricket. This IR interface is just like the IR Tower that most of you are familiar with from the LEGO® RIS 2.0. One of the exciting attributes of the Handy Cricket, however, is the action that contributed to its name.

A unique IR transmitter/receiver circuit built into the Handy Cricket enables communication between two or more Handy Crickets. Get it? Like the chirping of a cricket, the Handy Cricket is able to "chirp" IR signals at a 50-kilo data rate between each other. Oh, and just like a biological cricket, the Handy Cricket is a tiny sucker. The overall dimensions are just a bit under 2 1/2 inches per side.

Sure all of this hardware stuff is exciting, but the part of the Handy Cricket that leaves me salivating is the fantastic implementation of the Logo programming language, called "Cricket Logo," that is built into the Handy Cricket.

TINIPOD™

OK, you want smaller, then the TiniPod™ from New Micros of Dallas, Texas, can be inserted into any LEGO project with room to spare. But this type of manufacturing achievement is nothing new to New Micros. Although founded over 20 years ago, in 1987, New Micros set the trend for today's robot all-in-one microprocessor modules with a stand-alone, single-chip computer with a built-in programming language. The release of the slightly more than 1-inch square TiniPod marks a pinnacle in robot controller development.

Like its contemporaries, TiniPod sports 16 digital I/O lines that can drive servos and timers, a DSP56F803 MPU, 16-bit processor, 32-kilobyte by 16-bit Program Flash EEPROM, and 2-kilobyte by 16-bit RAM for data. Unlike other controller modules, TiniPod includes a high-level language known as IsoMax.™

No matter which high-level controller module you select for transplantation into your next NXT project, just make sure that you do your hack with some style. And if you're lacking in any inspirational style, then please consult *Katherine's Design Fun House* later in this book.

1 The Parallax BASIC Stamp development board.

2 Parallax BASIC Stamp 2.

3 NetMedia BASICX-24.

4 Acroname BrainStem General Purpose module.

A HACKER'S GUIDE TO SUPPLEMENTAL BRAINS

1

2

BASICX DEVELOPMENT STATION © 2000 NETMEDIA INC.
RESET
PROG
BX-24
BX-35
MADE IN TAIWAN R.O.C.
www.basicx.com
3

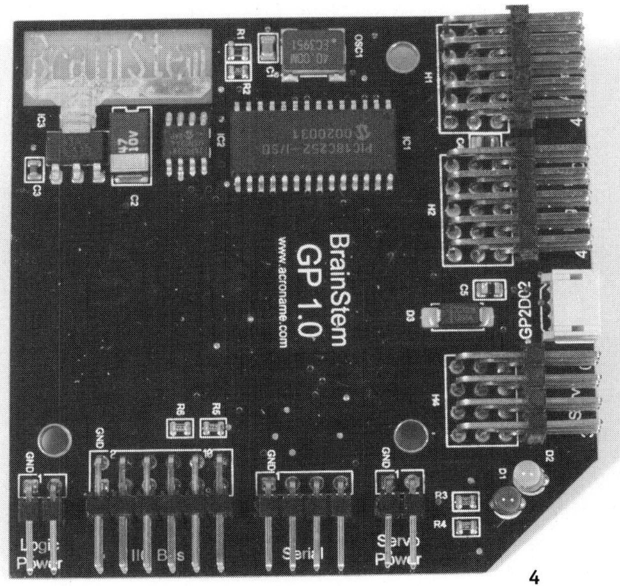
BrainStem
BrainStem
GP 1.0
www.acroname.com
Logic Power
Serial
Servo Power
GP2D02
4

5 Handy Cricket.

6 Handy Cricket with IR transmitter/receiver circuit and its IR programming interface.

7 Expansion bus for Handy Cricket.

8 An optional display for Handy Cricket.

5

6

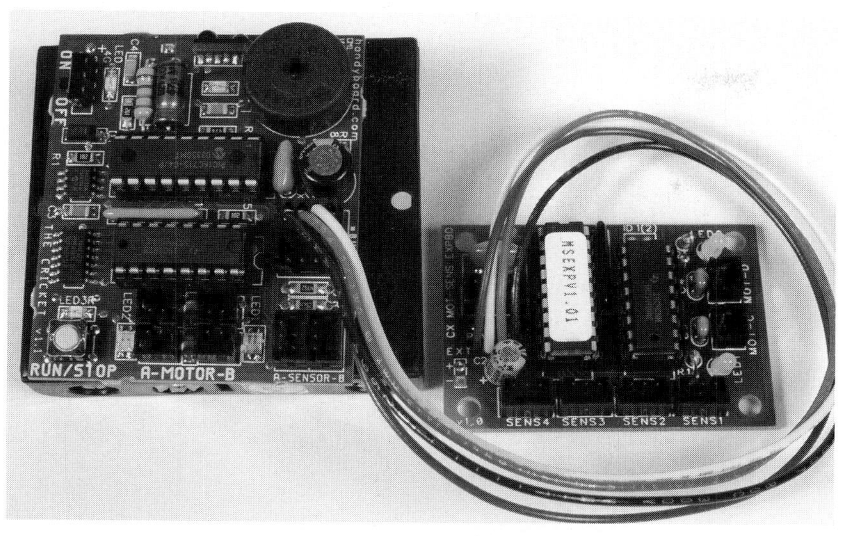
RUN/STOP
A-MOTOR-B
A-SENSOR-B
THE CRICKET
handyboard.com
SENS4
SENS3
SENS2
SENS1
MSEXPV1.01

7

RUN/STOP
A-MOTOR-B
A-SENSOR-B
LED3R
LDQ-N516RI
B03 5 26 15 P

8

9 Hmm, this looks familiar, the battery box is located on the underside the Handy Cricket.

10 Twin sons, separated at birth. The LEGO RCX (left) and the Handy Cricket (right).

11 OK, start humming the *2001: A Space Odyssey* theme song. The LEGO RIS IR tower (left) and the Handy Cricket IR programming interface (right).

12 It's all in this family...NXT (left), Handy Cricket (center), and RCX (right).

9

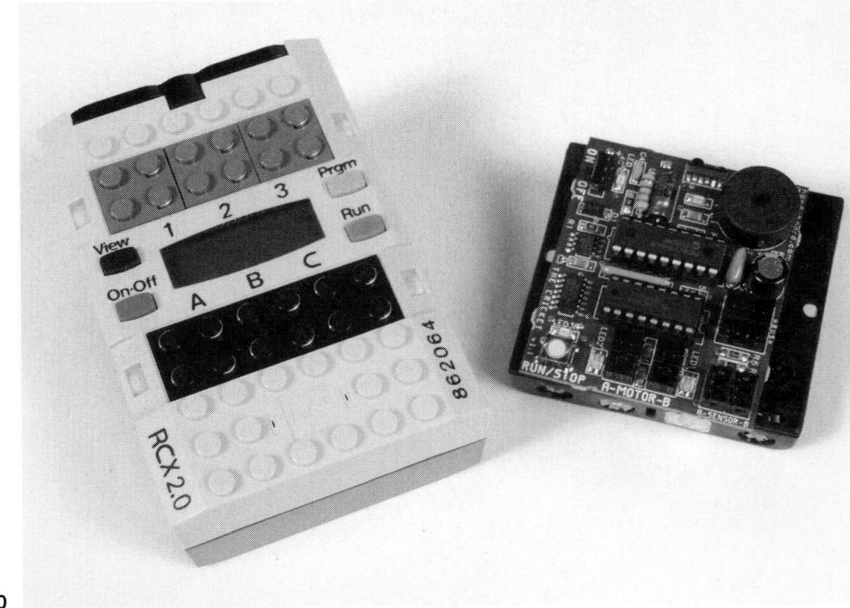

10

11

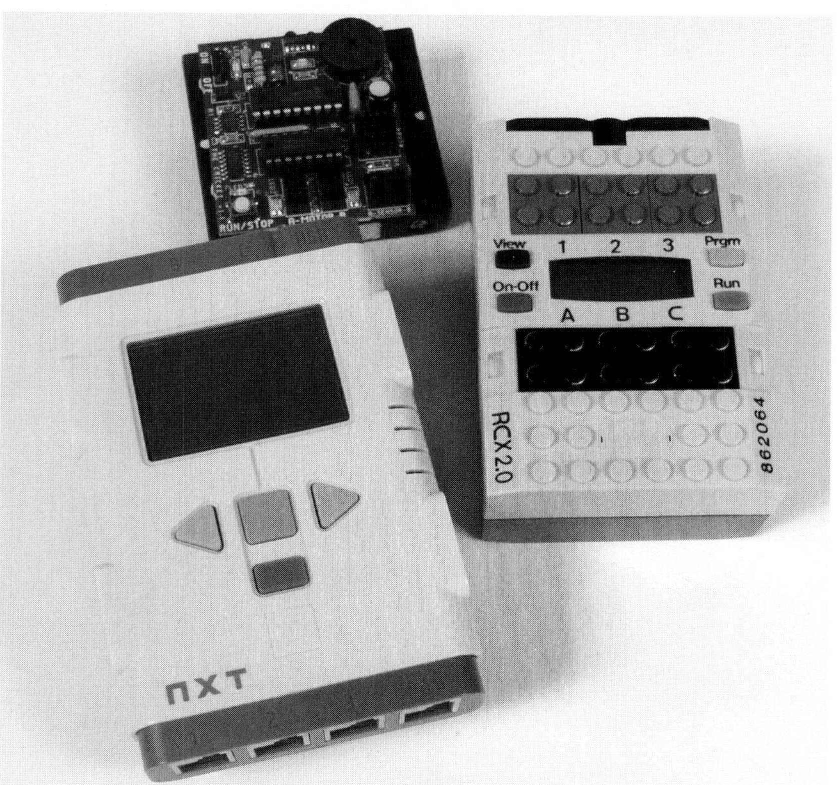

12

13 New Micros TiniPod.

14 Programming interface for TiniPod.

15 Unlike most other programming boards, the TiniPod module stands upright.

16 One of the best bargains in today's computer/robot brain maketplace...

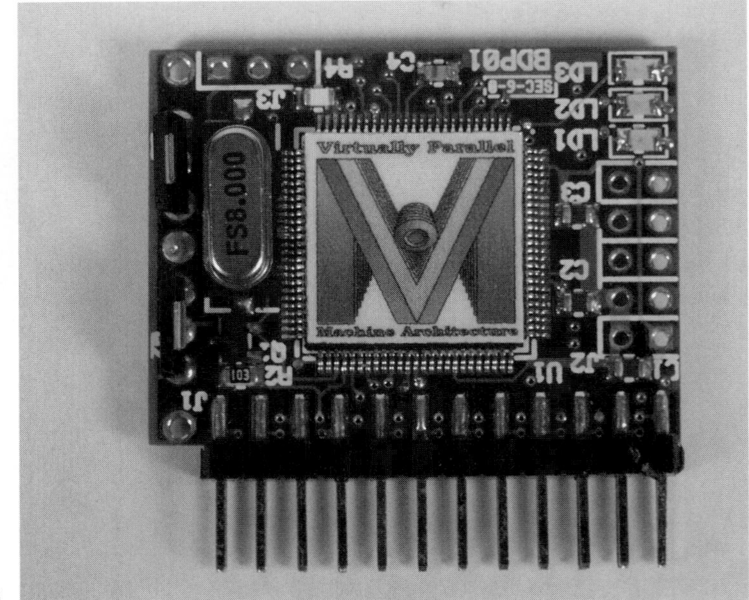

13

14

15

16

17 That is, if you can pry it out of this guy's hands.

18 Stay in the Atmel microcontroller fold with the Atmel Butterfly.

17

18

WELCOME TO THE ZIGBEE® FOLLIES

Believe it or not, there is a ZigBee Alliance. This global association of hardware manufacturers and technology providers is headstrong on creating yet another wireless "solution" for use in the home, commercial, and industrial venues.

Regardless of what you may think about ZigBee, this new technology is an interesting low-cost, low-power wireless standard that is gaining international acceptance for remote sensing.

If you're looking for a way to dip your toes into the ZigBee pool, Cirronet™ Inc. is now offering the ZigBee Basic Developer Kit. This low-priced development tool enables network designers to rapidly evaluate the ZigBee wireless standard within their specific application environment.

The ZMN24HPDK-B (Basic) Developer Kit includes one RS-232 wireless sensor modem and one 2.4-gigahertz sensor interface board, both equipped with the Cirronet's 2.4-gigahertz high-RF-power (100 microwatt) ZigBee OEM module, plus power adapters, cables, antennas, batteries, and demo software. You can use the kit to test range, speed, and operation of a sample application within your real-world application environment.

ZigBee Basic Developer Kit is now specially priced at an introductory offer of $199. Another ZigBee developer's kit is available from Freescale Semiconductor.

The Freescale Semiconductor Developer's Starter Kit (DSK) is great for experimenting with 802.15.4 ZigBee wireless technology. Consisting of two reference design boards equipped with accelerometer sensors, MC13192 2.4-gigahertz RF transceiver data modems, and MC9S08GT60 microcontrollers, the DSK includes all of the software, documentation, and even a trial version of Metrowerks CodeWarriorTM IDE that is needed for building and testing a ZigBee design. This kit is reasonably priced at under $200.

The DSK's major components include:

- MMA6200Q Series XY-Axis 1.5g Accelerometer. Able to measure small exertion from tilt, motion, positioning, shock, or vibration forces.
- MMA1260D Z-Axis 1.5g Accelerometer. A small package 1200 mV/g sensitivity along the z-axis.

- MC13192 2.4-gigahertz RF Data Modem. This packet modem is compliant with IEEE 802.15.4.
- MC9S08GT60 Microcontroller. A low-power, low-voltage 8-bit MCU.

Once you've got the transmitter attached to your NXT bot, your receiver/PC combo will be able to read four types of data; and, get this, these are really cool sensor readings for such a small footprint:

- **RAW SENSOR DATA**. The onboard MCU handles all of the big number crunching, but you can view voltage output, 8-bit A/D data, and axial g force (units of gravity).
- **ANGLE DATA**. Displays the x, y, and z orientation of the transmitter.
- **TILT DATA**. Just like that pinball game you could never beat, this reading represents the gravitational force (g) in each axis that is at an angle to the direction of rotation.
- **ORIENTATION DATA**. Turn the transmitter upside down and BINGO, the time of this event is recorded by the receiver.

What can you do with either of these developer kits? For starters, after you load some demo software, you will have a rudimentary data logger or HOBO. Now just drive your NXT robot to a desired sensing location and use your ZigBee system for recording real-world data. A veritable lobo HOBO for your LEGO bobo.

HOW TO HACK A HOBO TO YOUR NEXT ROBO PROJECT

1 The Freescale Semiconductor Developer's Starter Kit (DSK).

2 Everything you need to do some remote sensing. The DSK is equipped with two reference design boards with accelerometer sensors, MC13192 2.4-gigahertz RF transceiver data modems, and MC9S08GT60 microcontrollers.

3 This board becomes the remote sensing transmitter.

4 It doesn't matter which board is used for the transmitter and which one is used for the receiver.

1

2

J106
J102
J103
S106
ON
OFF
S105
J101
S104
S103
S102
S101

3

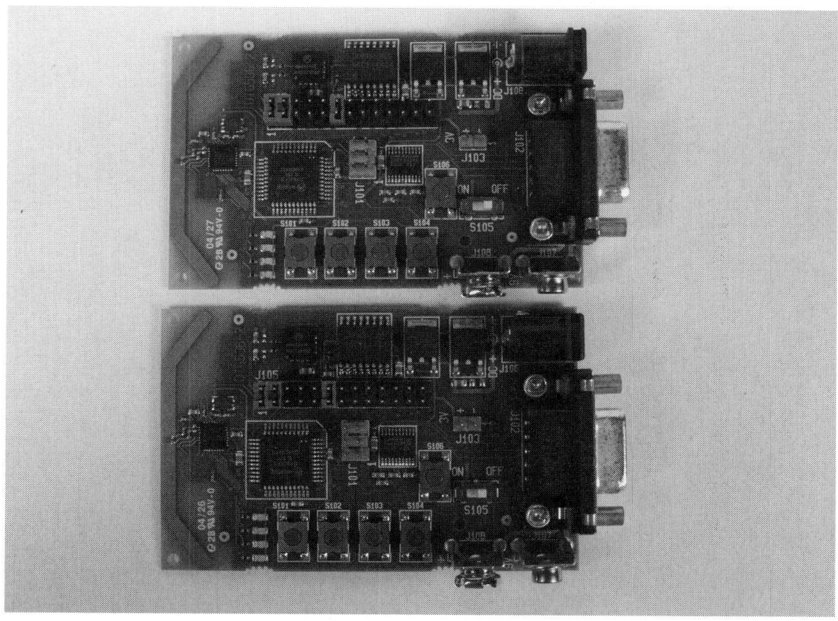
J103
ON
OFF
S105
J105

4

5 Only a little hacking is necessary for adding a remote sensing ZigBee device to an NXT project.

6 The easiest hack does not require any major league soldering, either. You can simply attach the DSK transmitter to your mobile NXT platform and record angle, tilt, and orientation data on your computer.

7 The simple demonstration software package supplied in the DSK is perfect for recording your pseudo-HOBO data from your NXT-powered robo platform.

5

6

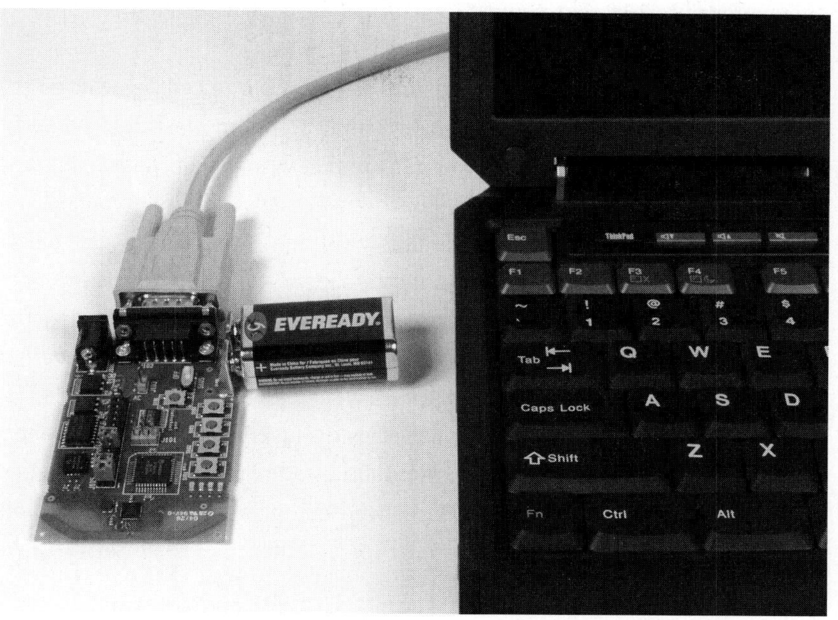
EVEREADY
Esc
ThinkPad
F1
F2
F3
F4
F5
Tab
Q
W
E
Caps Lock
A
S
D
Shift
Z
X
Fn
Ctrl
Alt

7

EL NXT

Are you ready to trick-out your NXT creatures? Then EL wire is a great product for modding any LEGO project.

Formerly known as electroluminescent wiring, EL wire can be attached to virtually anything. Wrap it around fabric, metal, wood, or plastic, EL wire will not short, overheat, or discharge when in contact with any of these materials.

Housed inside a plastic sheath, EL wire is actually a flexible copper wire that is coated with phosphorous. When low voltage is applied to this combination, the phosphorous glows. In order to supply the voltage to the copper wire, two transmitter wires are wrapped around the length of the EL wire. The external plastic sheath serves as a container for holding everything together and making EL wire a spiffy product that anyone can use for any purpose.

Powering the EL wire is accomplished by an inverter. The inverter generates a frequency which excites the phosphorous and makes a distinct colored glow at a specific intensity level. Most inverters operate in the 9- to 12-volt range and typically require an AC power source. There are some terrific low-power alternatives, however.

These lower voltage EL wires can accomplish their distinctive glow with as little amount of power as two AA-sized batteries. This low-power option comes at a price—reduced length. While their bigger AC brethren can drive 10 to 120 feet of EL wire, the pint-sized battery variety is limited to less than 3 feet.

There are some exciting "tricks of the trade" for making EL wire your own custom expression. Before you attempt to "customize" your EL wire, let me say two things. First, most EL wiring mods can be done with stock wire. You don't have to do any splicing or soldering, just plug your batteries in and create, bend, twist, and shape the wire. Second, this type of customizing work requires some special tools (e.g., wire cutters, wire stripper, and soldering iron) and a basic understanding of electrical circuits. If you feel uncomfortable with these requirements or don't have the necessary specialized talents, just use your EL wire "straight."

Most EL wire is sold in single-color lengths or spools. While this "monoglow" is great for a long, single-color run of wiring, you might want to mix your own palette. In this case, all you have to do is splice various colors of EL wire together. Just cut the EL wire, strip the ends, solder the two wires together, and seal the joint with some epoxy sealant.

Another creative trick with EL wire is to splice a length of regular non-EL copper wire to the inverter as an extender. This extender won't glow, but it will help in adding some extra space or separation between the inverter and the EL wire.

So the next time that you're glowing with pride over your latest NXT creation, your NXT bot can glow along with you.

A HACKER'S GUIDE TO EL AND LED LIGHTING

1 Electro-luminescent lighting is a great mod for any NXT project.

2 Most inexpensive EL wiring kits come with a self-contained, battery-powered inverter for powering the wiring.

3 This test "mockup" of the Hearst Headquarters design challenge, featured later in this book, worked wonderfully with EL lighting.

4 LEGO bricks can be modified to hold flashing LEDs.

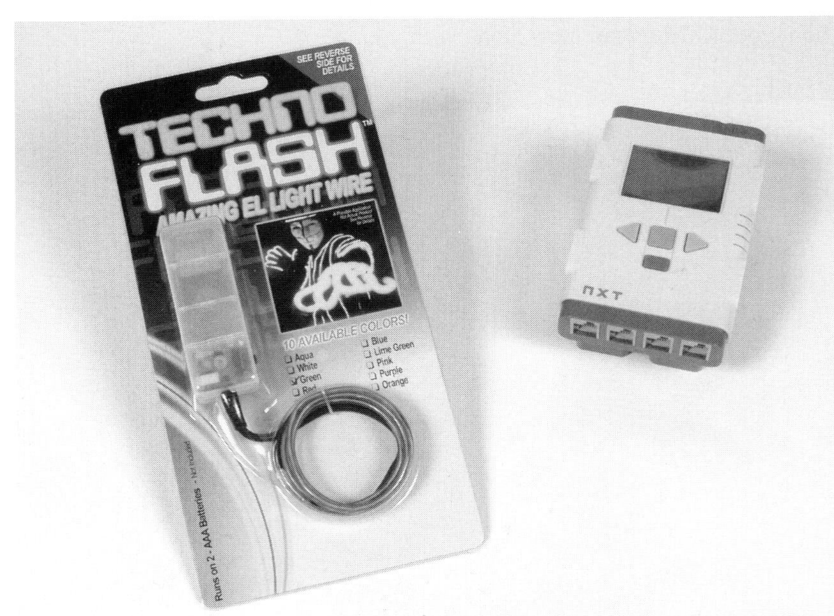

1

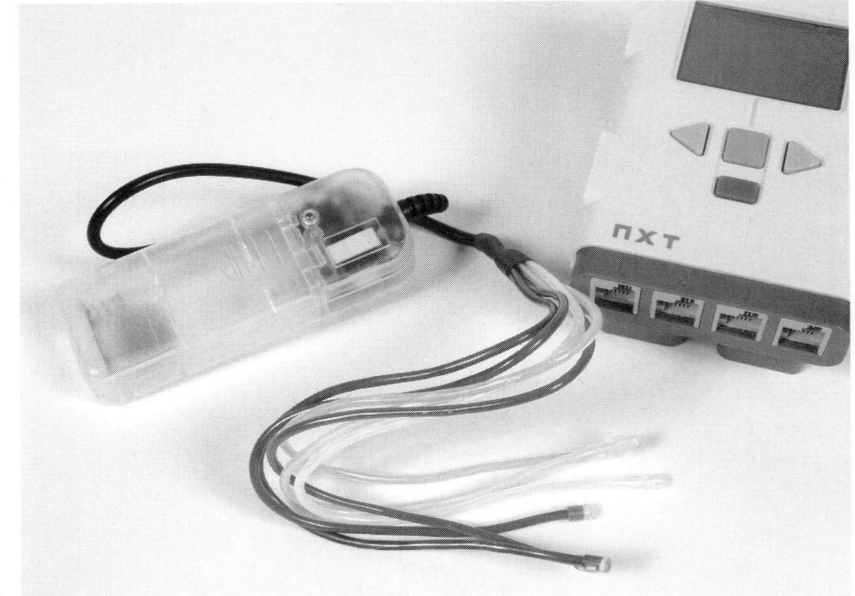

2

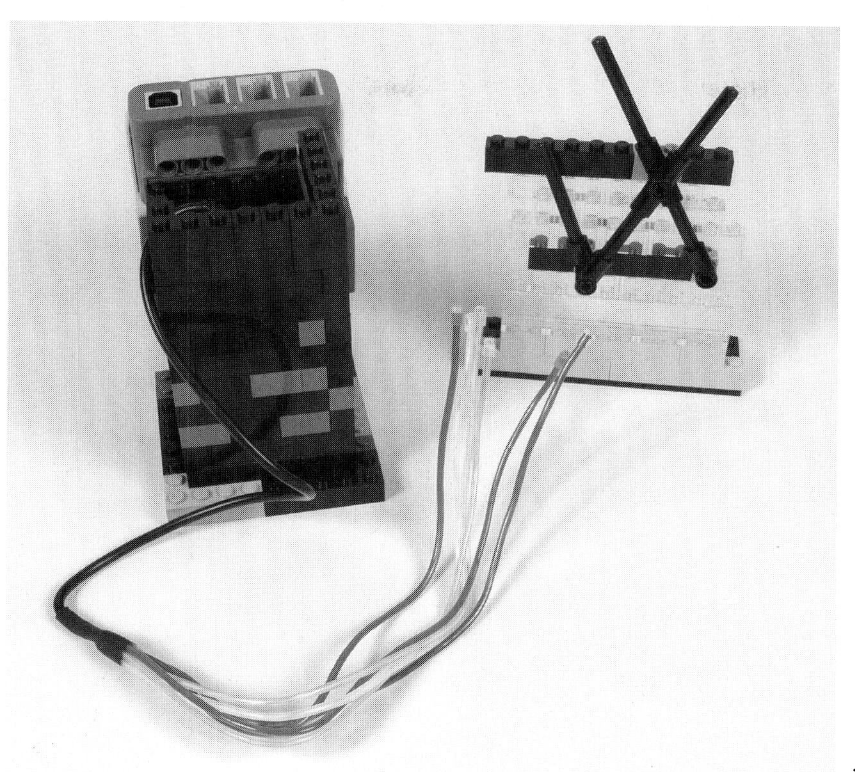

3

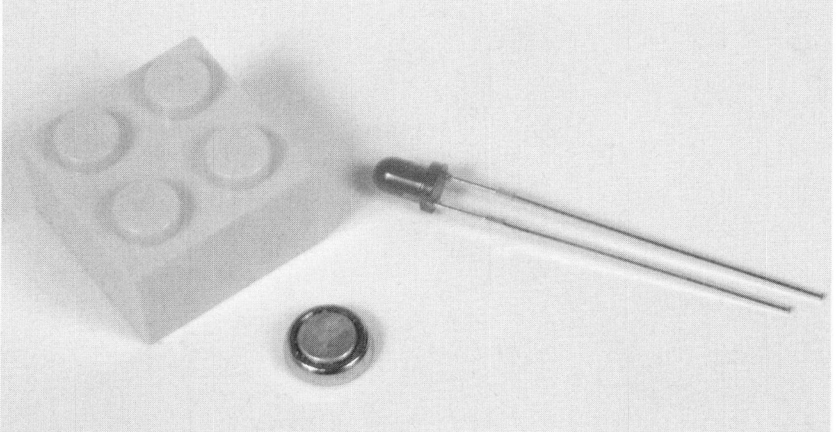

4

5 Very few parts are needed for hacking this lighting effect.

6 Use a hand drill for making a hole in the LEGO brick.

7 Only a couple of twists from a hand drill are needed for punching a hole through a LEGO brick.

8 Clean up the hole with a reamer.

5

6

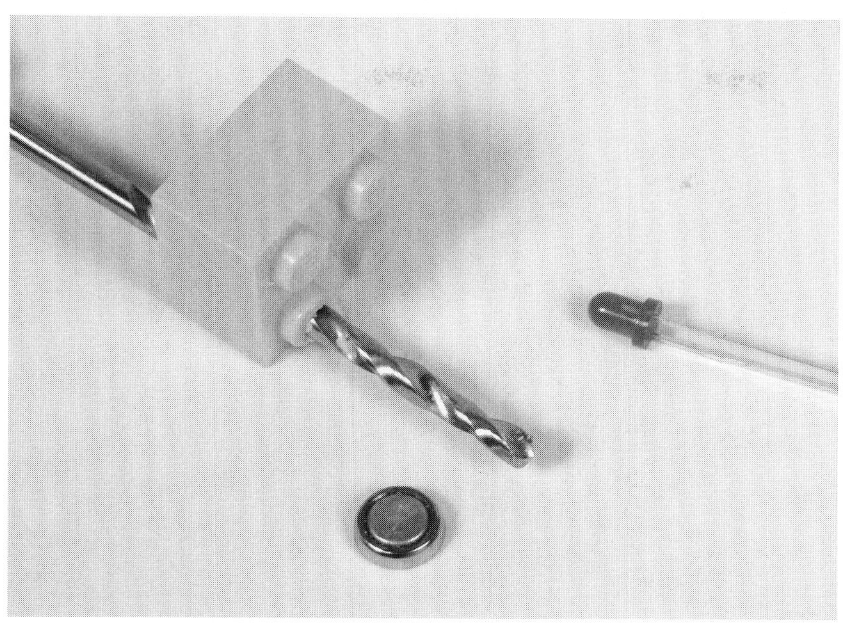
7

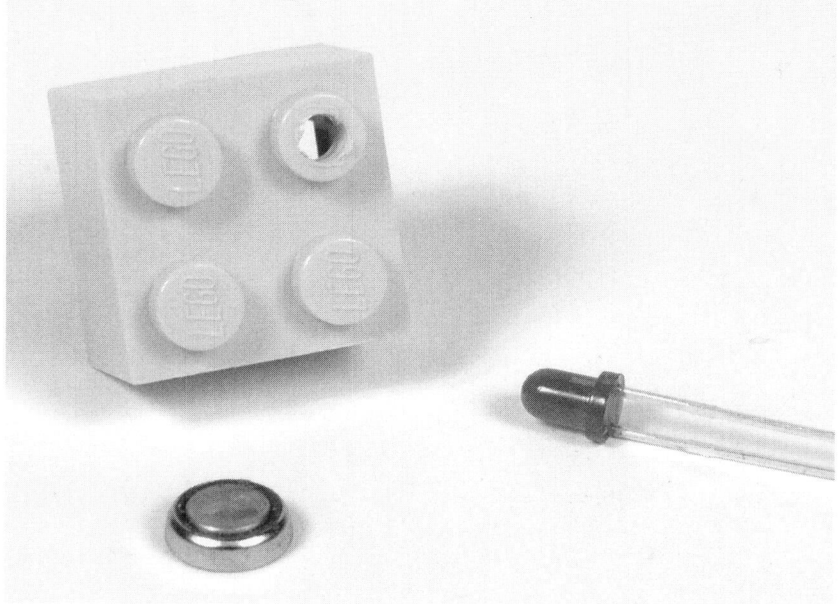

8

9 Add a small battery.

10 Ready for installation inside any LEGO project.

9

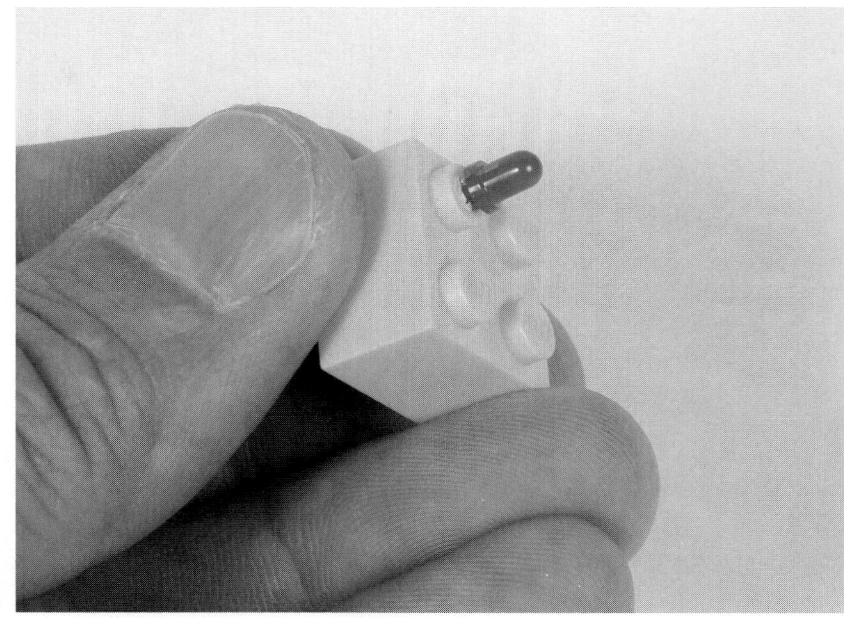

10

FROM FEET MADE OF TREAD

Inside the LEGO RIS robot design kit there was a terrific pair of "shoes" that would enable any robot to go anywhere. This all-terrain footwear was actually a pair of tank treads. Sadly, the NXT robot design kit is missing this pair of very adaptive "feet." One way to get around this omission is to hack your own set of tank treads.

Armor kits, armored fighting vehicle (AFV) kits, or just plain tank kits come in many different sizes (called scales), nationalities, and manufacturers. When hunting for a kit that can be used for providing a tracked chassis, keep in mind that most model tank kits are not motorized. This type of kit is for scale model builders who strive for accuracy over mobility. In my hacking of various motorized AFV model kits into viable robots (with varying degrees of success, mind you) I have come up with this short list of manufacturers who produce the best motorized tank kits:

Academy Hobby—Korean kits; best for general robot hacking; 1/48th-scale remote control mini tank series.

- Merkava MBT
- T-72 Russian Army MBT
- Challenger British MBT
- Leopard 2 A5 MBT
- M1A2 Abrams MBT
- Leclerc French Army MBT

Micro-X-Tech—Hong Kong kits; tiny, detailed kits requiring advanced hacking skills; 1/72nd-scale radio-controlled series (these kits can be difficult to find; you could consider the Hobbico 1/35th-scale radio-controlled combat armor tank set as an alternative—see Tower Hobbies for ordering these kits).

- M1A1 Abrams (different radio frequencies)

Tamiya—Japanese kits; outstanding quality for high-performance robot hacking; 1/35th-scale radio-controlled series (1/16th-scale series of AFV model kits are much bigger and much more expensive).

- Leopard 2 A5 Main Battle Tank
- King Tiger
- Tiger I

The armor kits from Academy Hobby represent one of the best bargains in robot building. Two motors complete with gearboxes are mounted in a neat, self-contained chassis. Even if you don't intend to build a tracked robot, this motor/gearbox combination can be used in a variety of other projects. Finally, this manufacturer's tank models do not require any cement for assembling. This important construction element is extremely important when you go to add, remove, or change batteries inside each tank chassis.

HOW TO HACK A PAIR OF TANK SHOES FOR YOUR NEXT NXT PROJECT

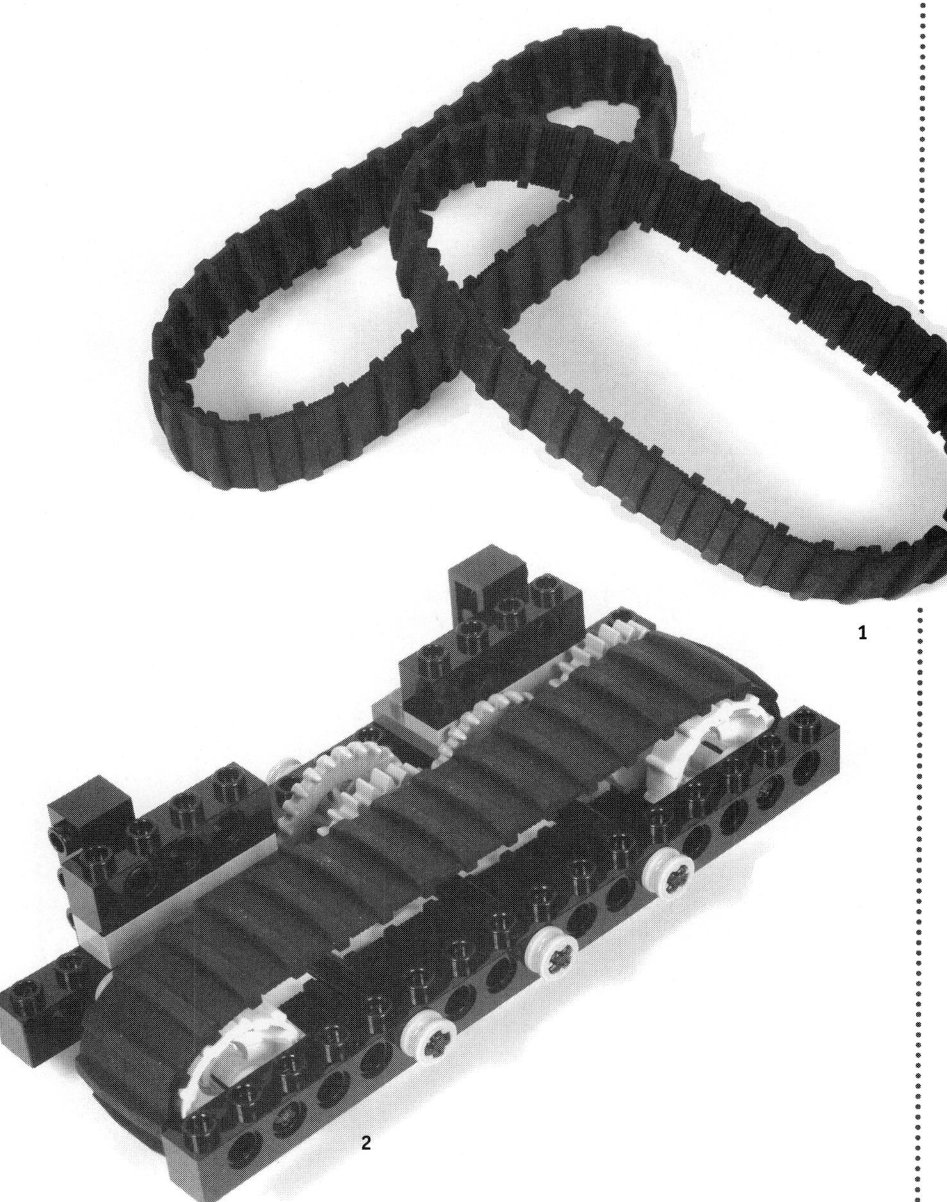

1 Remember these? Yup, treads are missing from the new LEGO MINDSTORMS NXT robot design kit.

2 One of the favorite mobility systems in the RIS was the treaded bot.

3 Tracked systems with RIS were easy to build and simple to power.

4 I discovered a better system for adding treads to your NXT kit during the research of an article that I wrote for *MAKE* magazine. Titled, "Panzeroids," this article showed how to build a robot from an inexpensive battery-powered tank kit.

5 These tanks are fun to build and drive around.

6 A sophisticated gearbox is pre-installed in these tank kits.

3

4

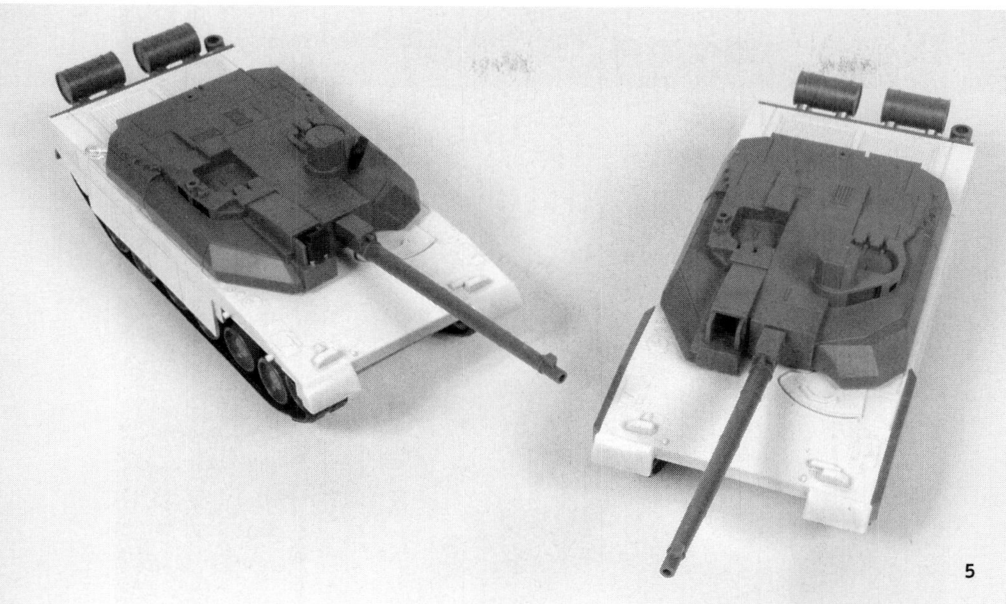

5

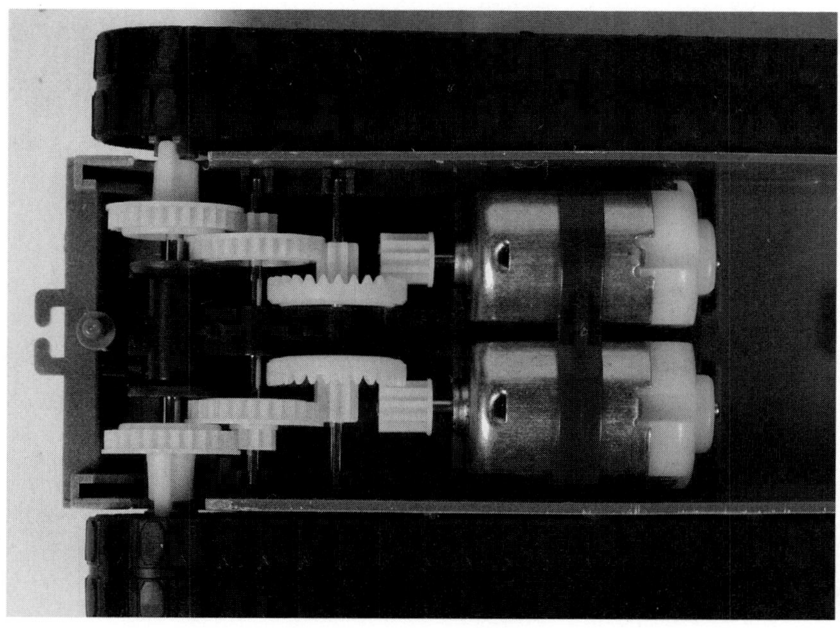

6

7 Just attach your own wiring to this gearbox and you can drive it with your NXT Brick.

8 Each tank kit comes with a controller. This controller will supply the wiring needed for connecting the tank's gearbox to the NXT Brick.

9 Take the controller apart.

10 Disconnect the wiring harness from the controller.

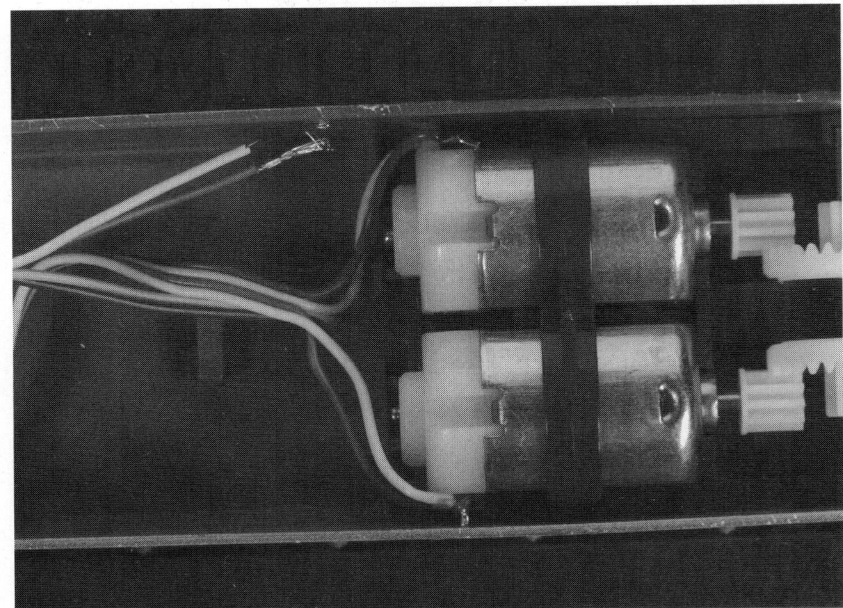

7

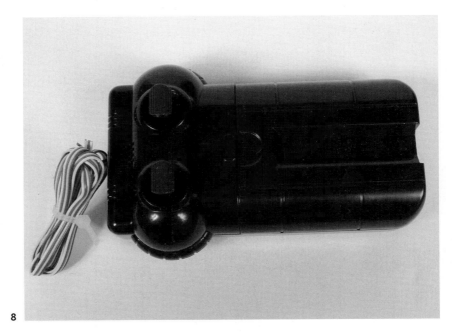

8

9

10

11 Attaching a treaded movement system to the NXT Brick is clumsy, at best.

12 Now you're talking. The NXT Brick can be easily attached to the tank kit's chassis with some ERECTOR set beams and fastners.

13 This bot is "good-to-go." Cabling has been removed for clarity.

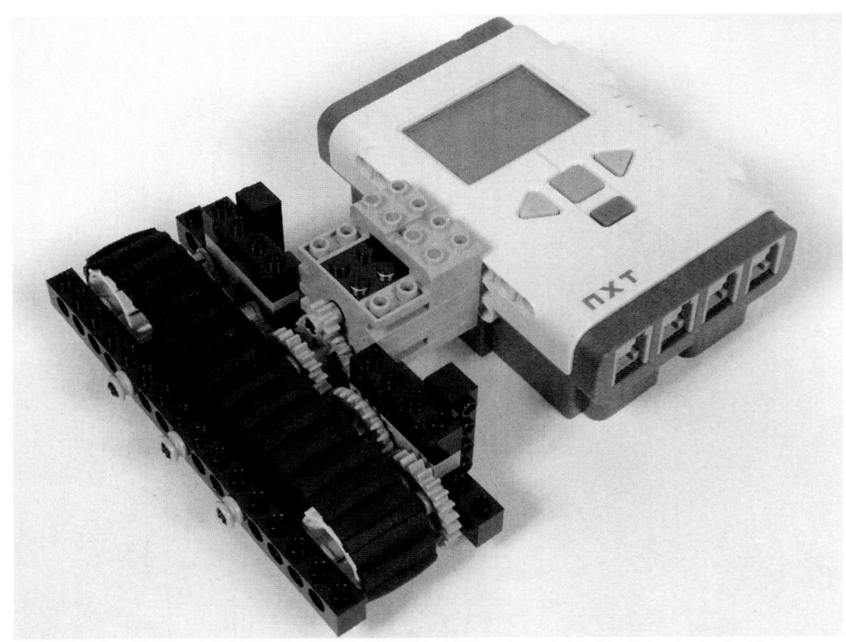

11

12

13

PUMP ME UP

Pneumatic pumps can be a great element for adding some life to an otherwise staid project. Over the years, pneumatic pumps and pistons have come and gone through the LEGO line of commercial kits.

Once again heaving up for hackers, the TECHNIC-flavored Mobile Crane (8421; although across the "pond" in the United Kingdom this kit is marketed as "Crane Truck") is the latest incarnation for the LEGO pneumatic pump/piston combination. In its own right, this kit is impressive (see Fig. III-5).

Although it's an eight-wheeled wonder, this truck features six-wheel steering. Furthermore, there are four deployable outriggers, so that you can recreate

your own Terminator chase scene between T-X and the Governor of California from *Terminator 3: Rise of the Machines*. Be prepared, though, your theatrical recreation will be big, real big—the Mobile Crane measures over 18 inches long.

In addition to the motor for extending the crane's telescoping boom arm and hook, nestled away underneath a lot of TECHNIC beams, is the real jewel from this kit. Inside this massive buildup of plastic beams is a pneumatic cylinder (along with an associated pump) that is used for lifting up the crane's arm. You can either elect to extract this pneumatic system for your own NXT projects or use the Mobile Crane kit for chasing ARNOLD all over the streets of Hollywood.

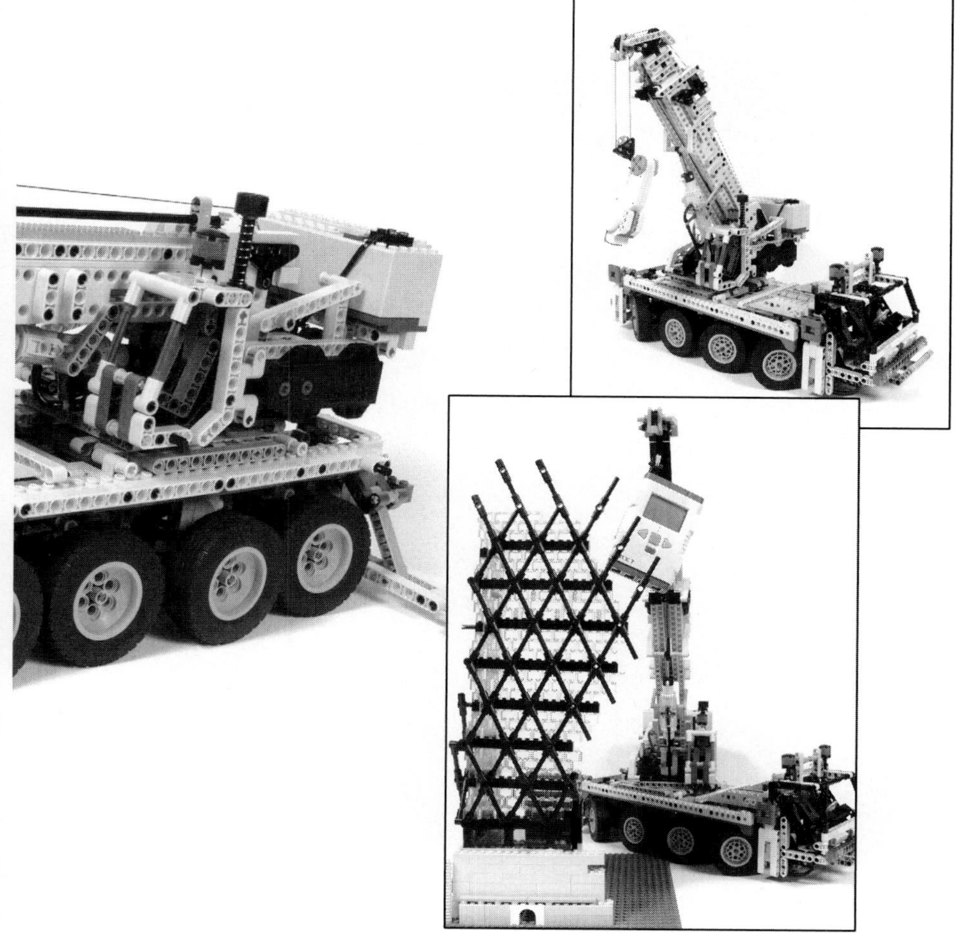

III-5 The LEGO Mobile Crane (8421) is a great source for two pneumatic pistons.

HOW TO HACK A PNEUMATIC PISTON PROJECT

1 You can find a couple of useable pneumatic pistons inside the LEGO Mobile Crane (8421) kit.

2 Use these pistons with care, they don't hold pressure very well. Of course, how many of us can?

3 A complete pneumatic system.

4 You will have to work at designing a suitable stand for holding your pump which, in turn, will power the pistons. This model is for demonstration purposes, only.

1

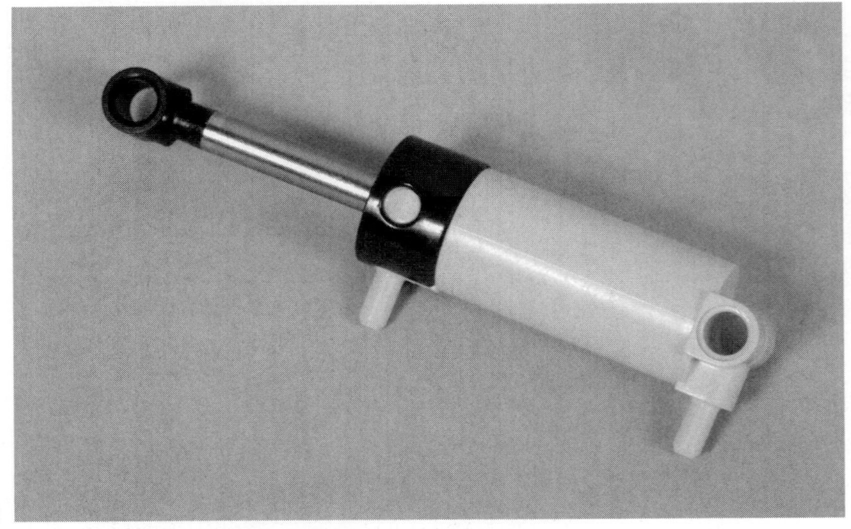

2

3

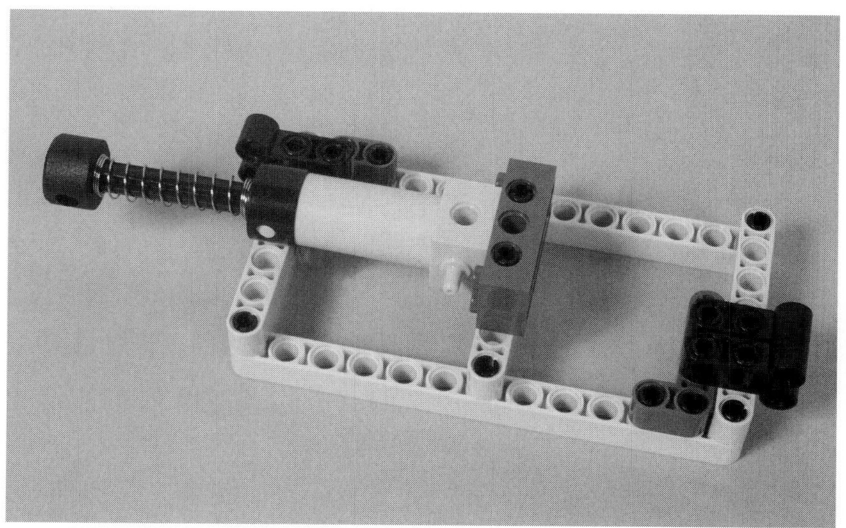
4

5 Attach an NXT motor for activating the pump. This model is for demonstration purposes, only.

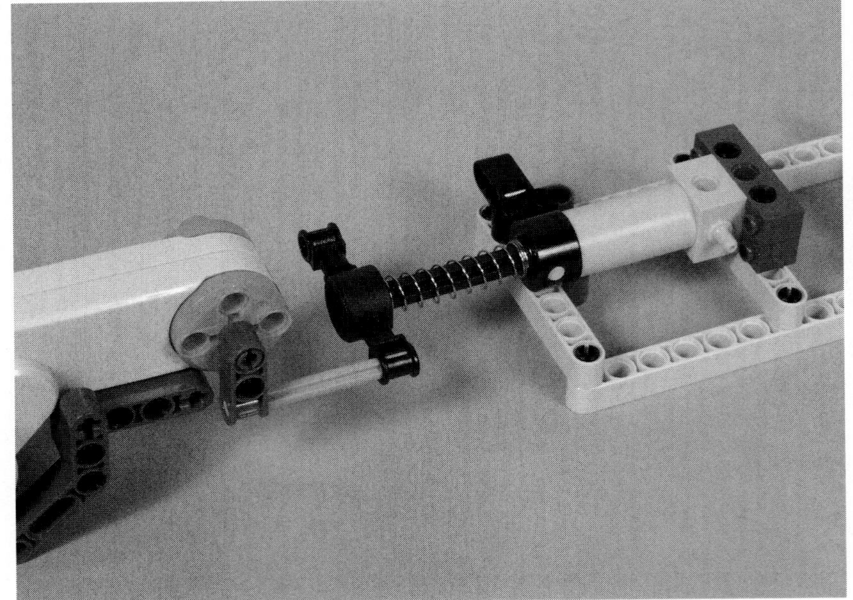
5

PICTURE THIS

Bitmapped graphics on portable handheld devices is nothing new. Yes, we all know about Portable Network Graphics (PNG) file format images being used for Mobile Information Device Profile (MIDP) programming with Java 2 Platform, Micro Edition (J2ME). Likewise, we've all dealt with the black and white PICT (colloquially known as PICTure graphics files) bitmaps inherent to Palm OS Programming projects. So what would possess the LEGO Group to saddle the NXT graphics screen with a Ricoh FAX Document format image?

Yup, the NXT Brick reads graphics files that are written in a .ric format. While this is a ridiculously obscure graphics format, luckily, there is a program that can convert your current multimedia file formats into the .ric format. Unfortunately, the NXT .ric format is *not* the Ricoh FAX Document format. Yes, they both share the exact same file extension, but the format is different. Got that?

During the reign of the MINDSTORMS Developer's Program, one of the intrepid members of this group developed a rudimentary graphics program for reading/writing/creating NXT .ric files. Whether this program will ever see the light of day for the "rest of us" remains to be seen. In the meantime, there is another program worth watching for handling NXT .ric files.

Konvertor by Logipole is a Windows-based graphics conversion program that is capable of converting your image files into over 1200 different multimedia file formats. Priced under $30, Konvertor will be your preferred tool for hacking your own graphics into the NXT. In fact, with Konvertor you *might* be able to put your own smile on your next robot.

Currently, Konvertor is able to read Ricoh FAX Document .ric files, but *not* the NXT flavor of .ric files. This program is worth watching, however. In fact, registered users of Konvertor can request new data file formats for conversion. So pony up your $30 and petition Logipole to add reading/writing capabilities for the NXT .ric format to Konvertor.

1 Locate your .ric files.

2 Launch Konvertor by Logipole and set up the program for reading Ricoh FAX Document format files.

3 Konvertor can be used for converting .ric files into .bmp graphics.

4 Navigate to the folder that holds your NXT .ric images.

HOW NOT (YET) TO READ/WRITE .RIC FORMAT GRAPHICS FILES

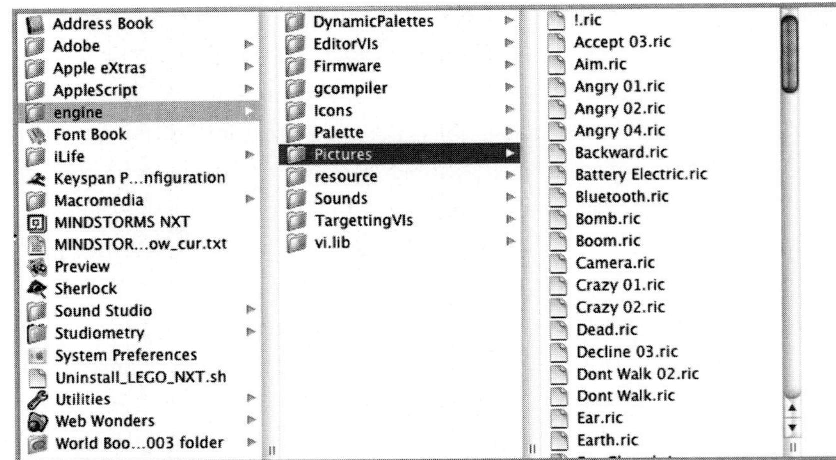

1

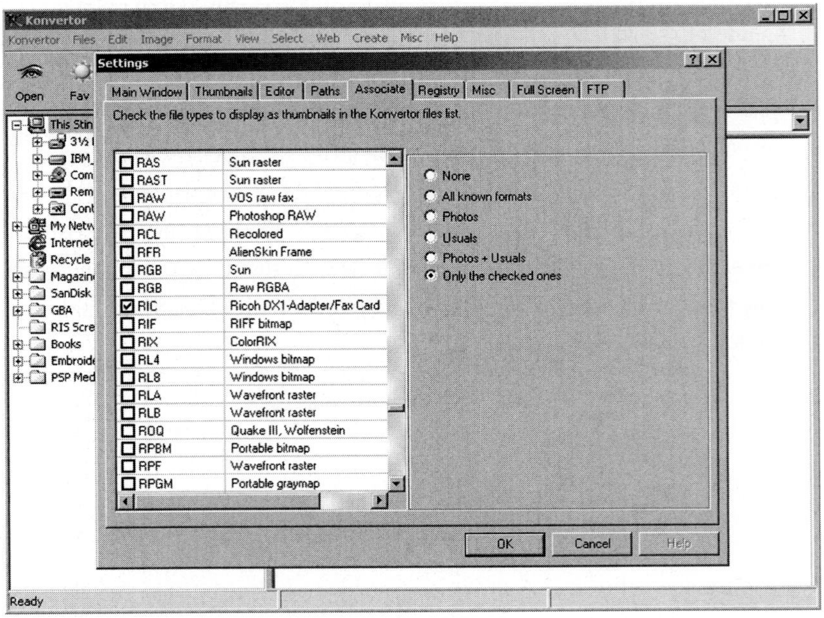

2

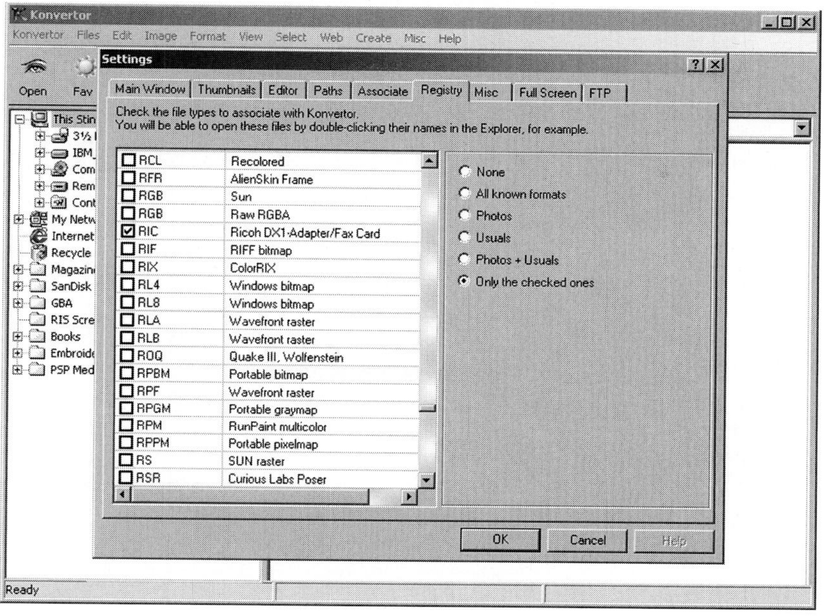
Konvertor
Konvertor Files Edit Image Format View Select Web Create Misc Help
Open Fav
Settings
Main Window Thumbnails Editor Paths Associate Registry Misc Full Screen FTP
Check the file types to associate with Konvertor.
You will be able to open these files by double-clicking their names in the Explorer, for example.
RCL Recolored
RFR AlienSkin Frame
RGB Sun
RGB Raw RGBA
RIC Ricoh DX1-Adapter/Fax Card
RIF RIFF bitmap
RIX ColorRIX
RL4 Windows bitmap
RL8 Windows bitmap
RLA Wavefront raster
RLB Wavefront raster
ROQ Quake III, Wolfenstein
RPBM Portable bitmap
RPF Wavefront raster
RPGM Portable graymap
RPM RunPaint multicolor
RPPM Portable pixelmap
RS SUN raster
RSR Curious Labs Poser
None
All known formats
Photos
Usuals
Photos + Usuals
Only the checked ones
OK Cancel Help
Ready

3

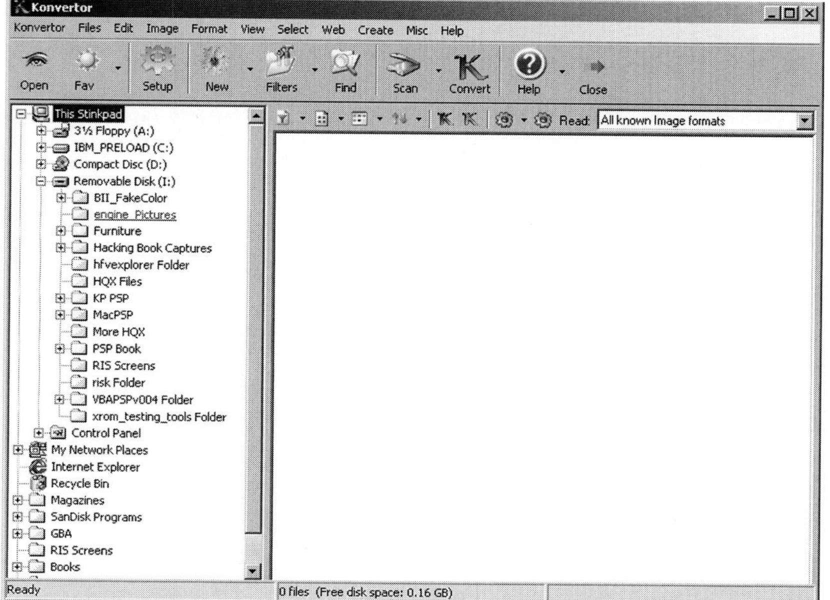
Konvertor
Konvertor Files Edit Image Format View Select Web Create Misc Help
Open Fav Setup New Filters Find Scan Convert Help Close
Read: All known Image formats
This Stinkpad
3½ Floppy (A:)
IBM_PRELOAD (C:)
Compact Disc (D:)
Removable Disk (I:)
BII_FakeColor
engine_Pictures
Furniture
Hacking Book Captures
hfvexplorer Folder
HQX Files
KP PSP
MacPSP
More HQX
PSP Book
RIS Screens
risk Folder
VBAPSPv004 Folder
xrom_testing_tools Folder
Control Panel
My Network Places
Internet Explorer
Recycle Bin
Magazines
SanDisk Programs
GBA
RIS Screens
Books
Ready
0 files (Free disk space: 0.16 GB)

4

5 **Currently, you can't read the NXT .ric images with Konvertor.**

6 **Nor can you convert these .ric files into industry-standard bitmaps.**

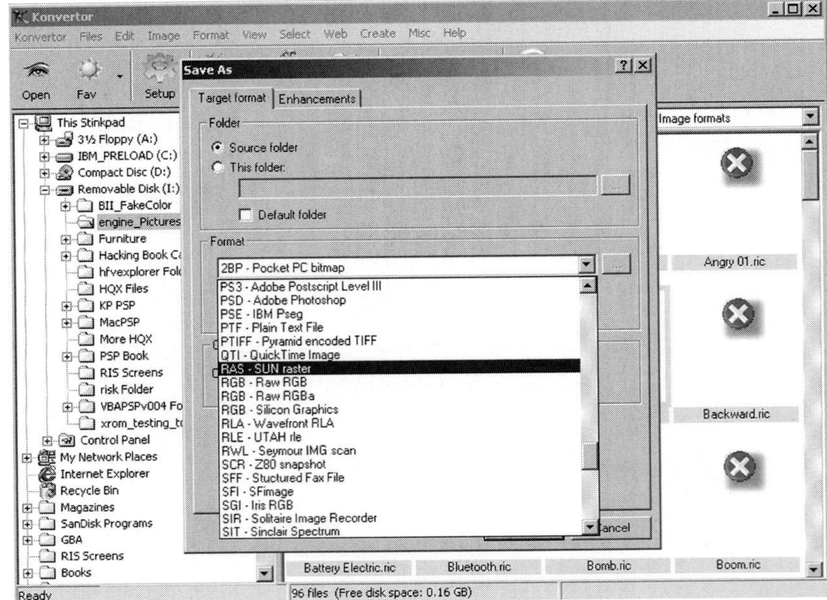

5

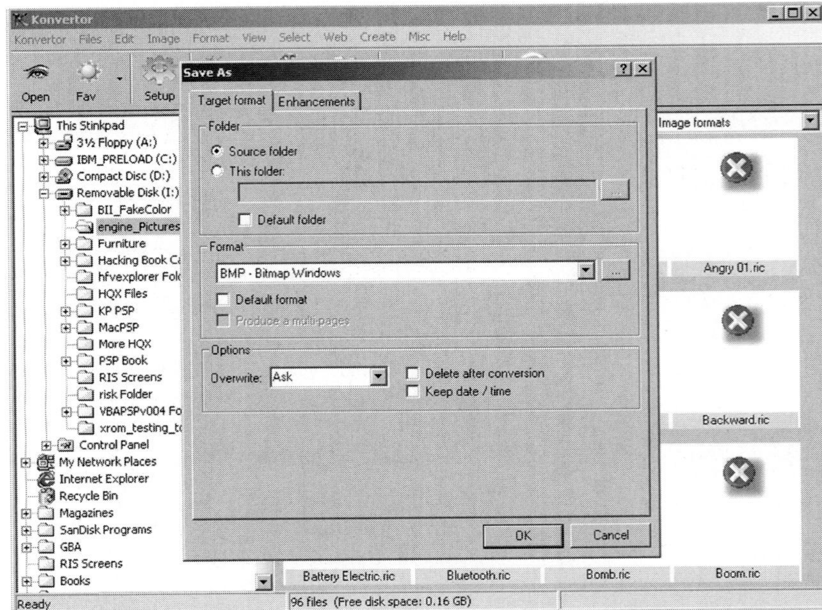

6

GETTING A BIT LONG IN THE BLUETOOTH

The Bluetooth specification is a low-cost, low-power radio standard that is used for wirelessly connecting devices. Global acceptance of this specification has helped incorporate limited-range wireless communication into everything from automobiles to Web players. Based on a nifty time-sharing architecture featuring frequency-hopping and tiny packet sizes, Bluetooth uses the 2.4-gigahertz radio band with a range of approximately 30 feet.

Apple PowerBook G4 portables were the first computers to offer Bluetooth 2.0+ Enhanced Data Rate (EDR). At that time, other computers were stuck with the older Bluetooth 1.x support. Bluetooth 2.0+EDR, while backwards-compatible with Bluetooth 1.x, is up to three times faster than the older standard. A maximum data rate of up to 3 Mbps is possible with Bluetooth 2.0+EDR. This throughput plus the peripheral nature of the connectivity feature has enabled some vendors to describe Bluetooth as "wireless USB."

Lucky for you, the NXT Brick is equipped with a CSR BlueCore™ 4 2.0 +EDR system. While you might not have heard of them before, Cambridge Silicon Radio (CSR) is one of the big players in the single-chip radio device market of short-range wireless communication. The CSR main offering is BlueCore. BlueCore is a fully integrated 2.4-gigahertz radio, base band, and microcontroller used in over 60 percent of all qualified Bluetooth v1.1 and v1.2 enabled products.

BLUETOOTH® COMMAND MESSAGES

Bluetooth communication is conducted through direct command "telegrams." Direct command telegrams are limited to 64 bytes in length exclusive of 2 bytes indicating the telegram's size. The first byte in a telegram is designated Byte 0.

BLUETOOTH COMMUNICATION TELEGRAM STRUCTURE*:

Byte 0 = Telegram Length, LSB
Byte 1 = Telegram Length, MSB
Byte 2 = Telegram Type
Byte 3 = Command
Byte 4 - Byte N = Data Bytes (padded, if necessary)

* Telegrams that are not delivering Bluetooth or USB communications begin their structure at Byte 2. In these other telegrams, Byte 2 is then renamed Byte 0.

TELEGRAM TYPE

0x00 = Direct Command Telegram, Response Required
0x02 = Reply Telegram
0x80 = Direct Command Telegram, No Response Required

COMMANDS

MessageWrite

Byte 2: Direct Command Telegram (0x00 = Response, 0x80 = No Response)
Byte 3: Command (0x09)
Byte 4: Inbox Number (0-9)
Byte 5: Message Size (<59)
Byte 6 - N: Message Data

Return

Byte 2: Reply Telegram (0x02)
Byte 3: Command (0x09)
Byte 4: Status Byte (0x00 = Success)

MessageRead

Byte 2: Direct Command Telegram (0x00 = Response, 0x80 = No Response)
Byte 3: Command (0x13)
Byte 4: Remote Inbox Number (0-9)
Byte 5: Local Inbox Number (0-9)
Byte 6: Clear Message (0x01 = Clear, 0x00 = Message Received, 0x40 = No Message Received)

Return

Byte 2: Reply Telegram (0x02)
Byte 3: Command (0x13)
Byte 4: Status Byte (0x00 = Success)
Byte 5: Local Inbox Number (0-9)
Byte 6: Message Size
Byte 7 - Byte 63: Message (Padded)

BLUETOOTH EXAMPLE TELEGRAM:

0x05, 0x00, 0x00, 0x13, 0x0B, 0x01, 0x01

where,

0x05 = Telegram length is 5 bytes in length
0x00 = Telegram length placeholder
0x00 = This is a Direct Command Telegram
0x13 = Using *MessageRead* command
0x0B = Read the Remote Message Mailbox Number 1
0x01 = Use the Local Message Mailbox Number 1
0x01 = Clear the Message When Finished

III-6 Pepper Pad 3 Web Player from Pepper Computer.

Naturally, you're going to want to find a Bluetooth-equipped device that you can hack into the NXT Brick. Such a device might be the Pepper Pad 3 Web Player (see Fig. III-6).

Although Pepper Pad from Pepper Computer, Inc. and Hanbit Electronics Co., Ltd. has been operating under the radar screen for the last couple of years, it's now ready for prime time with the release of a new Web player model. Dubbed Pepper Pad 3 Web Player, this latest incarnation of the familiar handheld entertainment device form factor features Wi-Fi , USB, and Bluetooth connectivity, a built-in Web camera, a 7-inch touch screen, integrated joystick/keyboard, a 20-gigabyte hard disk drive, and stereo speakers and microphone.

Measuring a scant 11-by-6 inches and tipping the scales at a svelte 2 pounds, the Pepper Pad 3 isn't some dainty consumer electronics device that must be safely ensconced inside a fancy leather case. Nope, the Pepper Pad 3 is a rugged, splash-resistant handheld that is equally at home in the kitchen, outside near the pool, or on the road in your car.

Don't confuse these beefy specs with a device that doesn't play well with others, either. The Pepper Pad 3 is also equipped with an infrared system that doubles as a universal remote for many different models of TVs and media players.

Powering all of this gusto is an embedded AMD Geode™ LX 800 microcontroller running a Linux Kernel 2.6 OS. Likewise, a handsome suite of software enables this worthy Web player to stream and download and play music, videos, movies, and photos through a suitable Wi-Fi network connection. Based on our extensive tests, the video playback capability of the Pepper Pad is outstanding, rivaling many portable DVD players. Furthermore, the integrated "prop" stand enables the Pepper Pad to play a movie to a much larger audience—hands free.

The Pepper Pad 3, with a suggested retail price of $699, can be purchased through Amazon.com. In a generous nod to current Pepper Pad costumers, current Pepper Pad owners can buy the new Pepper Pad 3 at a special discount price through a rebate offer.

Sprinkling a couple of Pepper Pad 3 Web players through your house can go a long way toward spicing up your family's access to hands-on entertainment. At a price that won't break the bank, either.

HOW TO HACK BLUETOOTH

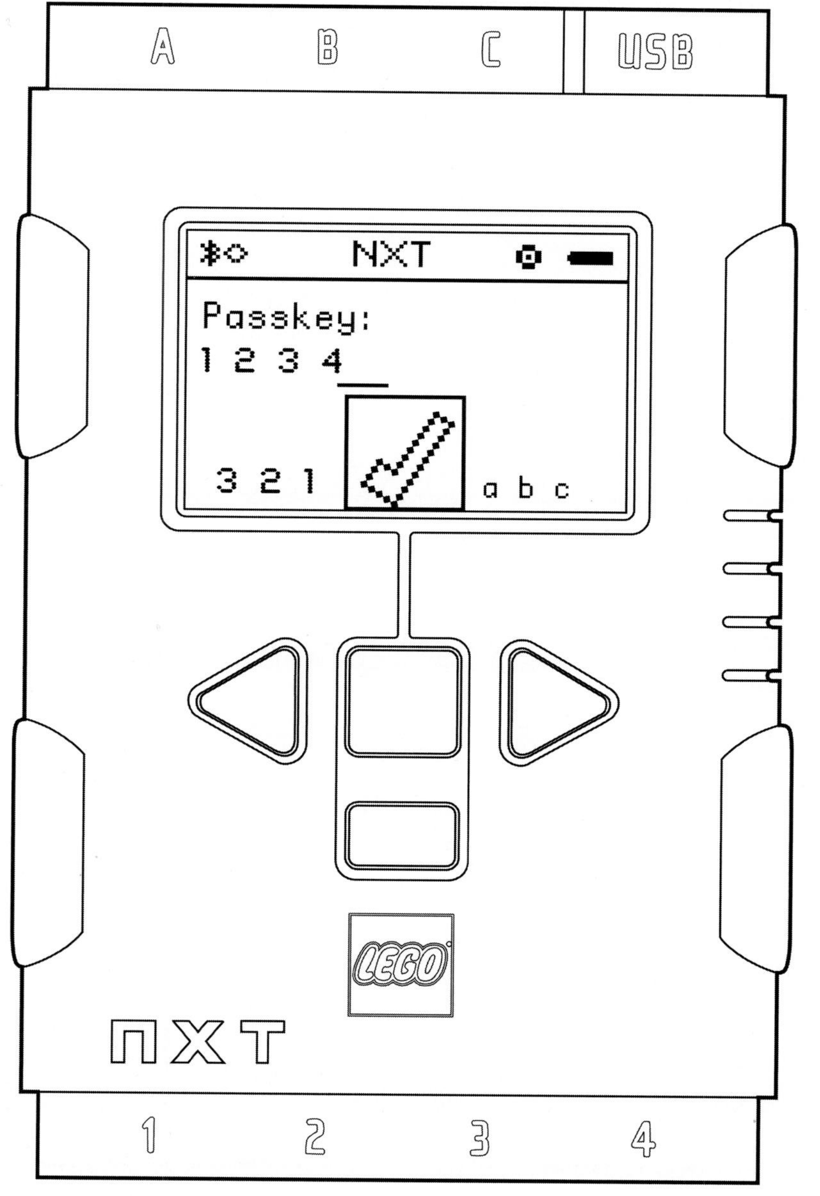

1

1 **Send your Passkey to a Bluetooth device.**

2 Bluetooth send program block inside the MINDSTORMS NXT software.

3 Bluetooth get or receive block.

4 Pepper Pad is an Internet appliance that can deliver a knockout punch to Microsoft's Origami project. Why? Pepper Pad is a rugged, portable touch-screen computer that uses a Linux OS and costs less than $850. Oh, and did I say that it's splash-resistant, too? Finally, pool-side computing is a reality.

5 All of the ports on Pepper Pad are covered with flexible rubber plugs.

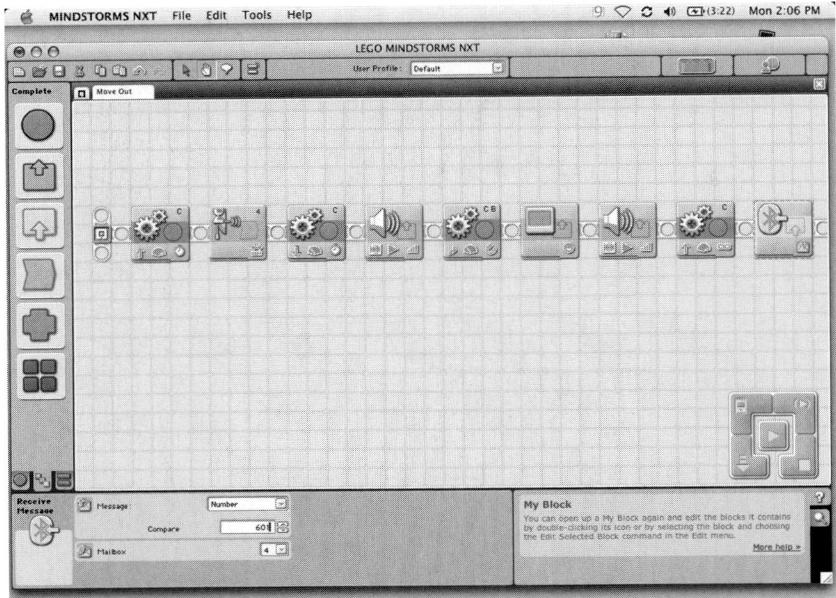

2

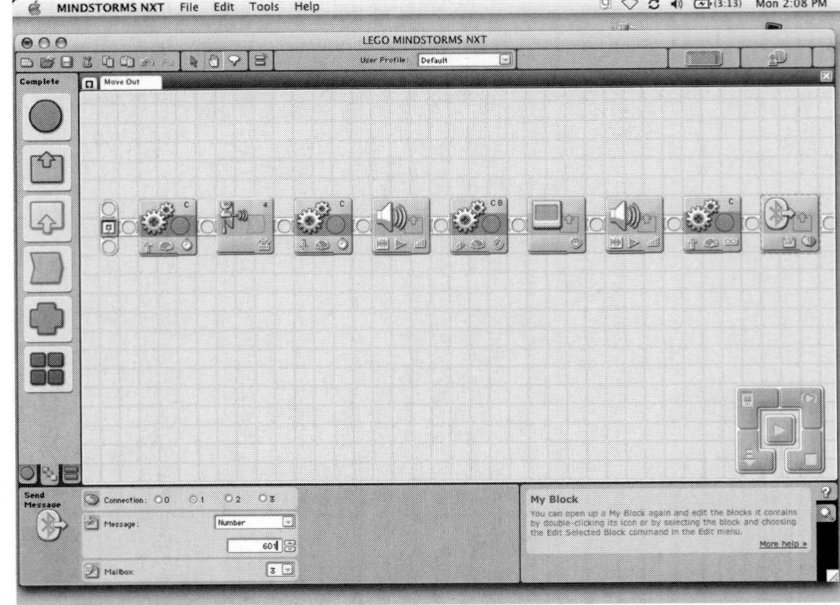

3

4

5

6 USB, audio in, audio out, and an external monitor port make Pepper Pad a portable that is equally at home on your desk, in the kitchen, or on a kid's bed.

7 An IR transmitter inside Pepper Pad can be configured to operate some popular TVs.

8 A built-in stylus controls Pepper Pad. Alternatively, for the touch screen–challenged, you can use a scroll wheel or arrow keys for input.

6

7

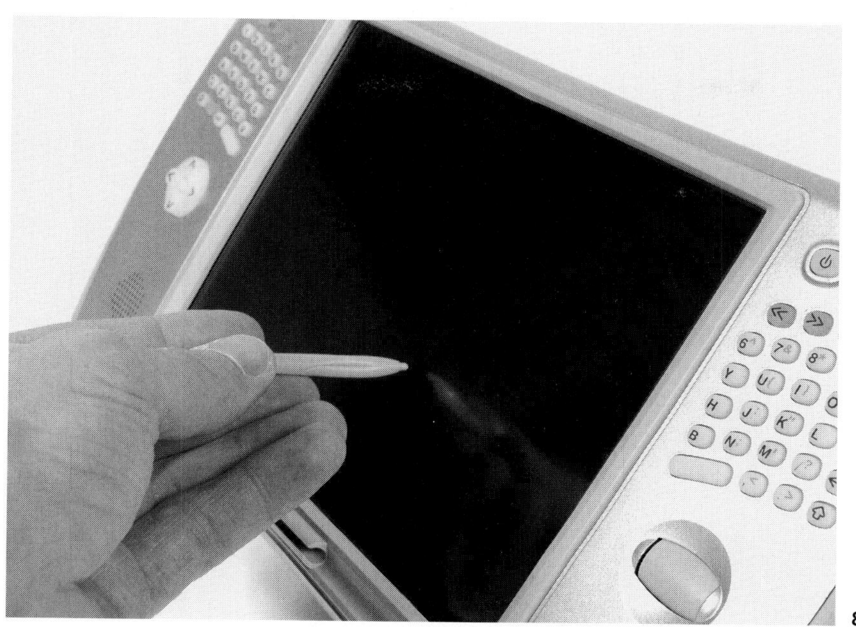

8

PART IV: KATHERINE'S DESIGN FUN HOUSE

LET'S STUDY the elements that constitute good design. OK, just what is "good design"?

That definition can be as elusive as trying to nail down a definitive statement about "what is beauty." For example, if you ask legendary robot designer Mark W. Tilden (the builder of a new breed of bot, *Robosapien*™) to define good design you would get this answer:

> My style of robotics is what they call "bottom-up," which is where you build things from simple components and see if you can push them to do something exotic. This is different to the usual "top down" approach which is to force a fast laptop computer to roll around in the hopes it doesn't get a disk crash.
>
> Top down robot development is being spearheaded by big names like Sony, Honda, Toyota, etc., and all the more power to them. Their stuff is wonderful, but are unlikely to get out of their media circus.
>
> I never could get into top down methods because it required huge costs and time. Few researchers ever cozy up to bottom-up because they claimed the behaviors were too simple.

It's a brawn verses brain issue. My Robosapien is just the latest in a long series of "evolved" designs that have shown some very complex behaviors indeed. Starting out as a simple 2 transistor roller in 1988, 15 years later Robosapien 1.0 has less than 30 transistors controlling 8 motors in a humanoid form. Doesn't fall, moves fast, looks around, picks up things, lasts for hours on batteries, and without a single line of code. It's the latest of hundreds (of bottom-up designs) that I've built.

Pity, as I've always been interested in the bizarre and unusual designs possible. Biology only comes in a small variety of round, tubular, flat, and bilaterally symmetric arrangements, whereas robotics allows us to build any design outside of convention or scale restrictions.

I've always believed there's a biological basis to aesthetics, that we admire the way nature builds because its function is so linked to the form. An aesthetic basis for robotics is harder to justify, seeing as how the majority of industrial robot machines would hardly win any beauty contests.—from *Personal Communication* (March 2005)

Smart guy, that Mark Tilden. Now compare Tilden's definition with this one espoused by renowned American architect Louis Sullivan:

All life is organic. It manifests itself through organs, through structures, through functions. That which is alive acts, organizes, grows, develops, unfolds, expands, differentiates, organ after organ, structure after structure, form after form, function after function. That which does not do these things is in decay! This is a *law*, not a word! And decay proceeds as inevitably as growth—functions decline, structures disintegrate, differentiations blur, the fabric dissolves, life disappears, death appears, time engulfs—the eternal night falls. Out of oblivion into oblivion—so goes the drama of created things—and of such is the history of the organism.

First I would dissolve for you this wretched illusion called American architecture, and then cause to awaken in your mind the reality of a beautiful, a sane, a logical, a human, living art of your day; an art of and for democracy, an art of and for American people of your own time.—from *Kindergarten Chats* by Louis Sullivan (ed. Claude Bragdon; Scarab Fraternity Press, 1934) [Reprinted as *Kindergarten Chats and Other Writings* by Louis Sullivan; Dover Publications, 1980]

Gee, that's pretty heady stuff. I personally like the "form follows function" Sullivan attribution. Contrast this view with one voiced by Sullivan's most celebrated protégé, American architect Frank Lloyd Wright:

> Style center group stations would grow natural in this way and television and radio, owned by the people, broadcast cultural programs illustrating pertinent phases of government, of city life, of art work, and programs devoted to landscape study and planting or the practice of soil and timber conservation; and, as a matter of course, to town planning for better houses. In short, these style stations would be inspired hives of creative energy all bearing on the character of modern industry wherever industry touched the common life. Without hesitation or equivocation let's say that architecture would, necessarily, again become the natural backbone (and architects the broad essential leaders) of such cultural endeavor.
>
> Such active work units in design, were they truly dedicated and directly applied to the radical culture of indigenous style and the building of our city, would at last stimulate popular growth as light stimulates the growth of a garden.—from *The Living City* by Frank Lloyd Wright (Mentor Book, 1958)

Maybe architecture is too, well, metaphysical for your taste. How about fashion design? Donna Karan might be speaking to you:

> It wasn't that I didn't want to be a designer. I just wanted the experience of being home with my child. I grew up with a working mother. And it's tough. Talk about what you resist, persists!
>
> Being a designer led me to the questions in life that I constantly have—the questions that I have to explore. What do I need? What do I desire? What do I hope for? Originally, I wanted simple, comfort clothes for me and my friends, such as a few black pieces.
>
> That's partly how I get satisfaction—supporting creativity. It's not necessarily about my creativity, but about putting the creativity of everyone together, and supporting creativity on many levels. I'm happy to be a catalyst for creativity.—from *Donna Karan New York* by Ingrid Sischy (Universe Publishing, 1998)

Finally, in terms that most of us can easily and readily comprehend, Steven Tolleson, founder of Tolleson Design, distills the essence of good design down to these three digestible bites:

Design is the latest in a string of jobs I've held since I began fending for myself.

I don't ever make the mistake of imagining that my design ideas are more valuable than anyone else's.

Support your colleagues; good design is hard to do.—from *Soak, Wash, Rinse, Spin* by Tolleson Design/Steven Tolleson (Princeton Architectural Press, 1999)

One thing that you might be able to glean from these excerpts is that no two designers think alike. Oddly enough, every designer addresses the same set of basic concepts when tackling a creative challenge. These basic design concepts are:

- **FORM**. This is your workspace, your definition, your canvas for solving your design problem.
- **ELEMENTS**. In a traditional sense, the elements of design are line, type, shape, and texture. Don't sell these elements short as not having any direct application in your NXT designs. Because they do—line, for example, is the visual organization of your robot. While type is the visual communication for your design, shape and texture are the visual skeleton and flesh for your next LEGO creation.
- **STRUCTURE**. There are five general ways for organizing the structure of your design elements. These structural organizers are called the principles of design. Balance, contrast, unity, value, and color will orchestrate your selected elements into a good design. By correctly handling these five general principles you can be assured that your NXT creation will generate a mood that is pleasing to your peers.

So no matter how *you* speak about your designs, your final design will speak very clearly to your audience. They will "know" good design when they see it. And you will hear about it—one way or another.

A FIELD GUIDE TO LEGO® MINDSTORMS® NXT

Field Marks for Identifying Species

NATURALIST'S NOTE: Rather than lurking around ancient Mayan ruins hopelessly rebuilding a handful of redundant robot carcasses, instead consider mounting massive explorations of primordial swamps, gothic cityscapes, and distant star realms. These are the true habitats of some unbelievable species. Capture of these beasts is rare, if not, impossible. Many times a visual identification is all that is possible. In the hopes that a fleeting glimpse can be made of one of these reclusive creatures, this field guide has been developed.

Beware that some of these beasts are nocturnal. Therefore, a positive identification might be impossible. Furthermore, carefully scrutinize the range maps for each species. These maps will help serve as a sanity check for ensuring that your presumed citing is indeed possible.

BIPEDALIA

Easily confused with Siththra. Not as moody. Can be downright good natured. Sometimes shy and hard to find. Turn over a few rocks and some might come scurrying out. Through colonial expansion has appeared on every continent—at least once.

RECOGNITION. Easy to spot "fifth" leg. This fifth leg is used for support when standing erect. Touch-sensor "feelers" provide quick reverse reaction when about to be squished under a shoe. Light-sensitive eyes elicit a defensive posture if caught in the middle of the floor when the room lights are turned on.

HABITAT. Ubiquitous; able to adapt to just about any environment.

HABITS. Doesn't seem to know when it is time to leave. Easily wears out its welcome.

YOUNG. Breeds like a rabbit.

RANGE. Found everywhere; leaves absolutely no rock underside unpopulated.

NUMBER OF SPECIES. One, but it's a big one.

COMMON NAMES. Also called a "bottom feeder" or a "lurking in-law."

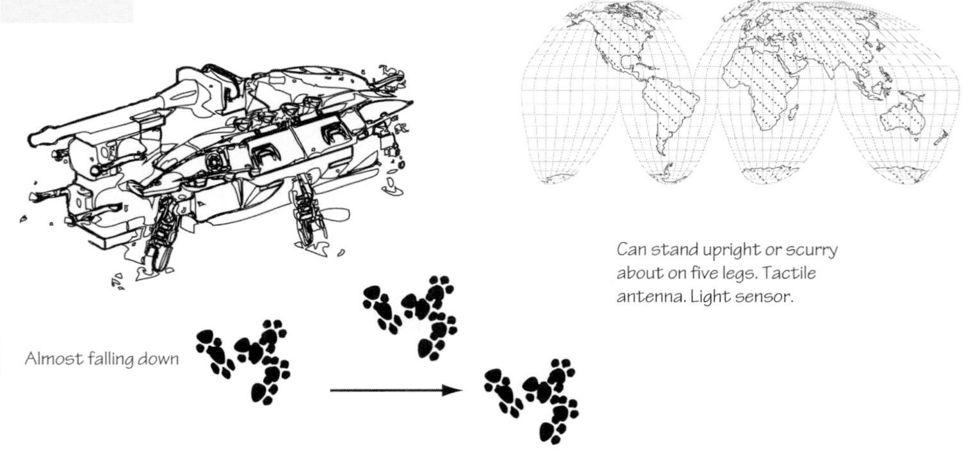

Can stand upright or scurry about on five legs. Tactile antenna. Light sensor.

Almost falling down

CARTERA

Our only gas combustion species. Restricted to most of the industrial nations. Appears to demonstrate an innate link to members of Dinotacea.

RECOGNITION. Smells like burning rubber. Moves with two or more rubber wheels.

HABITAT. Thrives on hard-topped open highways.

HABITS. Consumes unconscionable amounts of irreplaceable fossil fuels.

YOUNG. None; appears to be full-size at birth.

RANGE. Very common throughout the industrialized nations.

NUMBER OF SPECIES. There are hundreds of different species, some of which appear to have become extinct. Remarkably, extinct species are just as quickly replaced by new species.

COMMON NAMES. Well-known acronyms FORD ("Fix Often, Repair Daily" or "Found On Road, Dead") or FIAT ("Fix It Again, Tony").

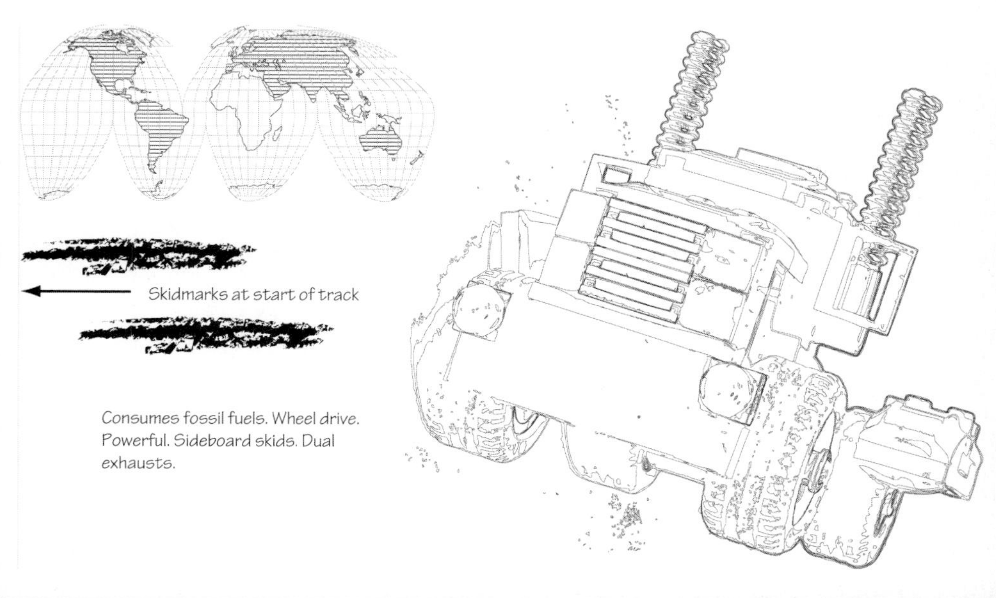

Skidmarks at start of track
Consumes fossil fuels. Wheel drive. Powerful. Sideboard skids. Dual exhausts.

DINOTACEA

Exact physical appearance of this species is a contentious issue. Can sometimes be seen in herds or solitary. Found in northern countries, but, oddly enough, requires warmer climates. Will feed on Bipedalia, if given the chance.

RECOGNITION. Stands erect. Snapping mouth with a love for eating flesh. Very sloppy eater. Thrives in warmer temperatures. Uses light sensor for seeking the maximum amount of sunlight.

HABITAT. Equally at home in a swamp or dense tropical forest. Occasionally seen in open plains running in packs. Woe be the lonely warm-blooded critter grazing in their path.

HABITS. Constantly moving in search of prey. Most vulnerable right after eating a large meal or during unseasonably cold weather.

YOUNG. Born as ugly as the adults and hungry right out of the egg.

RANGE. Wide-ranging in the northern hemisphere. Slowly penetrating into southern countries.

NUMBER OF SPECIES. Divided into smaller pack killers and larger solitary hunters, there are well over 100 different species.

COMMON NAMES. "Hail to the King, baby." Many lesser-known species have been reduced into a propellant fuel that is commonly pumped into Bipedalia. These variations are sometimes called "Pump Gougers" or "Tankosaurus."

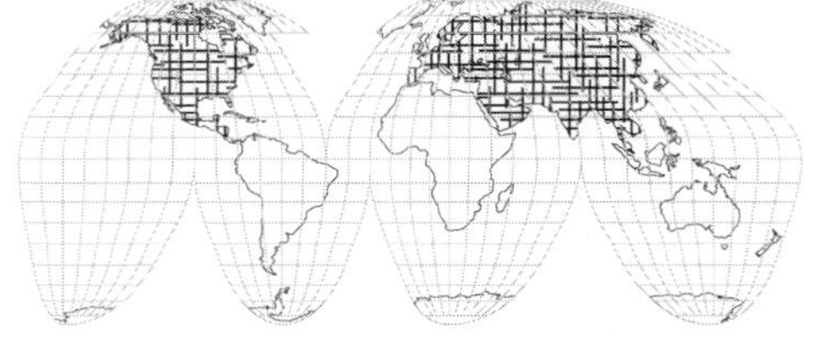

Biped with snapping jaws. Light sensor. Someday will be gasoline.

Crossover steps

BIONIPHA

A fantastic creature of mythical stature. Always angry. Able to handle awesome weapons with dexterity and power. Luckily, this species is restricted to a very small isolated spot in Antarctica.

RECOGNITION. Permanently angry. Always a weapon in one or more hands and, sometimes, in one or more feet. Red eyes. Accurate distance-measuring sensors. Oh, and did I mention that it's always mad?

HABITAT. Incredibly low temperatures. Rocky, barren, frozen ice-packed, open expanses. And you wonder why they are always mad?

HABITS. Kill or be killed. No time for recreation, until killing is considered an avocational pastime.

YOUNG. Thankfully, there are no young. Must be born full-grown.

RANGE. Limited to a small speck of frozen land located near Antarctica.

NUMBER OF SPECIES. There are rather large numbers of species (in excess of 45) with new species appearing annually.

COMMON NAMES. These murdering marauders should be treated with the utmost respect. Avoid using any kind of name calling, except for maybe "Sir."

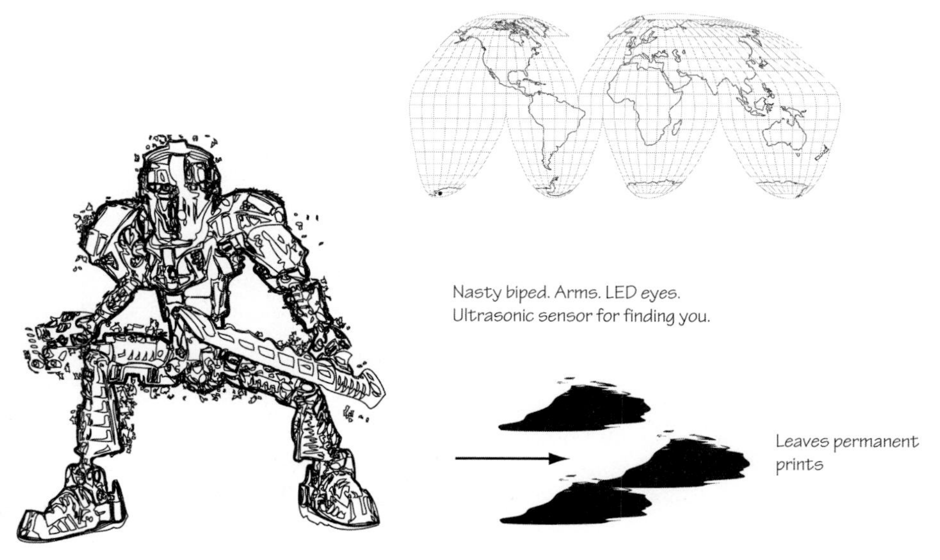

Nasty biped. Arms. LED eyes.
Ultrasonic sensor for finding you.
Leaves permanent prints

SITHTHRA

Inexplicably filled with remorse and sorrow. Extremely talented with a powerful light plasma sword-like object. This species is immensely popular and can be found worldwide.

RECOGNITION. Unusual symbiotic "rider" organism. Ultrasonic rangefinder used for locating, tracking, and chasing the dark side. Three toes per foot.

HABITAT. Appears to mainly manifest itself inside darkened auditoriums filled with adoring fanatics. Oddly, many of these fanatics will actually attempt to simulate or reproduce the appearance of the Siththra.

HABITS. Although raised by the forces of good, strangely compelled to follow the dark side.

YOUNG. An entire life cycle that crosses several generations and billions of light years.

RANGE. Worldwide distribution was ensured by the savvy marketing negotiation for DVD rights and limited-run theater engagements.

NUMBER OF SPECIES. At present there are six species, commonly known as episodes, but don't underestimate the lucrative lure of endless "return" episodes.

COMMON NAMES. Sometimes derided as "Lucas puke-as," more typically, however, succinctly identified with "Luke, I'm Your Father."

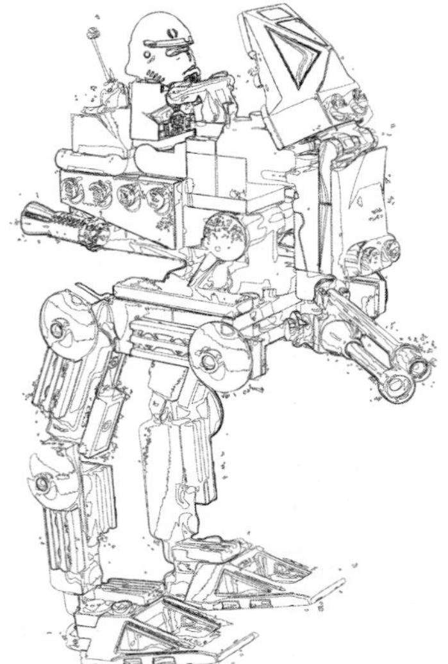

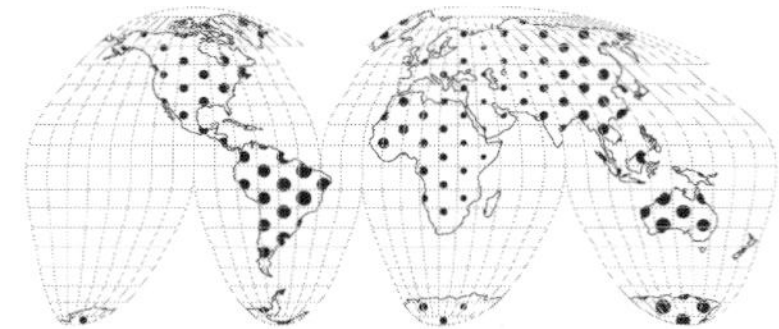

Biped with active sound sensor and ultrasonic rangefinder. LED beams located in front. Jockey rider.

Follows dark side

BATMANCETI

Always nocturnal in its habits. Usually found in areas of high criminal activity. Distribution is restricted to the northeastern United States. Features a partially masked head and face. When provoked, this species will eject a multifunction hard projectile.

RECOGNITION. Incredible night vision. Able to locate criminals in their own element. Watch for sudden stops. Prominent linkage with bats, including pointed ears.

HABITAT. Lives in caves. Older versions would emerge during either daylight or evening. Newer species, however, seem restricted to a Gothic noir lifestyle.

HABITS. Fights crime with obsessive passion. Seeks privacy, shuns publicity. Attempts to cast an air of philanthropy.

YOUNG. All young share an incredibly tough introduction to reality. At early age forced to watch the death of own parents.

RANGE. Limited to a very narrow island located along the Atlantic northeast of the United States.

NUMBER OF SPECIES. There have been at least four different species. Adamo westicus is considered the "father" of all current species.

COMMON NAMES. Originally called the "Caped Crusader," newer generations prefer the moniker "Dark Knight" for referring to this crime fighter.

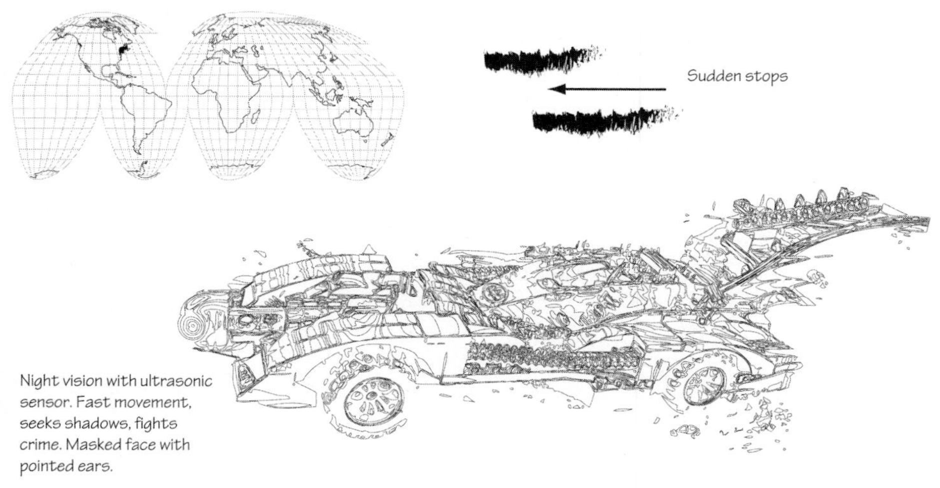

Night vision with ultrasonic sensor. Fast movement, seeks shadows, fights crime. Masked face with pointed ears.

IV-1 A shipping container.

IV-2 Dubbed the "box," these containers are a common sight along most of the nation's docks.

IV-3 Shipping containers, like this one, became a standard fixture along the Mississippi Gulf Coast in the wake of Hurricane Katrina during the summer and fall of 2005.

BUILDING INSIDE A BOTTLE

Who would even think of making their home inside a shipping container? Even better, what's a shipping container?

Just take a walk along the docks of any major shipping line and you will see lots and lots of steel (or, aluminum) boxes that resemble the trailers of an 18-wheeler's rig, minus the tires and axles. These are shipping containers (see Figs. IV-1, IV-2, and IV-3).

Uncovering the history of how these shipping containers rose to such prominence in the transportation of cargo is a difficult and sketchy venture. Luckily, Marc Levinson has assembled the definitive history of the shipping container in the apropos titled book, *The Box* (Princeton University Press, 2006).

Levinson was able to trace the origins of modern shipping containers to the odd partnership of trucking magnate Malcolm McLean and engineer Keith Tantlinger. Together, these two creative thinkers assembled approximately 200 aluminum Brown Industries' containers for a transportation experiment.

Tantlinger, as described in Levinson's book, was a shipping container expert who worked for Brown Industries. He had earned that reputation at Brown Industries by creating a unique 30-foot aluminum box that could be stacked on barges and strapped to tractor trailer trucks.

For McLean's shipping experiment, Tantlinger proposed a new container—a 33-foot-long aluminum container with a special steel track on the bottom. These steel tracks were used for attaching the containers to the deck of the transporting ship. Oh, and why were these new test containers 33 feet long rather than the proven Brown Industries' container length of 30 feet? Because

WALKING THE WALK

These architectural firms and museums are building modular dream homes which might be a starting point for your next LEGO construction project:

The Dwell Home—www.thedwellhome.com/winner.html

LOT-EK—www.lot-ek.com/

Resolution: 4 Architecture—www.re4a.com

Shigeru Ban Architects—www.shigerubanarchitects.com/

Walker Art Center—design.walkerart.org

TRITON
NCGU 288516

IV-1

TRITON
EC592
ENDOM
NCGU

IV-2

TRITON
LOWE'S
Low Prices
EVERY DAY!

IV-3

Tantlinger had calculated the available deck space on the test ship and found that it was divisible by 33.

Tantlinger's innovation didn't stop with the container, either. He also devised a method for lifting each container on and off a ship that didn't require longshoremen to attach hooks to the roof of each container. This innovation was the spreader bar—a center-expanding fork that would automatically engage hooks on each container *without* longshoreman intervention. Just one crane operator was needed to load and unload a container ship. In this manner, Tantlinger was thinking "outside of the box." I'm sorry, I just had to say that.

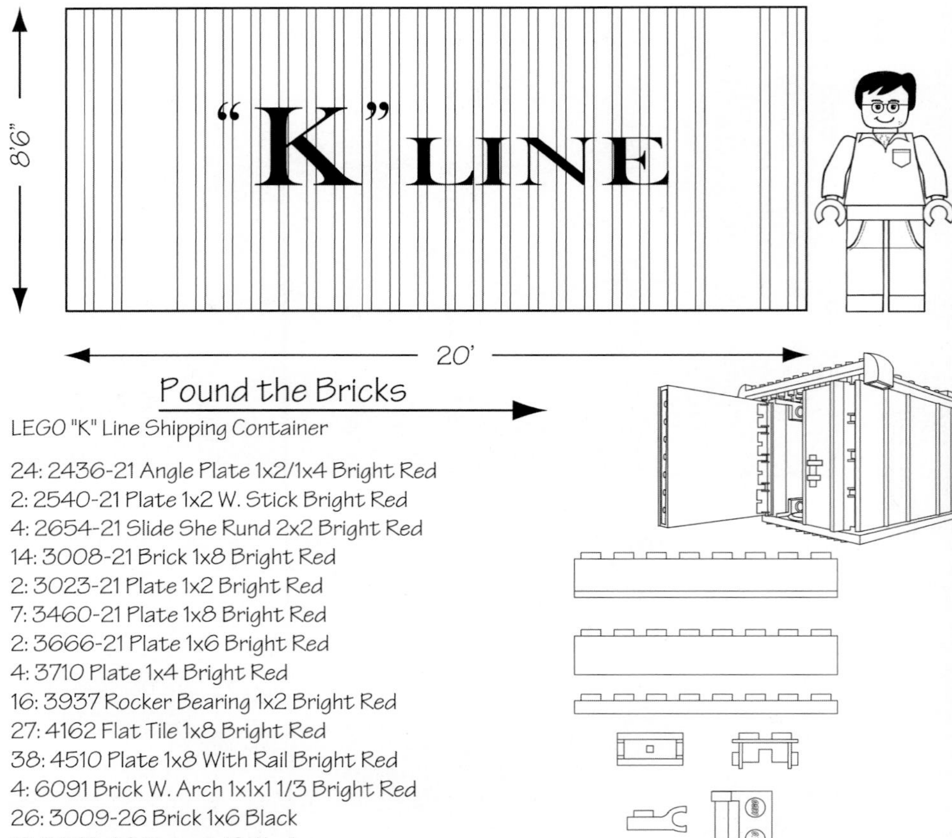

"K" LINE
8'6"
20'
Pound the Bricks
LEGO "K" Line Shipping Container
24: 2436-21 Angle Plate 1x2/1x4 Bright Red
2: 2540-21 Plate 1x2 W. Stick Bright Red
4: 2654-21 Slide She Rund 2x2 Bright Red
14: 3008-21 Brick 1x8 Bright Red
2: 3023-21 Plate 1x2 Bright Red
7: 3460-21 Plate 1x8 Bright Red
2: 3666-21 Plate 1x6 Bright Red
4: 3710 Plate 1x4 Bright Red
16: 3937 Rocker Bearing 1x2 Bright Red
27: 4162 Flat Tile 1x8 Bright Red
38: 4510 Plate 1x8 With Rail Bright Red
4: 6091 Brick W. Arch 1x1x1 1/3 Bright Red
26: 3009-26 Brick 1x6 Black
12: 3029-26 Plate 4x12 Black
52: 47905-26 Brick 1x1 w/2 Knobs Black

IV-4 A plan for a LEGO Dwelling Unit shipping container house.

On April 26, 1956, at the Port Newark dock in New Jersey, McLean loaded Tantlinger's-designed containers onto a Pan-Atlantic ship, a war surplus T-2 tanker named *Ideal-X*. After being loaded in less than eight hours, the ship set sail for Houston, Texas. When the ship arrived at Wharf II in Houston a shipping revolution had been born.

At first glance, you might think that shipping containers are the farthest thing from a viable building unit. But you'd be wrong (see Fig. IV-4).

In fact, some folks consider these ubiquitous cargo boxes to be an art form or, in the case of Gregory Colbert, suitable for holding art. So much so, that when Colbert was looking for the "perfect" museum to hold his 200-piece photo-

SHIPPING CONTAINERS

Founded in 1919, Kawasaki Kisen Kaisha, Ltd. or "K" Line is a global transportation company with a fleet of over 337 vessels carrying more than 70,000 containers.

20' Dry Freight Container

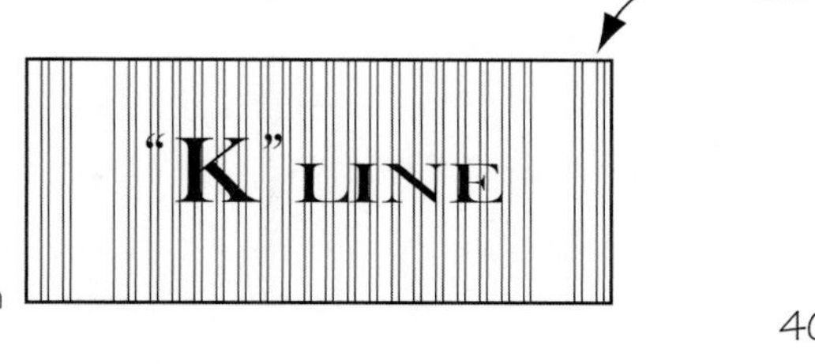

40' Dry Freight Container

A Guide to Shipping Containers		
		Notes Create K-Dwelling for LEGO® MINDSTORMS® NXT.
Sheet 1 of 1	dp	

IV-4

graphic show, "Ashes and Snow," he sought Japanese architect Shigeru Ban to build him one.

Using 148 shipping containers, paper tubes, and one million paper tea bags, Ban designed the 45,000-square-foot Nomadic Museum (see Fig. IV-5). Planned as a worldwide "touring" or traveling museum, the Nomadic Museum began its life on Pier 54 on the Manhattan waterfront adjacent to the Hudson River in New York City (see Fig. IV-6).

Opening in New York City on March 5, 2005, the museum was disassembled on June 6, 2005 and moved across the United States to the Santa Monica Pier in California. Between December 4, 2005 and February 28, 2006, the Nomadic Museum attracted thousands of visitors who were curious to see "what was inside that big stack of containers."

Following the successful engagement in California, the Nomadic Museum went international with a summer 2006 exhibition at the Vatican in Italy (although other unconfirmed sources have cited Tokyo, Japan as the next destination for this project). In spite of these three successful engagements, however, one important question remains: Why tea bags?

TALK THE TALK

You can't sound smart unless you know what you're talking about. These definitions should help you sound as smart as you look:

CONTAINER: a steel or aluminum box used for the intercontinental transportation of goods. These boxes generally adhere to a strict set of dimensions that make them interchangeable between different manufacturers.

DWELLING: a home or a hovel, take your pick.

MODULAR: a utilitarian prefabricated structure that enables customization through unique building design. Think LEGO brick equals unlimited design possibilities.

PREFABRICATED: a prebuilt unit that can be assembled into a modular dwelling. Wall panels, prehung doors, and floor and ceiling panels are prefabricated elements that rely on economical manufacturing processes and, in turn, provide faster, environmentally friendly building techniques.

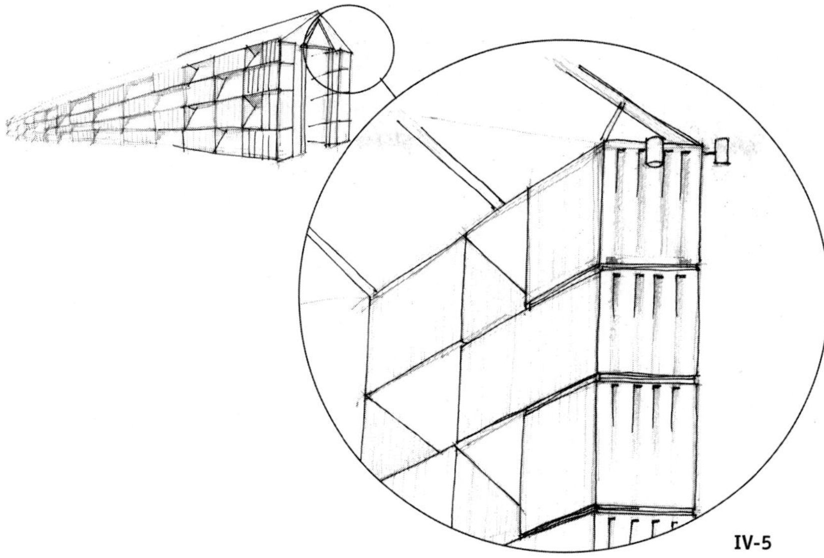

IV-5

NOMADIC MUSEUM - NEW YORK, USA, 2005

This was a temporary museum housing an exhibition of large-scale photographic works by artist Gregory Colbert from March to June in 2005.

It is constructed with paper tubes and 148 shipping containers. 20m-wide by 205m-long and 4500m²-surface.

IV-6

IV-5 A sketch of the Nomadic Museum.

IV-6 The Tokyo-based firm of Shigeru Ban Architects used shipping containers for "housing" the Gregory Colbert traveling photography exhibition "Ashes and Snow." This unique temporary structure was called the Nomadic Museum. (Screen capture courtesy of www.shigerubanarchitects.com/SBA_WOR/SBA_OTHERS/SBA_OTHERS_14/SBA_others_14.html)

FUN FACTS

The "K" Line VERRAZANO BRIDGE has an engine that weighs approximately 2,000 tons.

About 7,000 tons of ballast water had to be pumped into the hold so that the hull could be turned.

It took only 7 1/2 months to build the VERRAZANO BRIDGE at Ulsan Dockyard of Hyundai Heavy Industries Co., Ltd., South Korea.

The tea bags, devoid of tea leaves, were used to fabricate a large suspended curtain that was draped over the museum's entrance. This 34-foot-tall handmade curtain lent a soothing dreamlike aura which played against Colbert's mystical images.

Ban isn't the only architect championing the virtues of shipping containers as building materials. Kathy Prochnow attempted to convince Gulf Coast residents of the merits of modular container architecture as alternatives to Federal Emergency Management Agency (FEMA) trailers (see Fig. IV-7). Other architectural firms, like the New York City–based Resolution: 4 Architecture, have adopted their own spin on marketing custom modular prefabricated dwellings (see Fig. IV-8). None of these architects, however, have matched the level of expertise and design aesthetics as LOT/EK (see Fig. IV-9).

IV-7 A proposed housing solution to hurricane-threatened coastal areas that is derived from shipping containers.

IV-8 Your next modular dream home design can be customized by Resolution: 4 Architecture. These customization concepts look like a great design challenge for LEGO bricks. (Screen capture courtesy of www.re4a.com/modern-modular)

IV-9 LOT/EK is the leading "think tank" for shipping container design. In 2006, they created the Container Home Kit (CHK). (Screen capture courtesy of www.lot-ek.com)

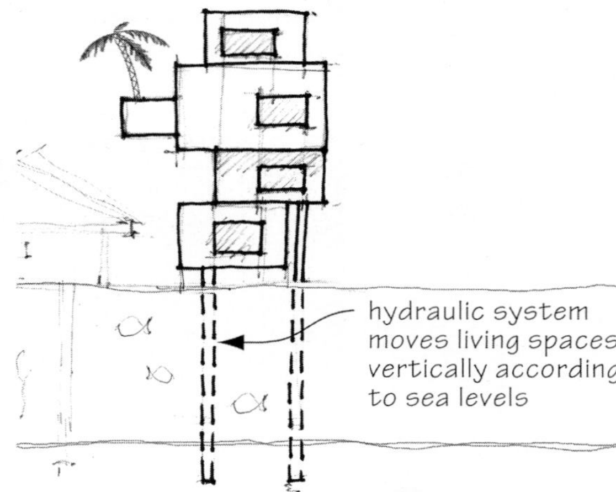
hydraulic system
moves living spaces
vertically according
to sea levels

IV-7

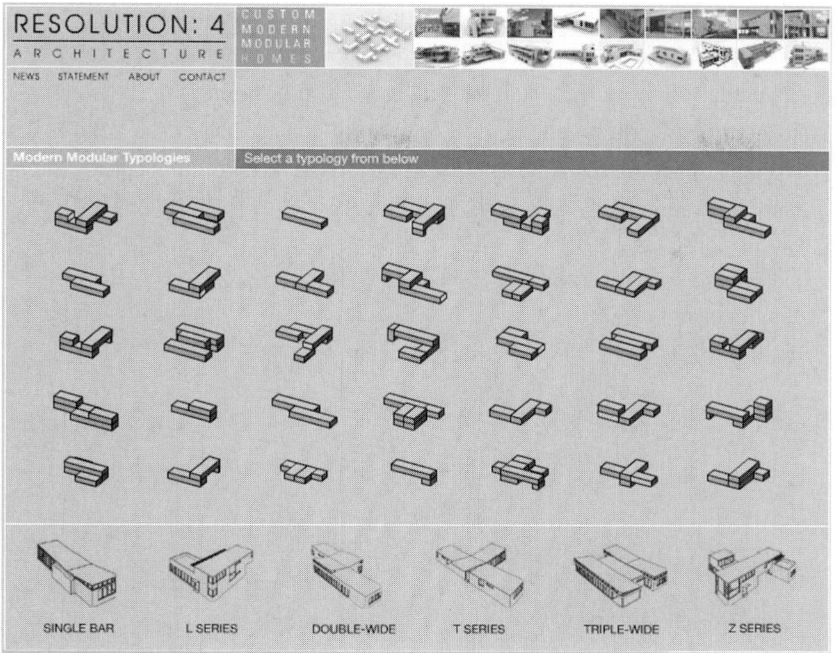

IV-8

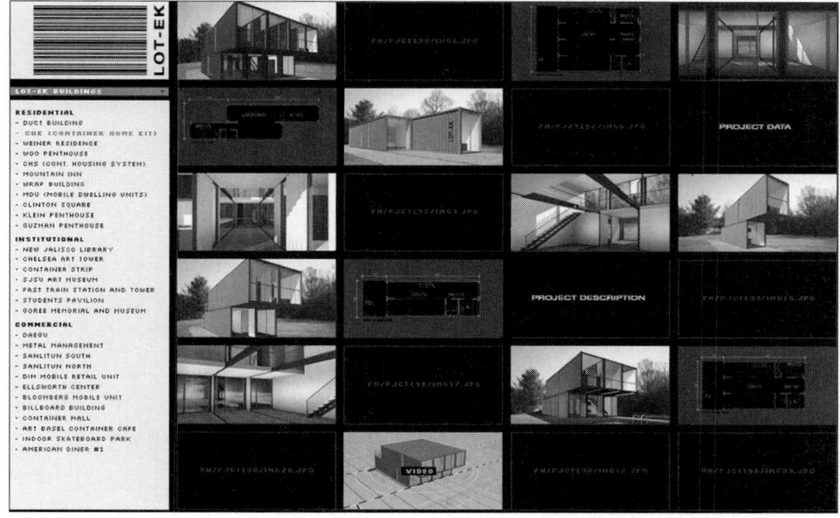

IV-8

LOT/EK (pronounced "low tech") is a New York City–based design firm that has created a buzz among architects with their reutilization of shipping containers as workable modular housing building materials. LOT/EK was founded by partners Ada Tolla and Giuseppe Lignano. These natives of Naples, Italy opened their New York studio after graduating from Columbia University.

The duo that is LOT/EK claims to be blurring the "boundaries between art, architecture, entertainment, and information." Remarkably, their dabbling in shipping container housing continues to strike a favorable chord among prospective homeowners. Home, Sweet Box (see Fig. IV-10).

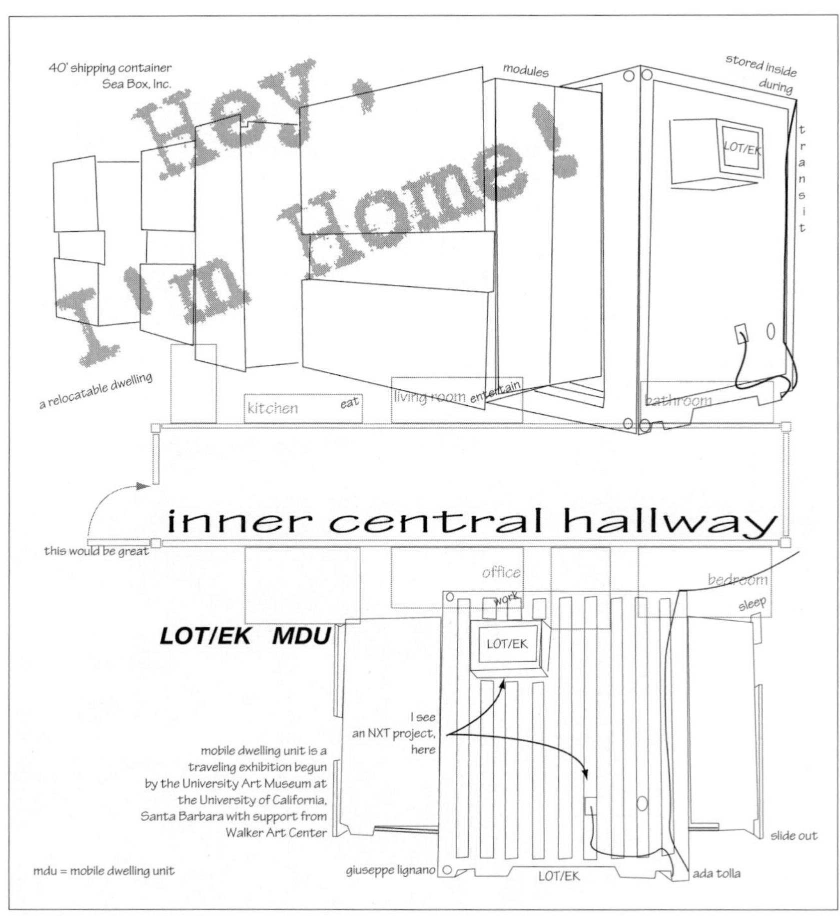

IV-10

IV-10 A sketch of the LOT/EK Modular Dwelling Unit (MDU).

HOW TO BUILD A LEGO DWELLING UNIT (LDU)

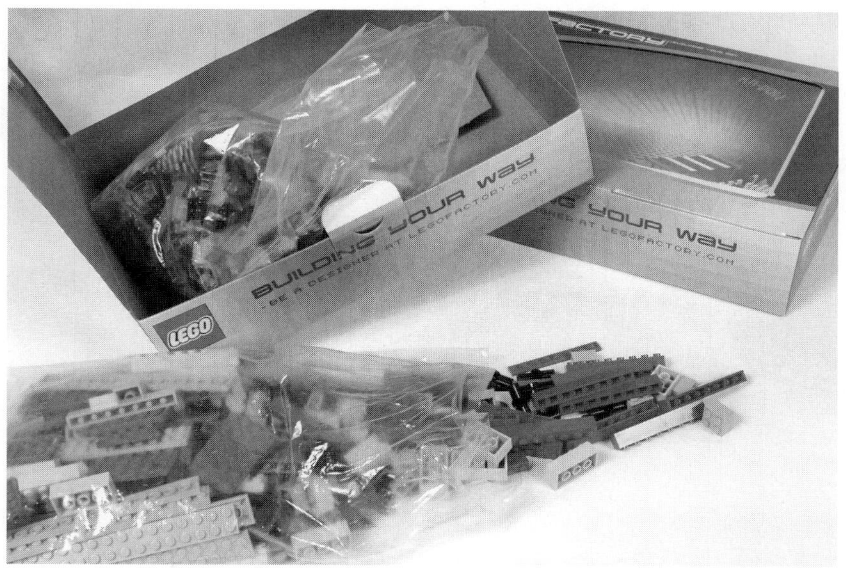

1

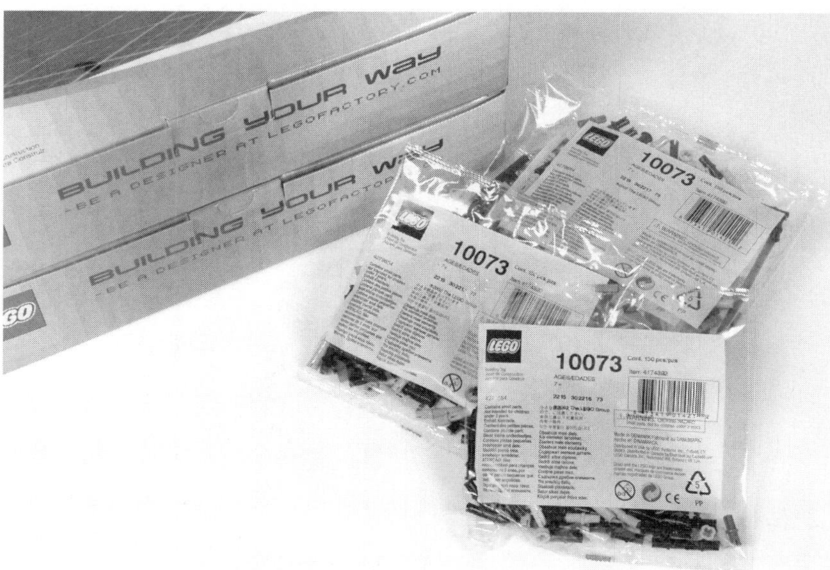

2

1 You can purchase the LDU as a LEGO Factory kit from my Web site: www.pco2go.com/lego.

2 You might want to add some connectors and axles to your purchase for being able to also handle the building of the Hearst Headquarters model, discussed elsewhere in this section. The LDU LEGO Factory also contains parts for building a sample of the Hearst Headquarters model.

3 The parts for building an LDU.

4 The general parts assembly for making the sides of the LDU.

5 Temporarily build a floor for holding the sides of the LDU. The two outer plates will be removed later.

6 Begin with one of the long sides.

3

4

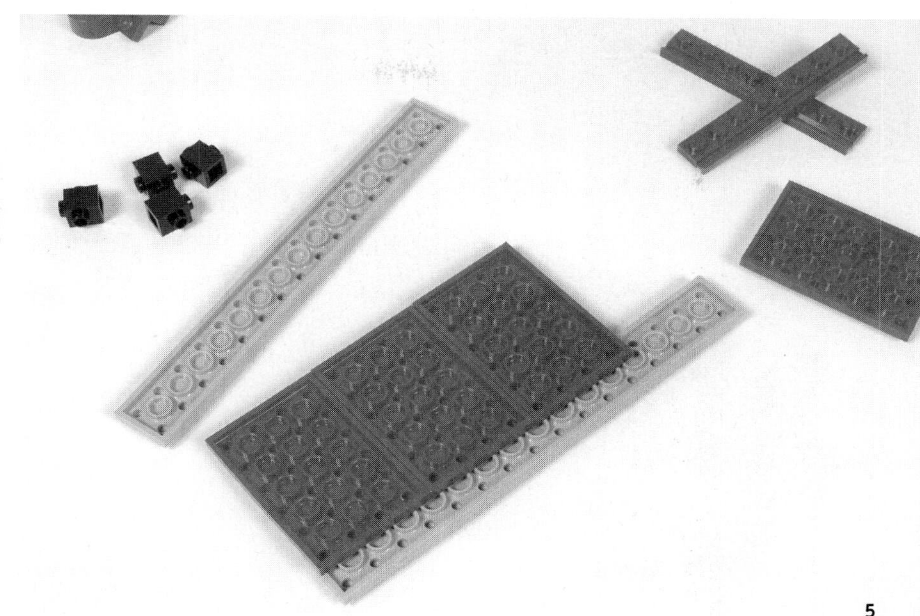

5

6

7 Assemble one complete side.

8 Press this side onto one of the long sides of the floor.

9 Remove the two outer plates from the floor. The LDU side has been removed for clarity.

10 Turn the corner with 1x1 bricks with studs on two sides.

7

8

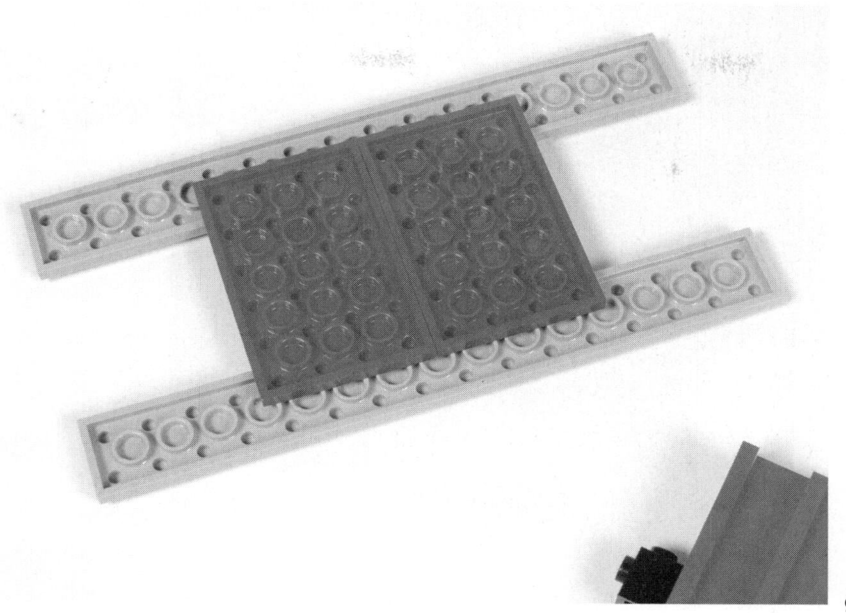

9

10

11 Get ready to turn another corner.

12 Begin the final long side.

13 Complete and ready for your choice of end treatment, roof, and floor.

14 First add a roof.

11

12

13

14

15 Finish the corners with 1x8 tiles.

16 Dress up the roof with 1x8 plates with rails.

17 Use tiles for the edge studs and 1x2 half-arches for each corner.

18 The open end can be finished with doors, windows, clear openings, etc. Act like an architect and be creative.

15

16

17

18

19 Home, Sweet Box. A completed LDU.

20 Combine more than one LDU.

21 Remove exterior wall sections and add some interior floor space.

22 Now hoist in some NXT equipment and you're ready for living the life of Riley.

19

20

21

22

23 **How about an LDU security system? Looks good, doesn't it?**

23

LDU* BRICK LIST

NUMBER OF BRICKS: 1260

1 - 2377 Wall Element 1x2x2 W. Window White
1 - 3665 Roof Tile 1x2, Inverted White
2 - 6182 Arch 1x4x2 White
30 - 3001 Brick 2x4 Brick Yellow
9 - 3002 Brick 2x3 Brick Yellow
12 - 3007 Brick 2x8 Brick Yellow
2 - 3031 Plate 4x4 Brick Yellow
6 - 3958 Plate 6x6 Brick Yellow
25 - 6141 Round Plate 1x1 Brick Yellow
1 - 2921 Brick 1x1 W. Handle Bright Red
3 - 3005 Brick 1x1 Bright Red
48 - 3008 Brick 1x8 Bright Red
1 - 30133 Scarf Bright Red
1 - 3020 Plate 2x4 Bright Red
1 - 3022 Plate 2x2 Bright Red
1 - 3023 Plate 1x2 Bright Red
2 - 3024 Plate 1x1 Bright Red
2 - 2412 Radiator Grille 1x2 Bright Red
16 - 3032 Plate 4x6 Bright Red
1 - 3046 Corner Brick 2x2/45 Inside Bright Red
2 - 3069 Flat Tile 1x2 Bright Red
2 - 3623 Plate 1x3 Bright Red
2 - 3660 Roof Tile 2x2/45, Inverted Bright Red
2 - 3665 Roof Tile 1x2, Inverted Bright Red
1 - 3666 Plate 1x6 Bright Red
1 - 3794 Plate 1x2 W. 1 Knob Bright Red
1 - 3853 Window Frame 4x3 Bright Red
1 - 4070 Angular Brick 1x1 Bright Red
16 - 4162 Flat Tile 1x8 Bright Red
1 - 73312 Front Door 4x5 Right Bright Red
58 - 4510 Plate 1x8 With Rail Bright Red
26 - 50746 Roof Tile 1 X 1 X 2/3 Bright Red

* LEGO Dwelling Unit

8 - 6091 Brick W. Arch 1x1x1 1/3 Bright Red
1 - 6141 Round Plate 1x1 Bright Red
9 - 6636 Flat Tile 1x6 Bright Red
2 - 3020 Plate 2x4 Bright Blue
3 - 3068 Flat Tile 2x2 Bright Blue
3 - 3069 Flat Tile 1x2 Bright Blue
1 - 3624 Mini Cap Bright Blue
2 - 3660 Roof Tile 2x2/45, Inverted Bright Blue
1 - 3710 Plate 1x4 Bright Blue
1 - 73200 Mini Lower Part Bright Blue
5 - 2555 Plate 1x1 W. Upright Holder Black
60 - 3004 Brick 1x2 Black
103 - 3005 Brick 1x1 Black
5 - 30374 Light Sword - Blade Black
60 - 32034 Angle Element, 180 Degrees [2] Black
1 - 3626 Mini Head Black
2 - 3666 Plate 1x6 Black
40 - 3700 Technic Brick 1x2, Ø4.9 Black
120 - 3705 Cross Axle 4m Black
1 - 4083 Hanger 1x4x2 Black
1 - 44676 Banner 26 Deg. M. 2 Holdere Black
30 - 47905 Brick 1x1 W. 2 Knobs Black
40 - 6541 Technic Brick 1x1 Black
1 - 3471 Spruce Tree H64 Dark Green
6 - 3795 Plate 2x6 Dark Green
8 - 3023 Plate 1x2 Transparent
71 - 3024 Plate 1x1 Transparent
279 - 3065 Brick 1x2 Without Tap Transparent
1 - 3855 Glass 23, 75x28, 5x2 Transparent
1 - 4862 Pane For Wall Element Transparent
4 - 6141 Round Plate 1x1 Transparent
2 - 3023 Plate 1x2 Transparent Red
3 - 3024 Plate 1x1 Transparent Red
1 - 6141 Round Plate 1x1 Transparent Red
1 - 6141 Round Plate 1x1 Transparent Yellow
1 - 6141 Round Plate 1x1 Transparent Fluorescent Green

1 - 49668 Plate 1x1 W/Tooth, Pc Transparent Brown

30 - 4864 Wall Element - 1x2x2 Transparent Brown

1 - 173200 Mini Lower Part Earth Blue

1 - 3624 Mini Cap Dark Red

1 - 2555 Plate 1x1 W. Upright Holder Medium Stone Grey

6 - 3001 Brick 2x4 Medium Stone Grey

6 - 3003 Brick 2x2 Medium Stone Grey

1 - 3070 Flat Tile 1x1 Medium Stone Grey

32 - 3004 Brick 1x2 Medium Stone Grey

3 - 3008 Brick 1x8 Medium Stone Grey

3 - 3010 Brick 1x4 Medium Stone Grey

6 - 4162 Flat Tile 1x8 Medium Stone Grey

8 - 4282 Plate 2x16 Medium Stone Grey

1 - 3032 Plate 4x6 Dark Stone Grey

1 - 3795 Plate 2x6 Dark Stone Grey

1 - 83447 Mini Head 'no. 126' Bright Yellow with Glasses

1 - 176382 Mini Upper Part Earth Blue with Driver Outfit

1 - 76382 Mini Upper Part White with Mailman Outfit

THE BUCKY STOPS HERE

He was born in 1895 in Milton, Massachusetts as Richard Buckminster Fuller. But the world grew to know him better as Bucky. Bucky was a consummate tinker, inventor, engineer, and teacher who had a penchant for walking to a different drummer. You can get a glimpse of his unique train of thought by studying Fuller's scholarly works.

In 1950, for example, Fuller set up an outline for a course in Comprehensive Anticipatory Design Science. Later, he taught this course at MIT in 1956 as part of the Creative Engineering Laboratory. The students who took the course—engineers, industrial designers, materials scientists, and chemists—represented research and development corporations from across America.

Bucky's course syllabus was written before his Dymaxion Map reached its final, icosahedral phase and followed the publication of his seminal theory titled

IT'S A DOME NATION

If you're interested in bringing structural integrity into you life, there are several dome manufacturers who can help you get out of the corner that you've painted yourself in to.

AMERICAN INGENUITY, INC.— A basic 40-foot diameter dome "kit" with 2176 square feet of living space costs $24,644. "Extras" like dormers, entryways, cupolas, and skylights are available at an additional cost.

OREGON DOME, INC.— If you're more of a "window shopper," then consider a plan book purchase prior to raising your own dome. This modest $15 investment can go a long way toward ensuring that you are ready for a life in the round.

PACIFIC DOMES, INC.— Conversely, this company's basic 36-foot kit costs $15,300. These are more "casual" shelters with a fabric covering or skin that is stretched over a tubular steel frame.

TIMBERLINE GEODESIC DOMES— In a true do-it-yourself mentality, these dome kits use a unique connector system that mates with precut, predrilled, color-coded lumber. Hmm, that sounds *very* familiar. Doesn't it? I just can't seem to put my finger on it, however. Oh well, now I can't remember the name of that "other" connector-based building system.

LEGO OF MY TENSEGRITY

Inside his fabulous book *Geodesic Math and How to Use It*, Hugh Kenner provides a neat model that demonstrates the tensional integrity found in a geodesic dome. In an admiring nod to Kenner, see if you can adapt his elegant model to the LEGO beam elements that are available in your NXT kit. Try using three 11m beams (roughly 3 3/8 inches long) and nine pieces of string measuring about 2 1/2 inches long (roughly two-thirds the length of the 11m beams). Tie the string pieces through the holes in each end of the beams. Make sure every end hole of the beams has three pieces of string tied to it. You will have a total of six vertices. Tighten the last string and *voilà*, tensegrity.

Synergetics I and 2 by 25 years. This is a theory which, at its outset in 1927, promised to develop international effectiveness of an individual's initiation despite the prevailing economic constraints. The name of Fuller's theory was a *Industrially Realizable, Comprehensive, Anticipatory Design Science*. As expressed in the words of "Bucky Speak," "if adopted by one individual and proved to be valid, it then must become comprehensively operative even though none other than the initiating individual ever consciously and formally adopted it. It would, however, emerge into popular adoption only through emergency."

Whew; that's some heady stuff.

Back on Earth, two of the more celebrated Bucky designs which gained some modest popularity are the Dymaxion Car and the geodesic dome—both of which were awarded patents (U.S. Patent 2,101,057, patented December 7, 1937 and U.S. Patent 2,682,235, patented June 29, 1954, respectively). [Note: Also, please refer to this subsequent dome patent for building construction, U.S. Patent 2,881,717, patented April 14, 1959.]

In each of these designs, the public was simultaneously perplexed and engaged by the "space age" look of a three-wheeled, bullet-shaped automobile and a wall-less, lightweight, igloo-shaped building structure. And while the Dymaxion Car faded into oblivion, the geodesic dome has slowly been accepted by the public—albeit as an interesting, novelty structure (see Fig. IV-11).

Built in a Bridgeport, Connecticut factory by the 4D Company, the Dymaxion Car soon captivated the interest of the automotive industry. Sporting a top speed of 120 mph, a streamlined monocoque body, and a rearward sighting periscope, the Dymaxion Car was design epiphany, but a marketing failure. Built

IV-11 Epcot Center at Walt Disney World in Orlando, Florida.

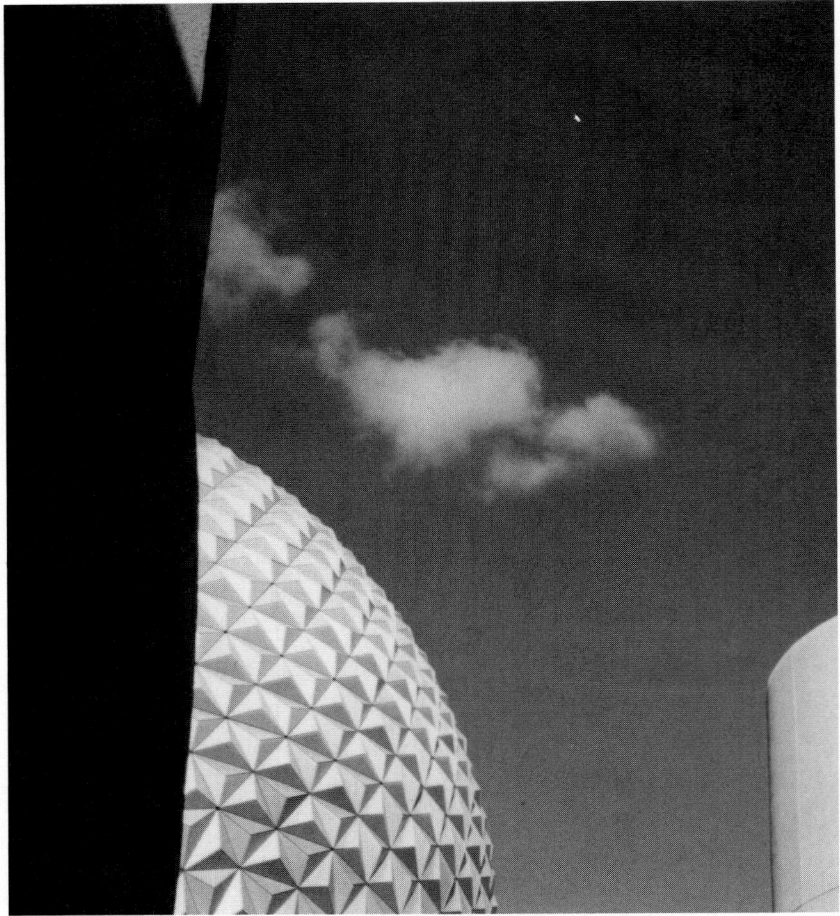

IV-11

by Starling Burgess and financed by Anna Biddle, the 4D Company was only able to build three prototypes of this streamline beauty (err, that's the car, not the lovely Ms. Biddle).

Although the radical innovations in the Dymaxion Car have yet to be integrated into our modern automobiles, it was the unconventional rear steering mechanism that proved to be its undoing. In fact, a fatal test drive in one of the prototypes has been attributed to a steering error—not a mechanical error, but an operator error. Even the master tinker himself was not immune to driving mishaps with the Dymaxion Car. Fuller rolled one of the prototypes, but escaped any injury.

Only one of the Dymaxion Car prototypes exists today. You can see car #2 at the National Automobile Museum in Reno, Nevada.

HOW TO BUILD A MODEL OF BUCKY'S DYMAXION CAR

1 An NXT servo motor.

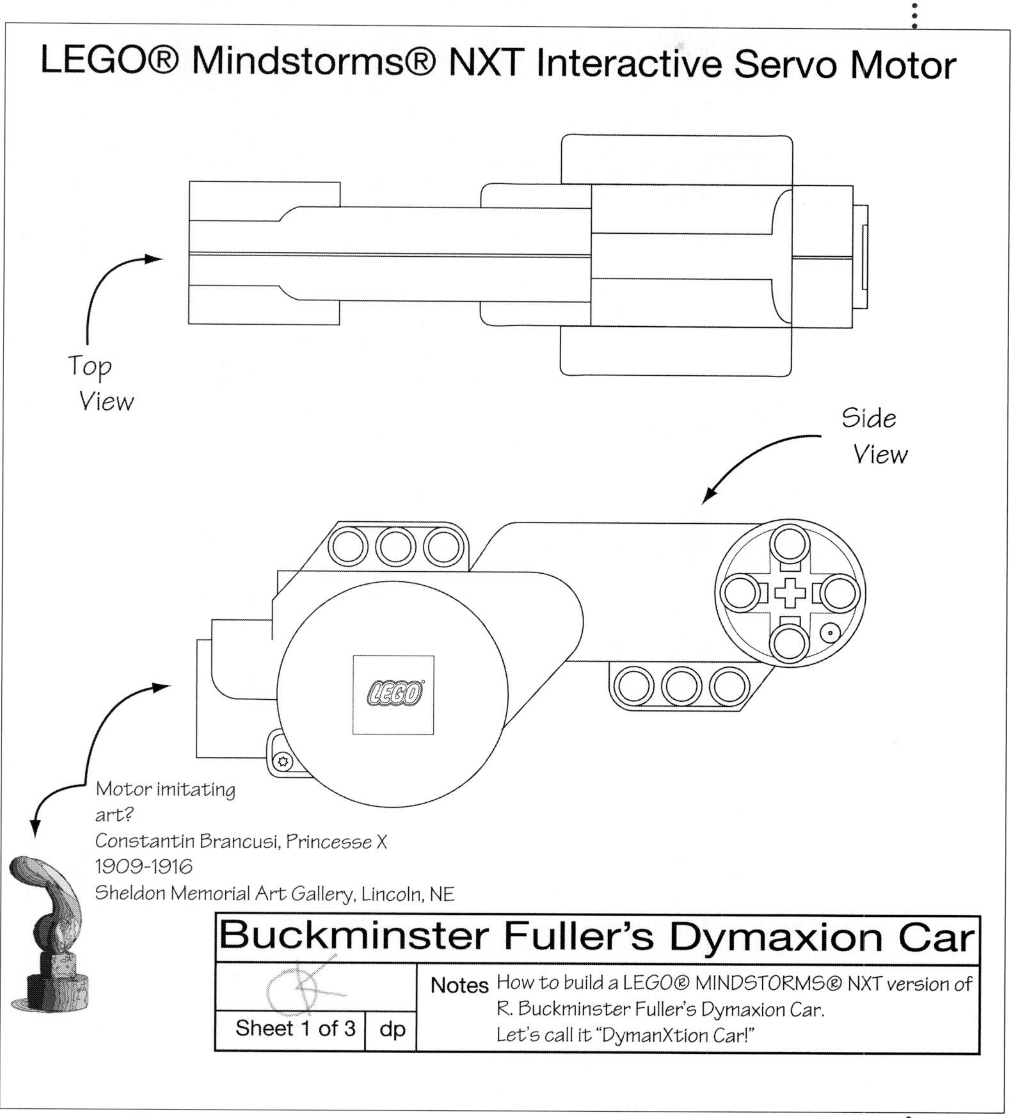

LEGO® Mindstorms® NXT Interactive Servo Motor
Top View
Side View
LEGO
Motor imitating art?
Constantin Brancusi, Princesse X
1909-1916
Sheldon Memorial Art Gallery, Lincoln, NE
Buckminster Fuller's Dymaxion Car
Sheet 1 of 3
dp
Notes How to build a LEGO® MINDSTORMS® NXT version of R. Buckminster Fuller's Dymaxion Car.
Let's call it "DymanXtion Car!"

1

2 A plan for the complete Dymaxion Car.

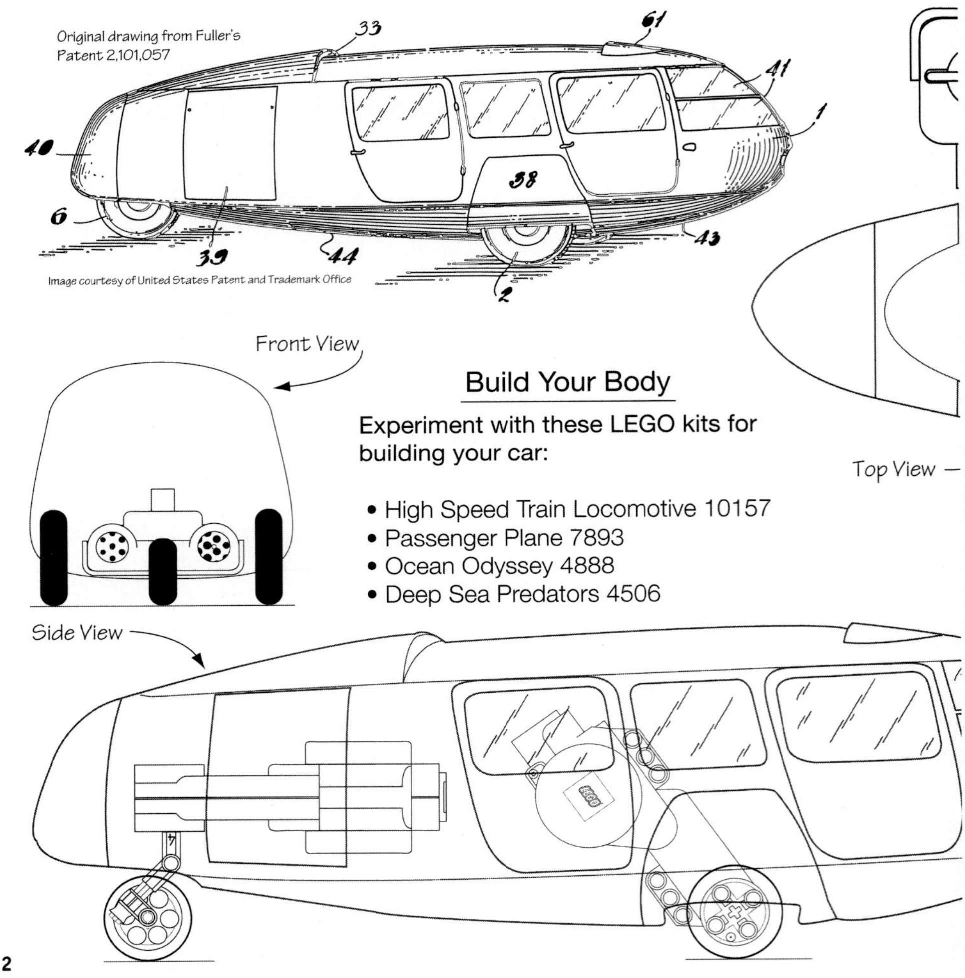

Build Your Body

Experiment with these LEGO kits for building your car:

- High Speed Train Locomotive 10157
- Passenger Plane 7893
- Ocean Odyssey 4888
- Deep Sea Predators 4506

2

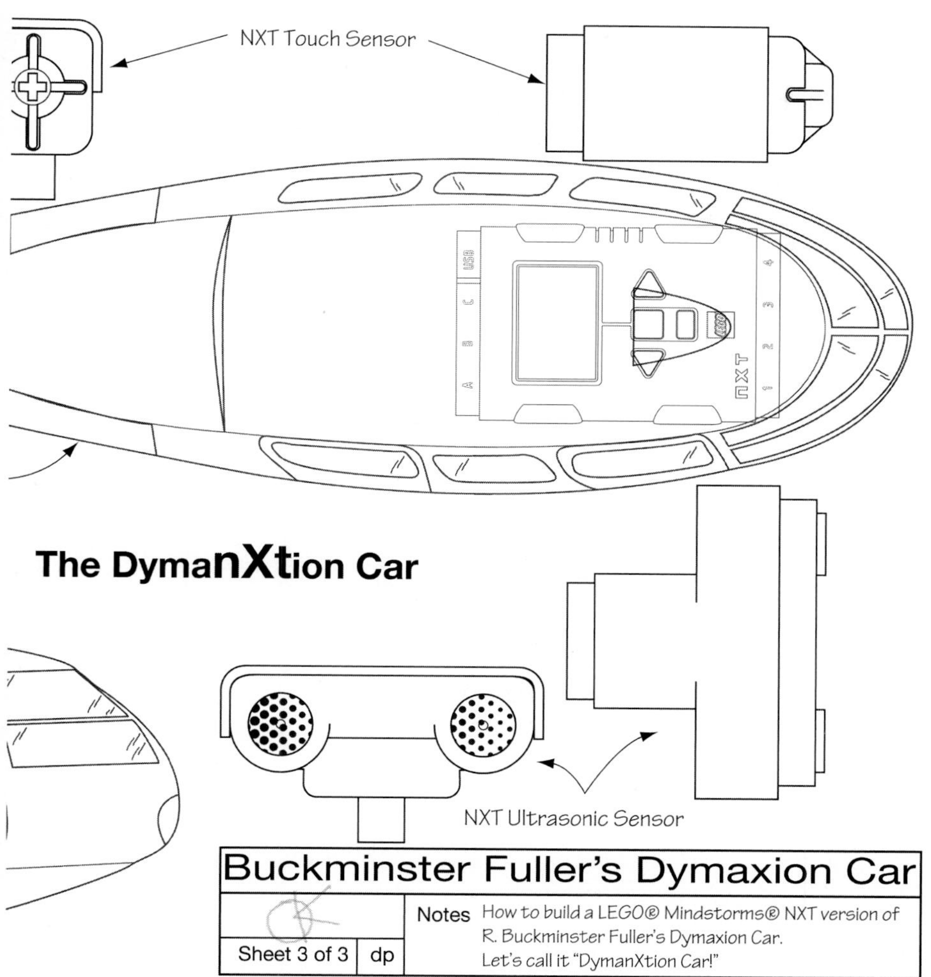
NXT Touch Sensor
The DymanXtion Car
NXT Ultrasonic Sensor
Buckminster Fuller's Dymaxion Car
Notes How to build a LEGO® Mindstorms® NXT version of R. Buckminster Fuller's Dymaxion Car.
Let's call it "DymanXtion Car!"
Sheet 3 of 3
dp

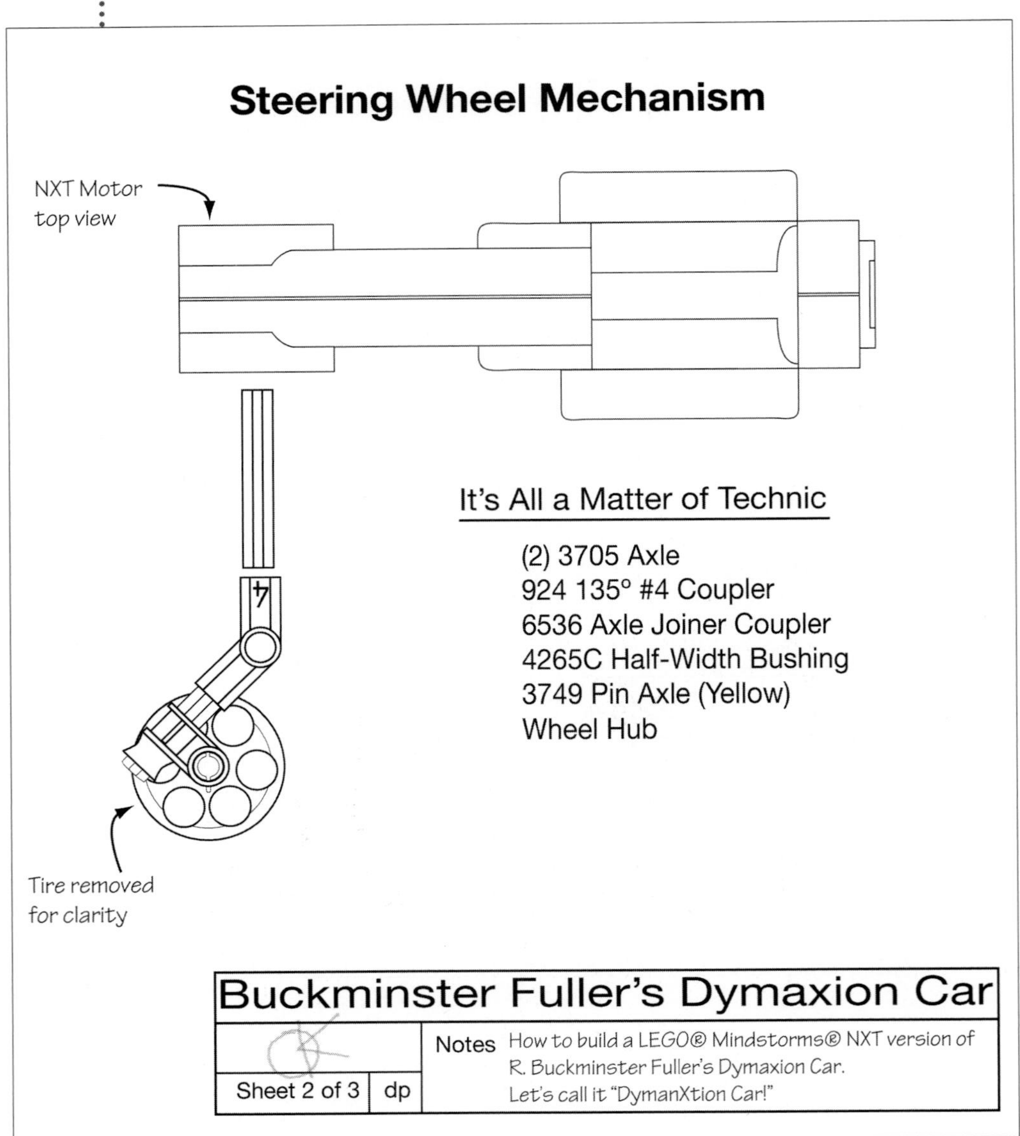
Steering Wheel Mechanism
NXT Motor
top view
4
Tire removed
for clarity
It's All a Matter of Technic
(2) 3705 Axle
924 135° #4 Coupler
6536 Axle Joiner Coupler
4265C Half-Width Bushing
3749 Pin Axle (Yellow)
Wheel Hub
Buckminster Fuller's Dymaxion Car
Notes How to build a LEGO® Mindstorms® NXT version of
R. Buckminster Fuller's Dymaxion Car.
Let's call it "DymanXtion Car!"
Sheet 2 of 3
dp

3

4

3 A plan for a steering mechanism.

4 First, we're gonna have to hoist a motor into place.

5 The parts for building a steering mechanism.

6 Add an axle and angled coupler.

7 Add an axle joiner coupler and another axle.

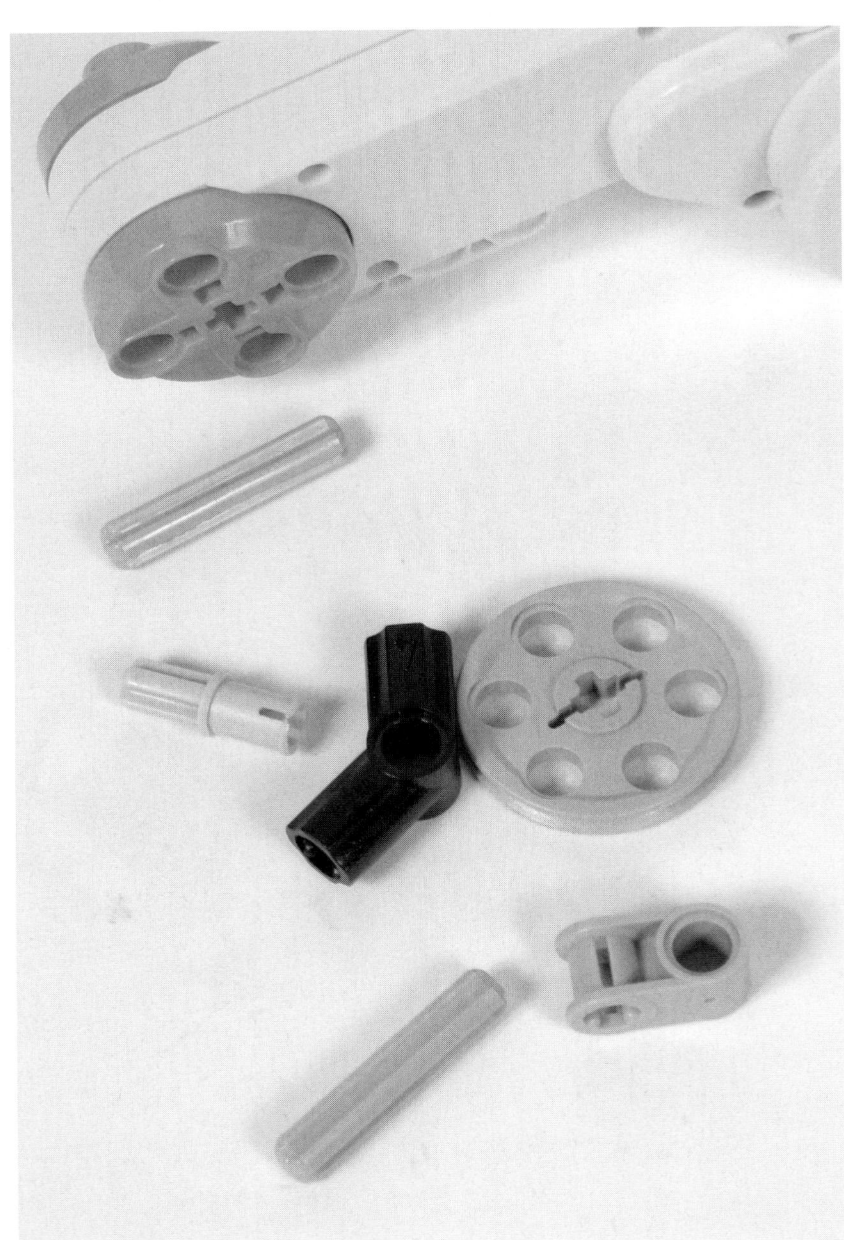

5

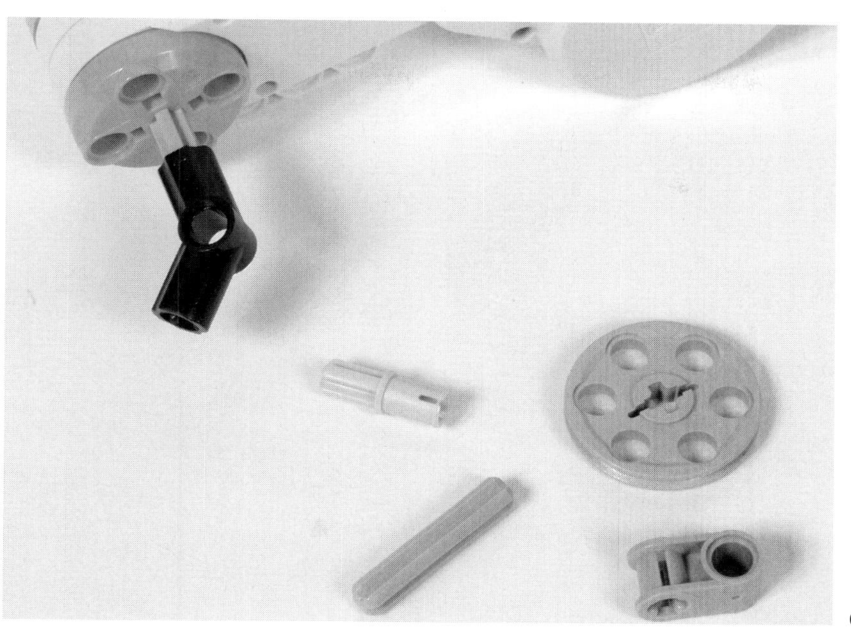
6

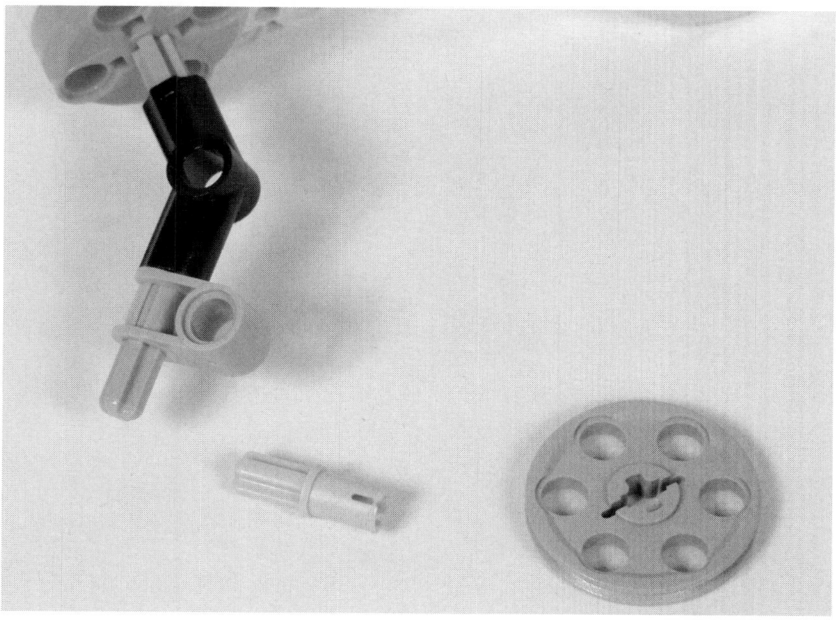
7

8 Insert a connector without friction into a pulley.

9 This pulley will become your steering tire.

10 Finished and ready for a rubber tire.

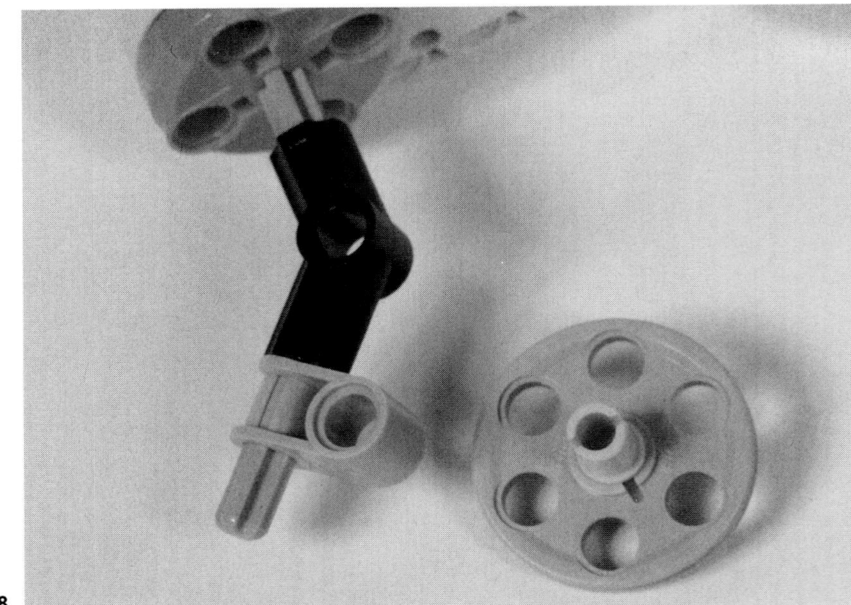

8

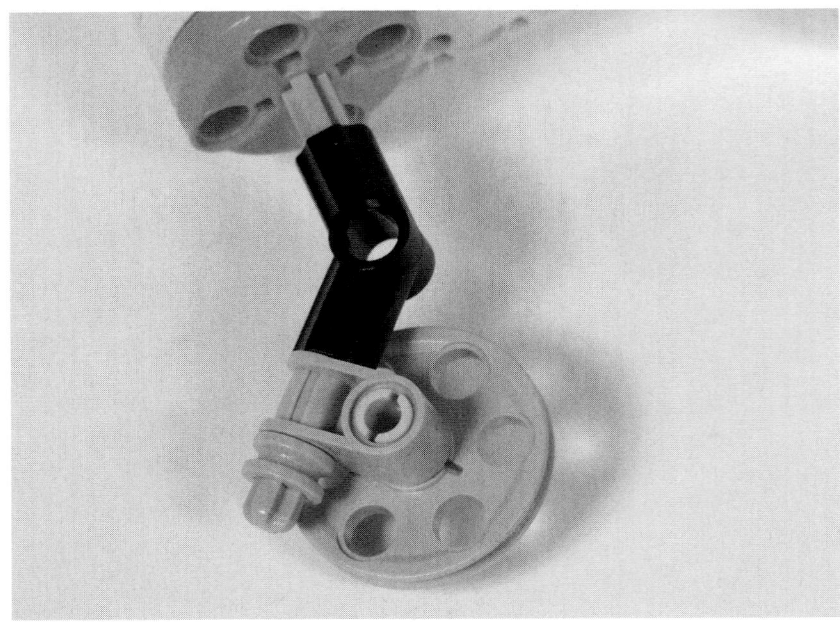

9

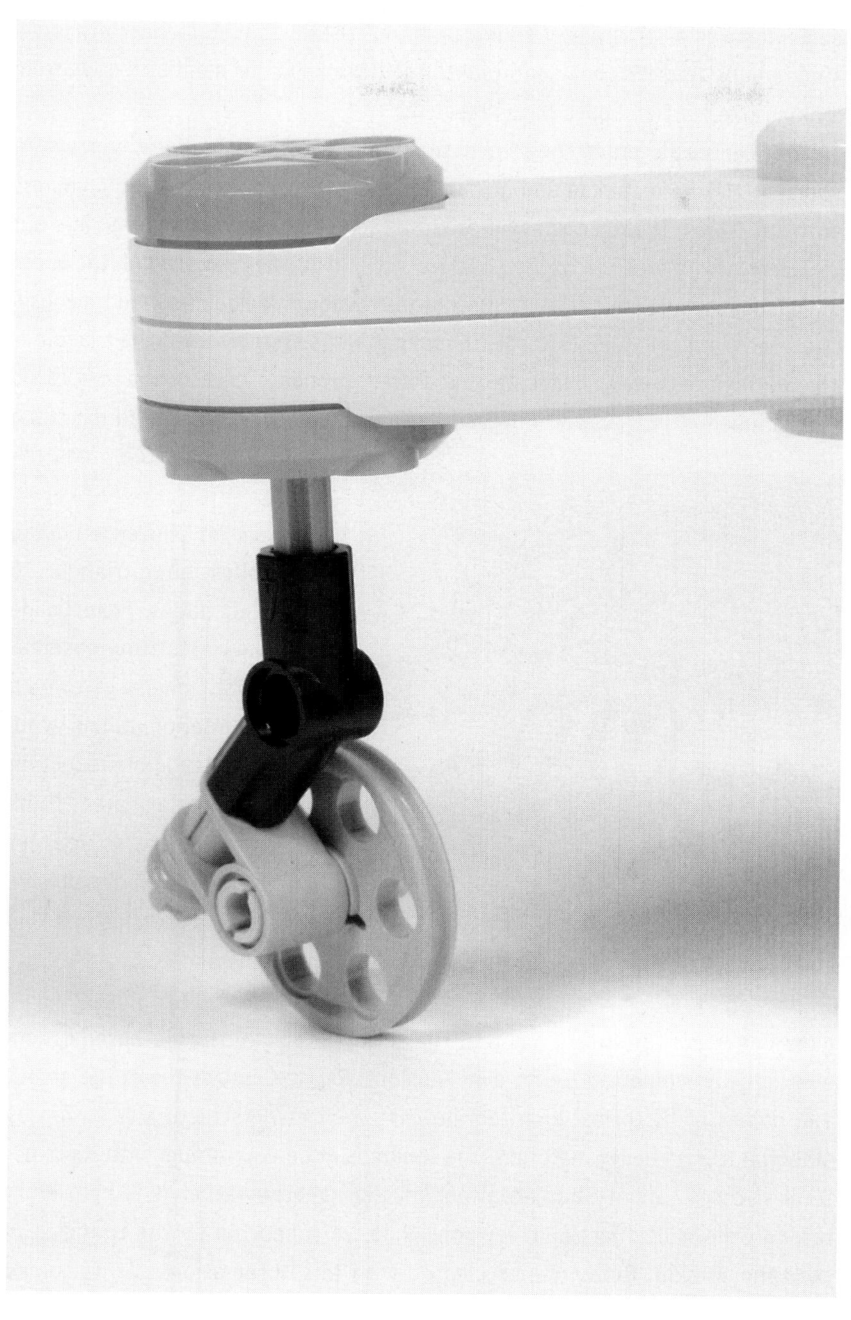

10

Following in similar unconventional steps (or, is that tracks?) as the Dymaxion Car is the geodesic dome. Backed by some serious math, the geodesic dome is a structurally sound building system that unfortunately looks kinda goofy. And it's these odd looks that have kept our neighborhoods from looking like Martian space bases (see Fig. IV-12).

You can easily prove the structural soundness of the geodesic concept to yourself. Make a rectangle and a triangle from the same material and compress both structures. The rectangle will collapse (actually, as Hugh Kenner points out in *Geodesic Math and How to Use It*; University of California Press, 2003, the upper member of the rectangle is compressed while the lower member is simultaneously stretched), but the triangle is able to endure the pressure. In fact, the triangle is able to tolerate twice as much pressure as the rectangle. The structural resistance that is inherent in triangles intrigued Bucky and he devised a method for lacing a sphere with intertwined triangles.

IV-12 Botanical Garden at Des Moines, Iowa.

Once he started studying the application of triangles to spherical shapes, Fuller made a couple of startling observations. First, spheres give you a lot more interior volume while providing a tangible reduction in exterior surface area. Think of this discovery as you get more from less. Or, more inside space for less outside building materials and cost.

Fuller's second, and even more important, discovery was that the structure of the triangle-based sphere was actually supported by its own tension. You see, unlike the stone arches and domes of historical architecture that were made structurally strong by being built from heavy materials, the sphere is at an equilibrium with its structural stresses. Gravity has no effect on the sphere—rather it is held erect through the multidirectional tensions that are supported by the triangles in the sphere's skin. Bucky was so pleased with this observation that he coined the term *tensegrity*, or, tensional integrity.

And that, ladies and gentlemen, is what makes the geodesic dome tick.

HOW TO BUILD A GEODESIC BUCKY DOME

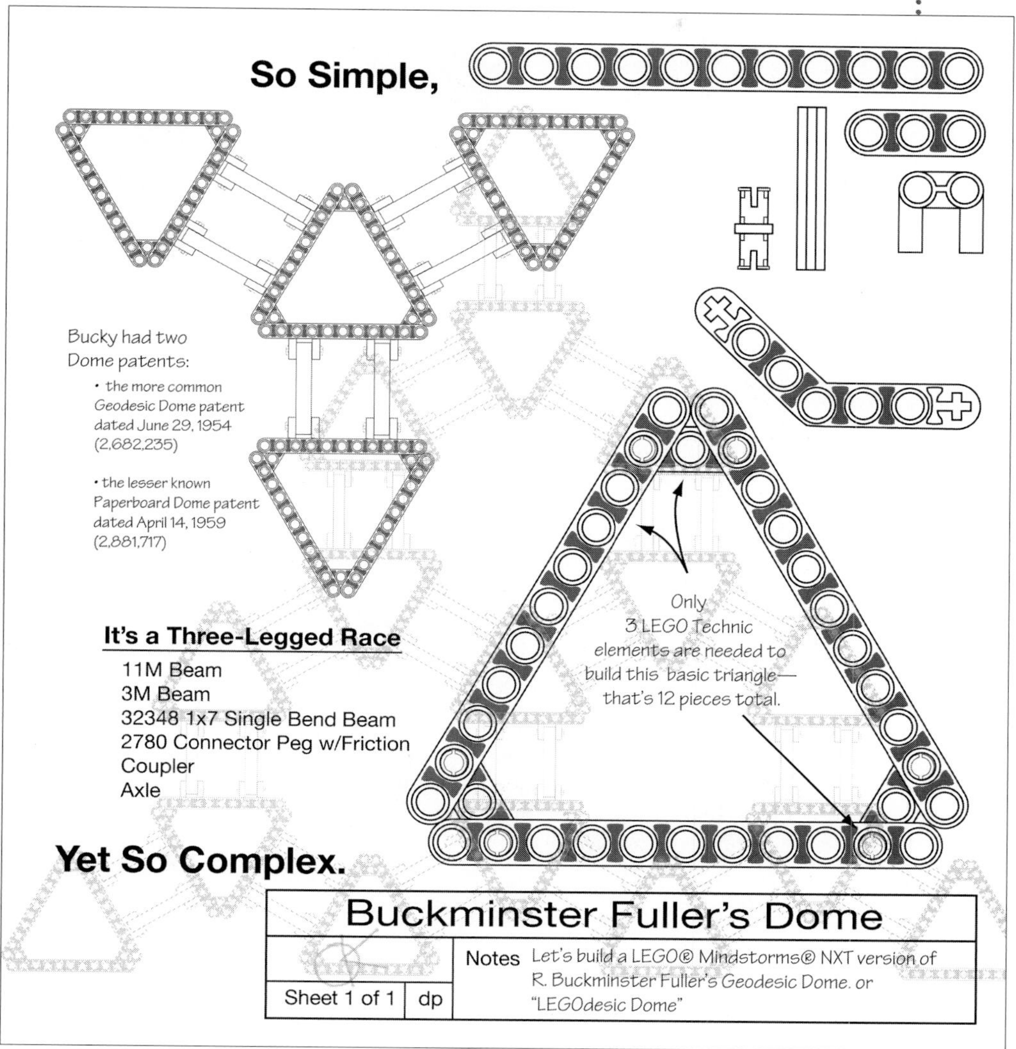

So Simple,
Bucky had two Dome patents:
• the more common Geodesic Dome patent dated June 29, 1954 (2,682,235)
• the lesser known Paperboard Dome patent dated April 14, 1959 (2,881,717)
It's a Three-Legged Race
11M Beam
3M Beam
32348 1x7 Single Bend Beam
2780 Connector Peg w/Friction
Coupler
Axle
Only 3 LEGO Technic elements are needed to build this basic triangle—that's 12 pieces total.
Yet So Complex.
Buckminster Fuller's Dome
Notes Let's build a LEGO® Mindstorms® NXT version of R. Buckminster Fuller's Geodesic Dome. or "LEGOdesic Dome"
Sheet 1 of 1
dp

1 A plan for a Geodesic dome.

2 Only a handful of parts are needed for building a sample dome section.

3 First form a triangle. Use 3m beams for connectors.

4 Attach the beams with two-stud friction connectors.

5 The beams will move slightly before the entire dome structure is finished.

2

3

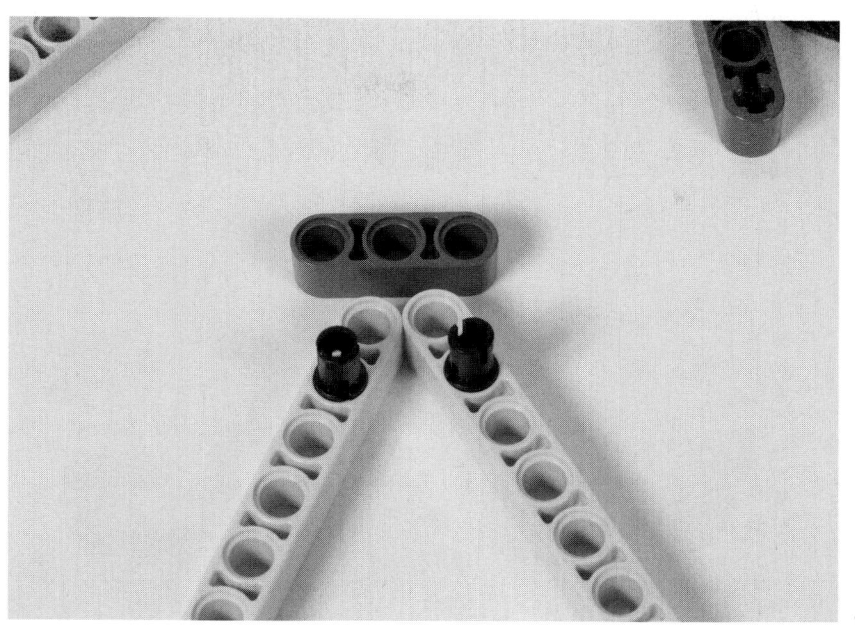

4

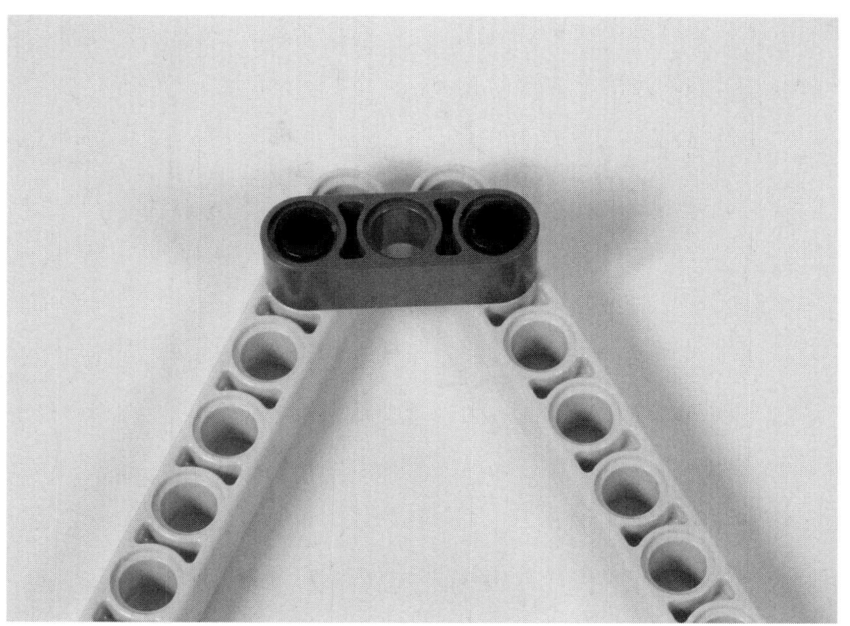

5

6 **Make two more connections.**

7 **A finished triangle.**

8 **Two completed triangles.**

9 **Use angled beams for joining triangles.**

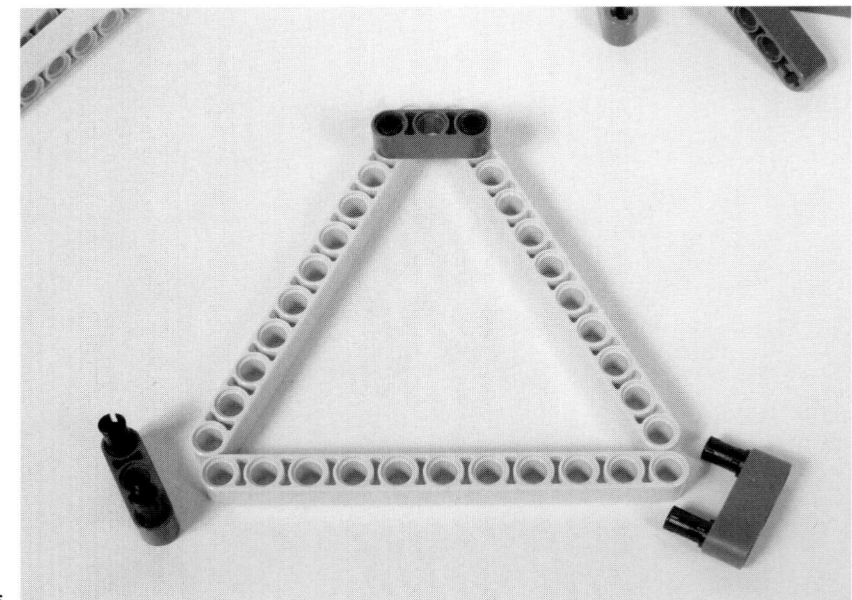

6

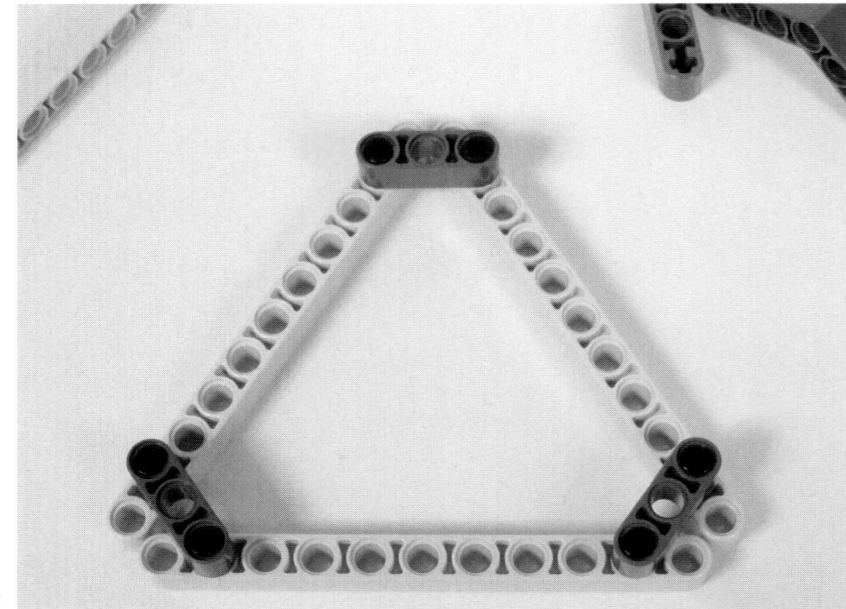

7

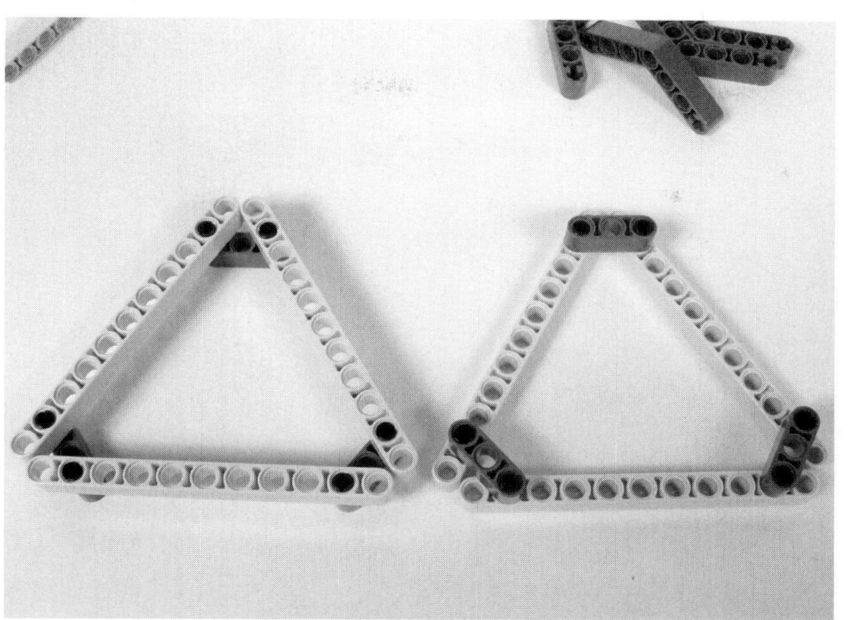

8

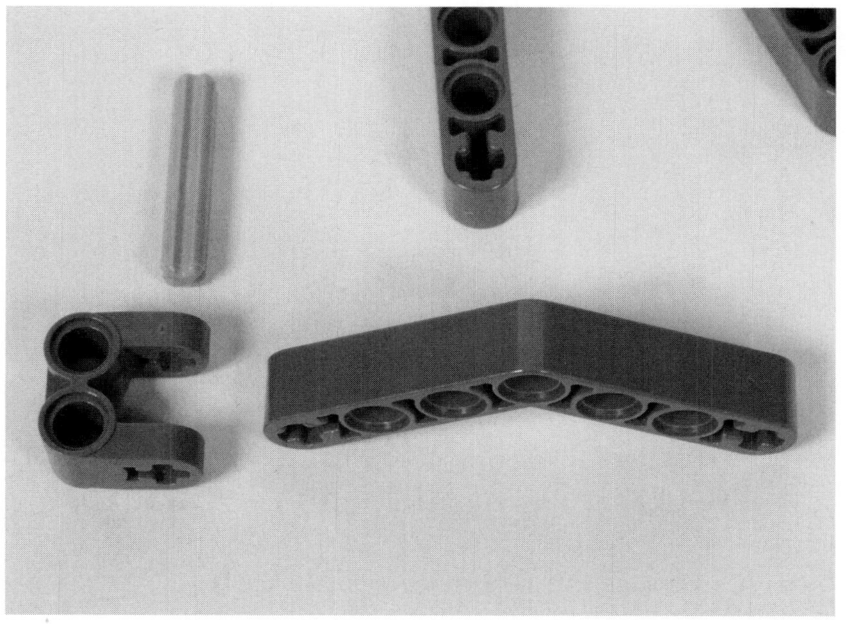

9

10 Fix one two-connector axle joiner to one end of the angled beam.

11 Connect the assembly to one triangle.

12 Add connectors to the axle joiner.

13 Attach this assembly to the triangle. You can make your connections to either face of a triangle.

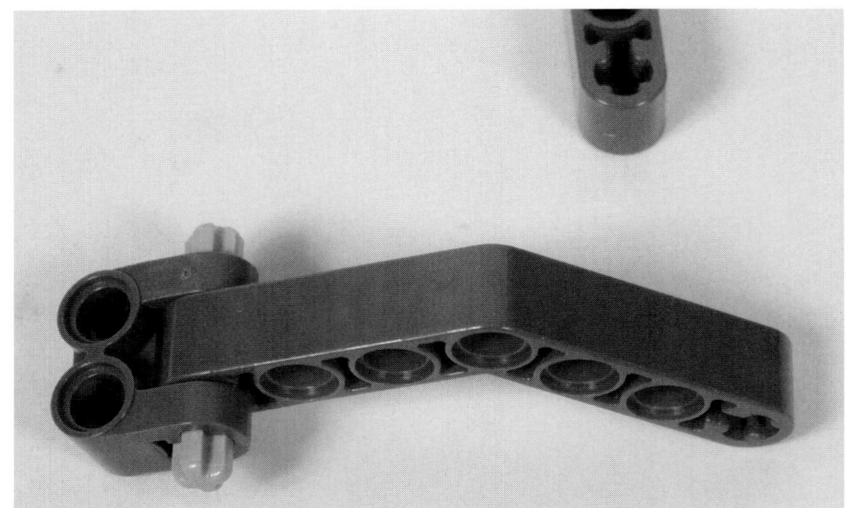

10

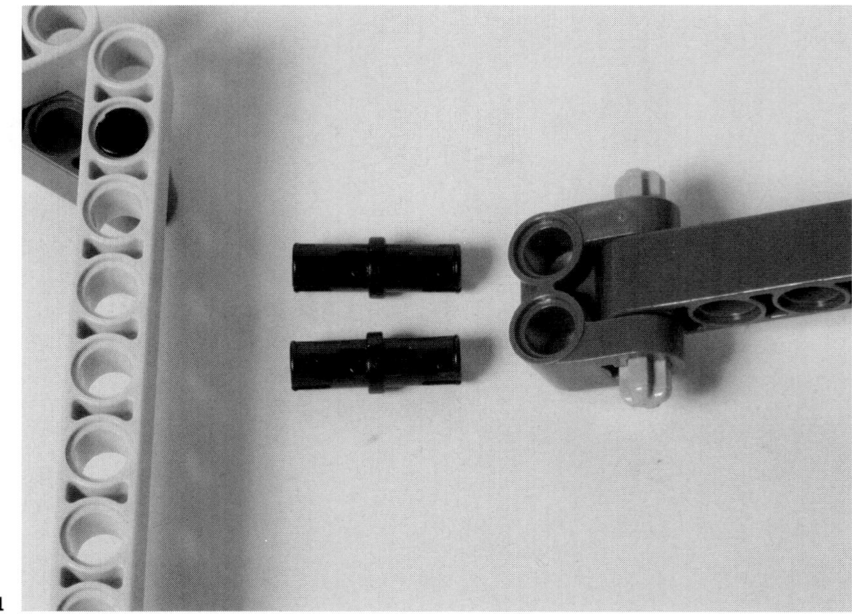

11

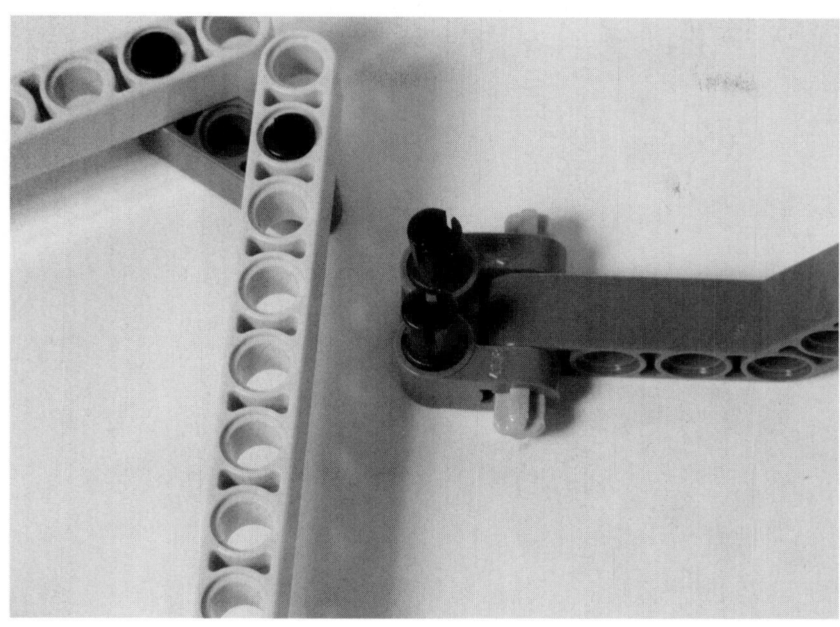

12

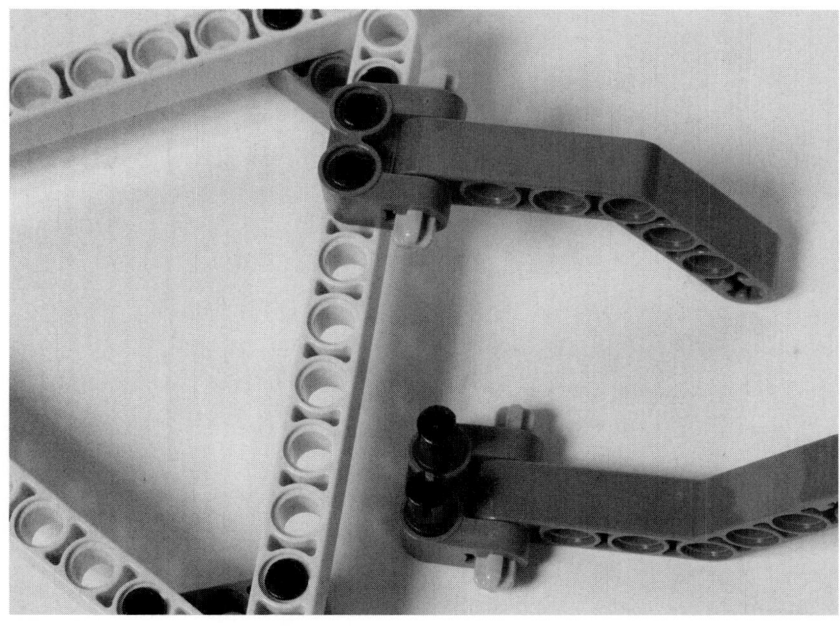

13

14 Just make sure that you are consistent with your connections. In other words, both connectors should be attached to the same face of a triangle.

15 Prep the other end of your angled beams.

16 Attach the second end of the angled beams to the same face of another triangle.

17 A completed joint.

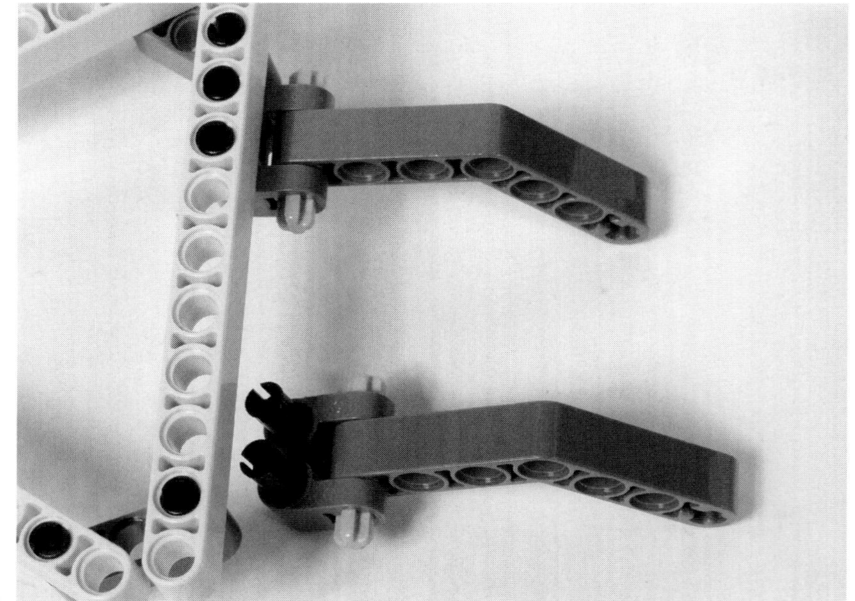

14

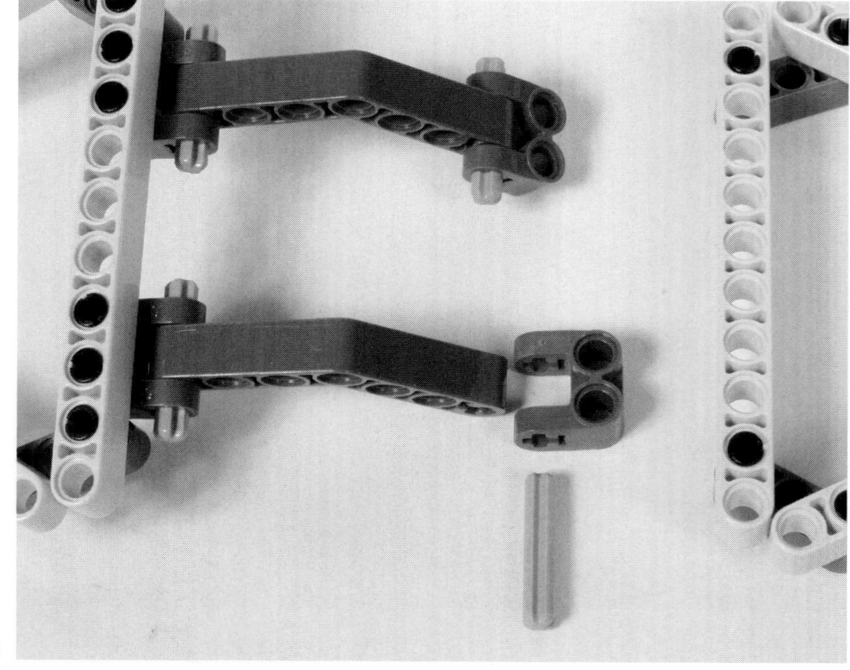

15

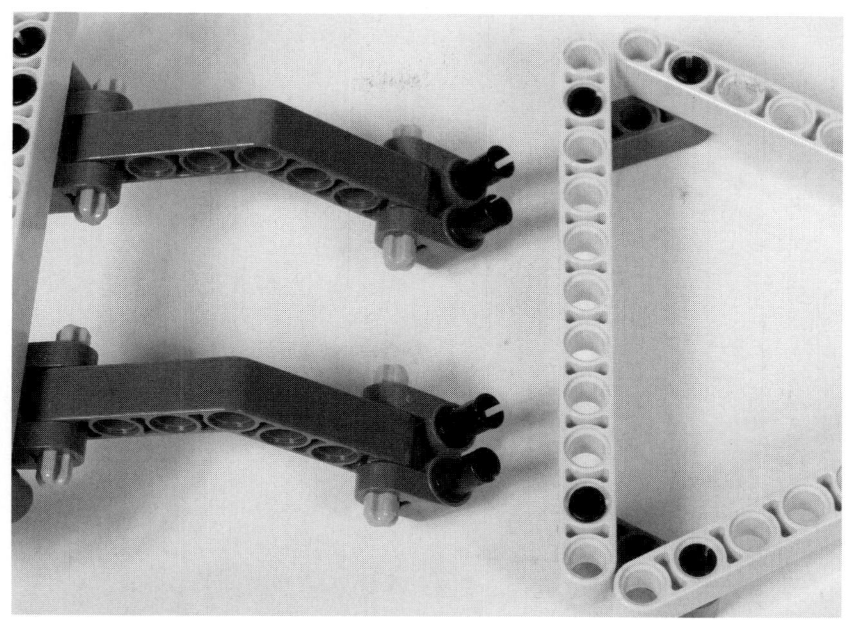

16

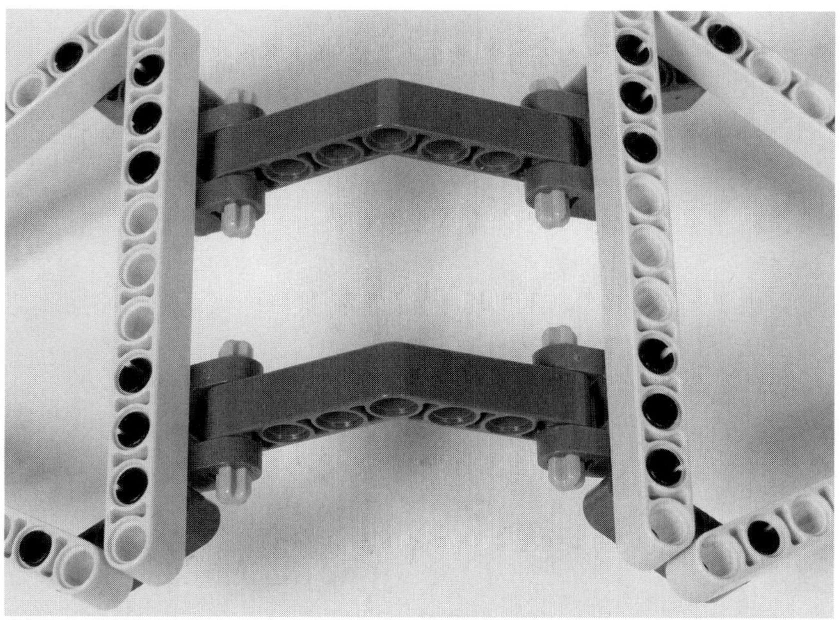

17

18 Two triangles joined together

19 The underside of the completed triangle pair.

20 Now make some more triangles.

21 If you run out of two-connector axle joiners, make your own from two axle joiners.

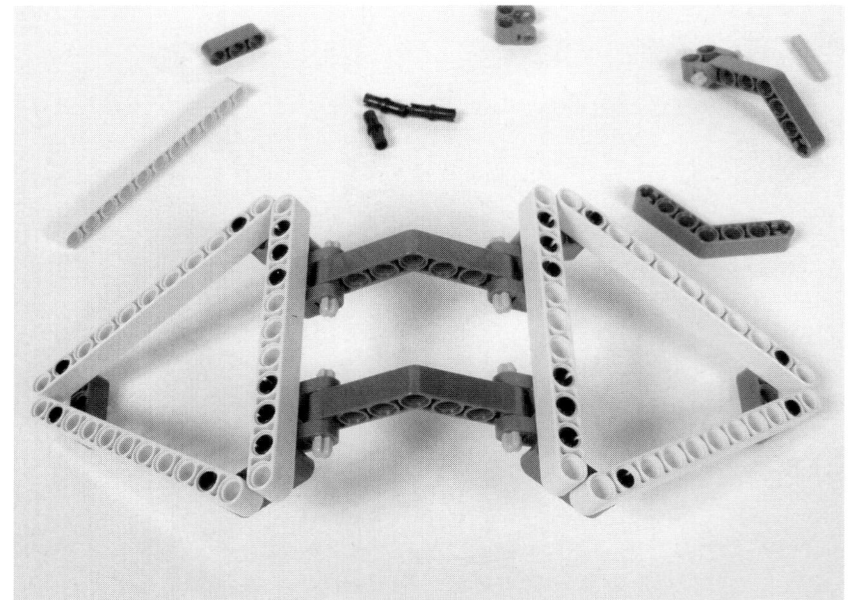

18

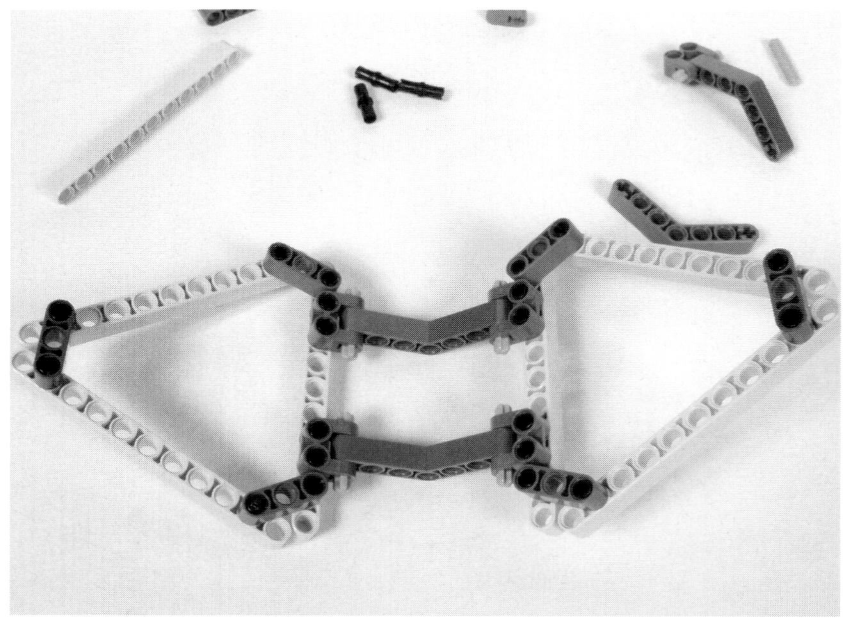

19

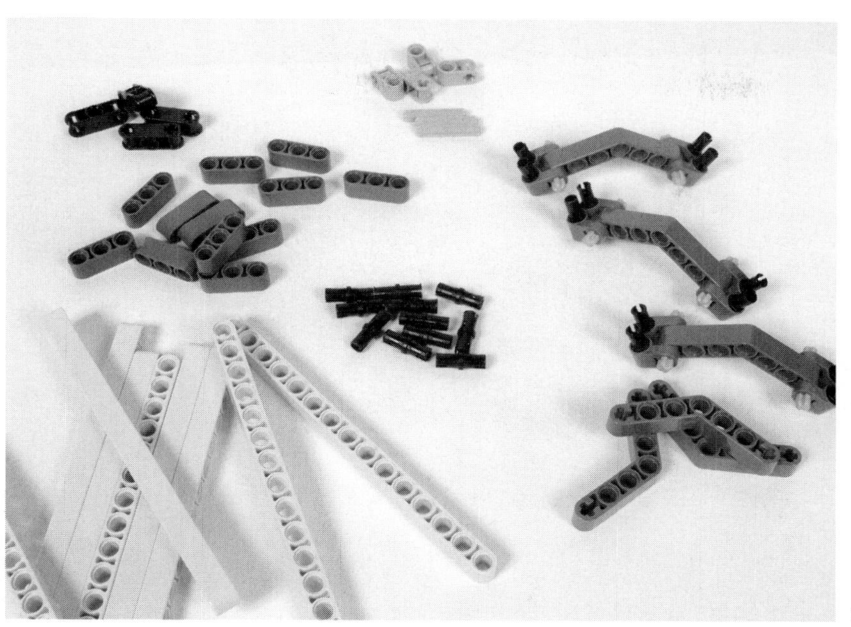

20

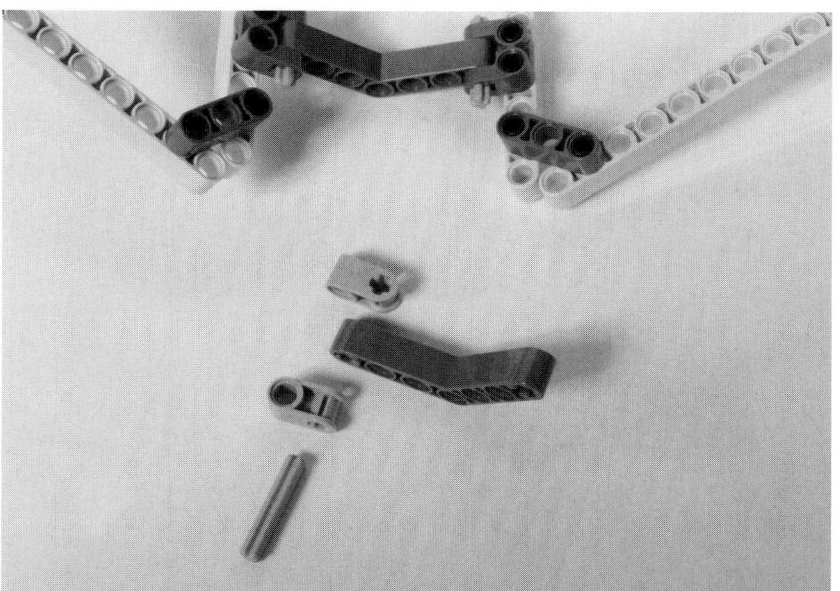

21

22 They might not be as pretty, but they will work.

23 Even these connectors can be used for fixing angled beams to your triangles.

24 Various angled beam connectors.

25 Although they look like dead bugs on the ground, these angled beam connectors can be substituted for the axle joiner variety.

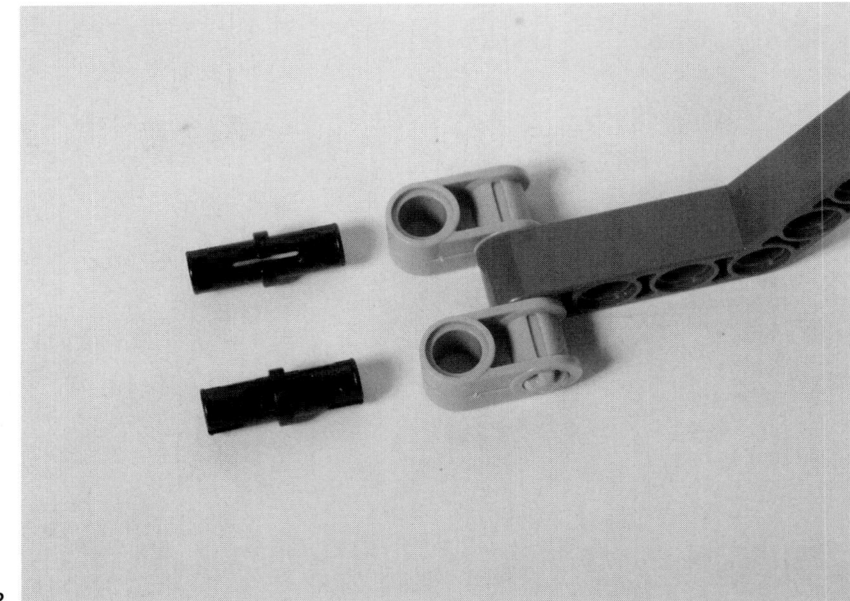

22

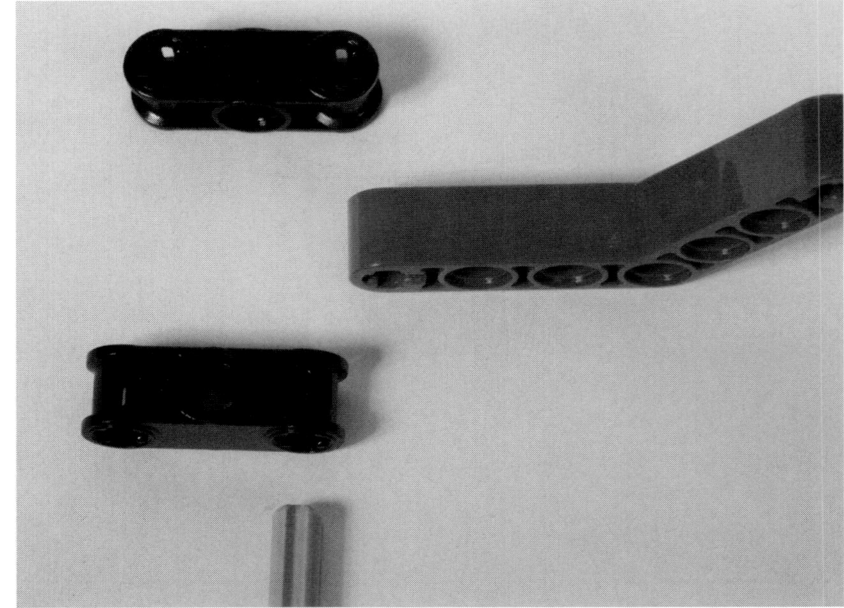

23

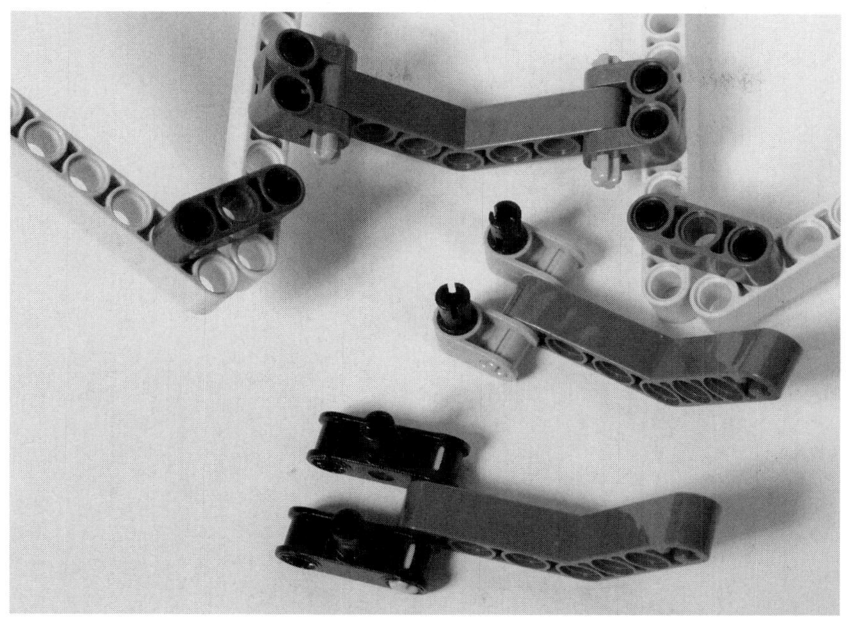

24

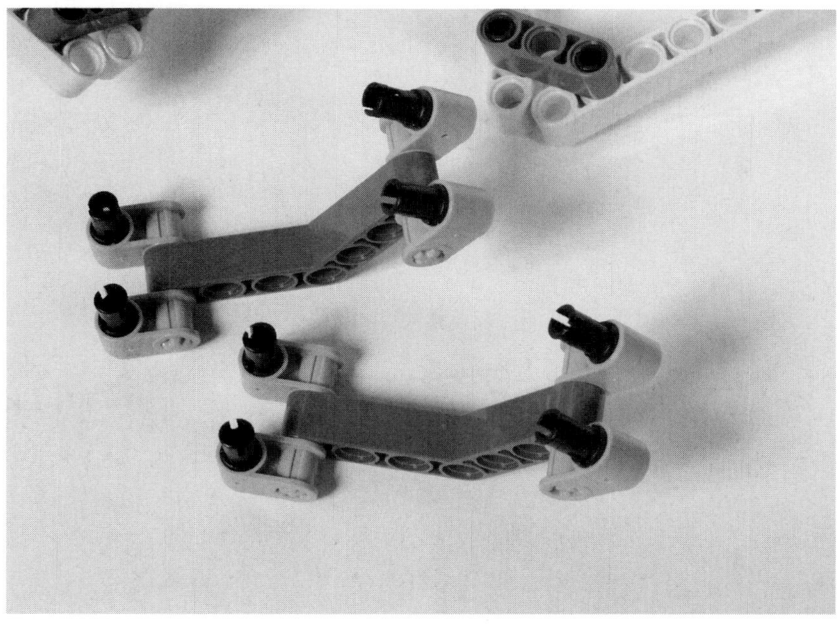

25

26 The spacing between your triangles will be slightly different for each variety of angled beam connector.

27 Three finished triangles.

28 Now start adding more triangles.

29 If your triangles aren't structurally sound enough for your tastes, you will have to add some extra bracing.

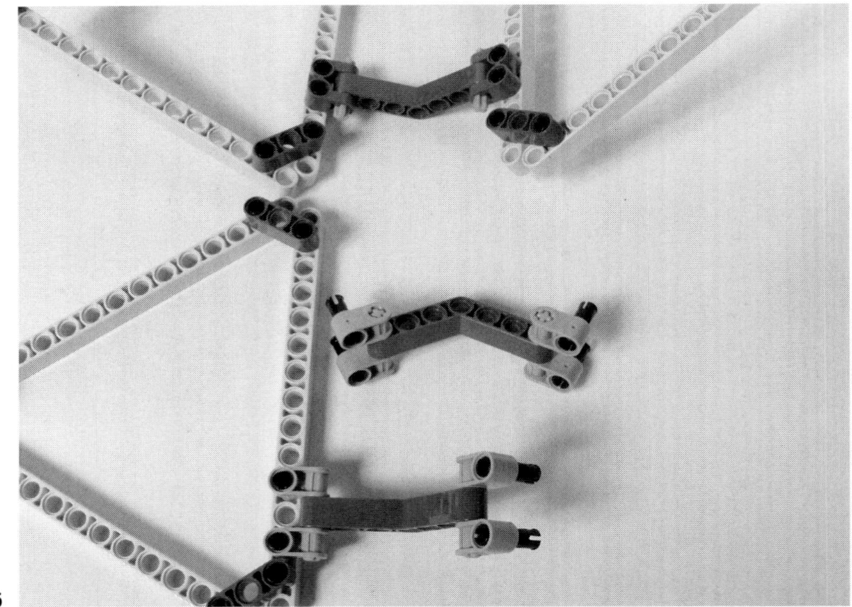

26

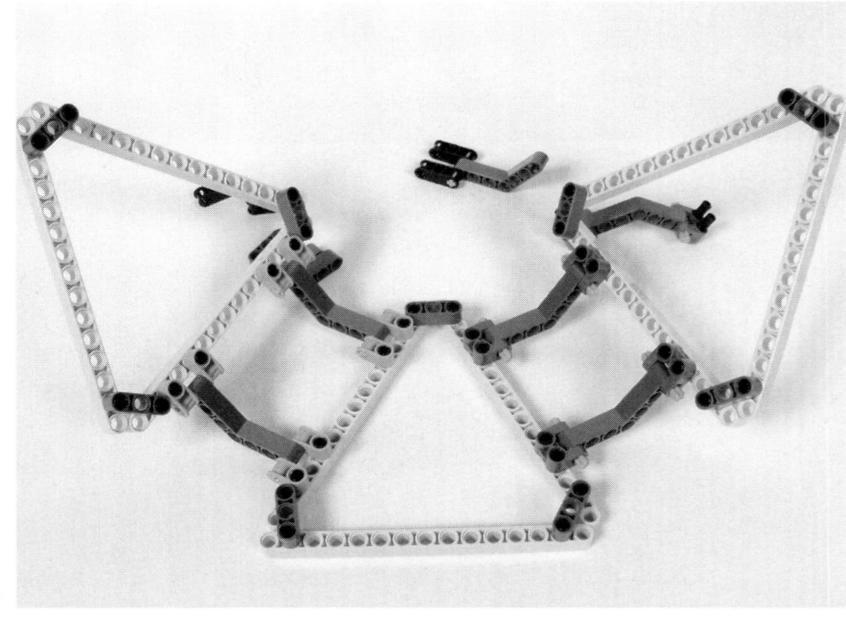

27

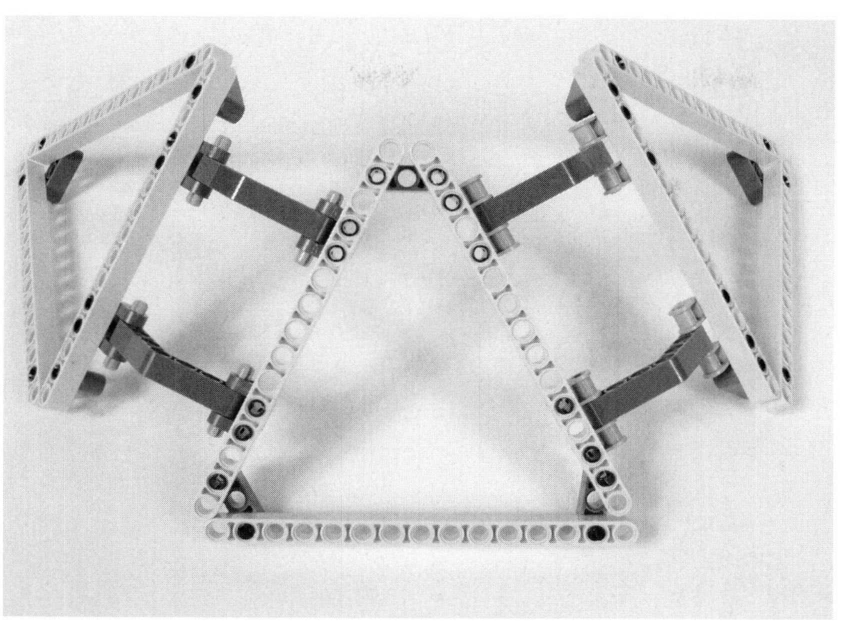

28

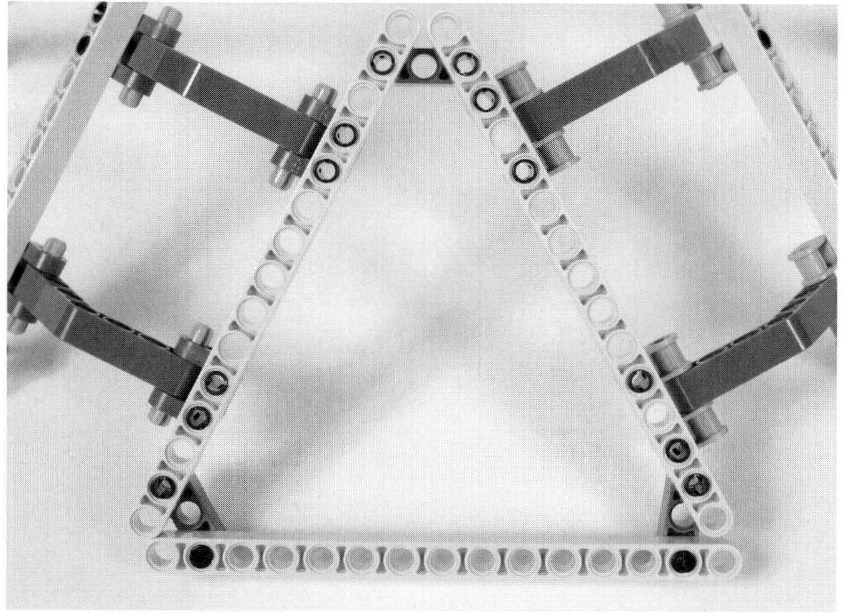

29

30 Note the connector spacing difference between these single-connector axle joiners and the two-connector versions in 31.

31 Compare the connector spacing on this angled beam connector with the one in 30.

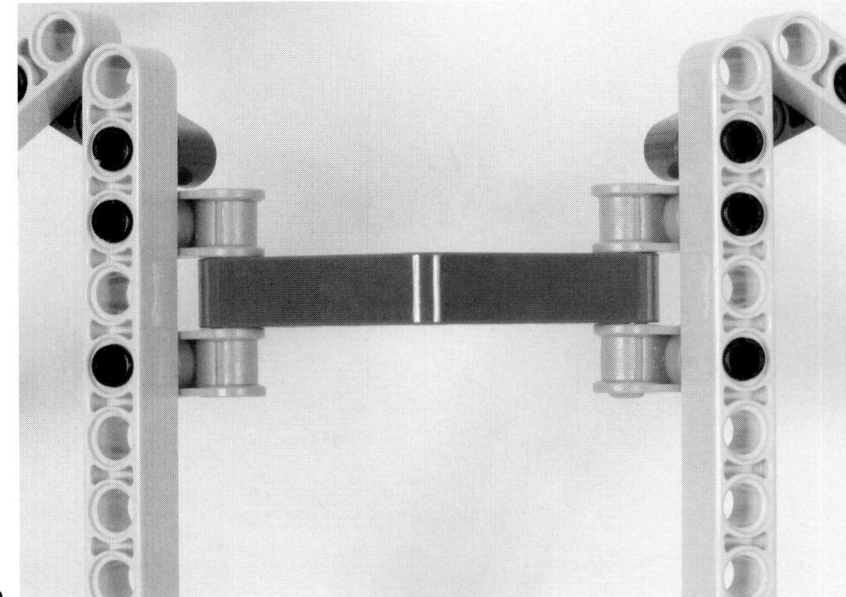

30

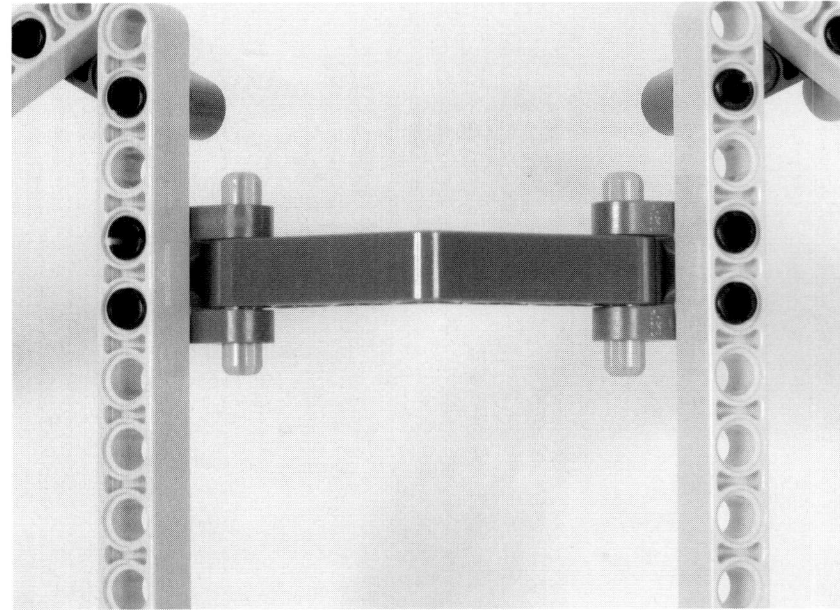

31

I DON'T KNOW, IS IT JUST ME?

Cubism looks so simple, so geometric, so, I don't know... so *arty*. Does that make sense? Yeah, it's arty, alright.

Don't feel badly, even well-known art critics have a hard time understanding, as well as explaining, a Cubist painting. In fact, über critic and author, Robert Hughes opined in his seminal treatise on modern art, *The Shock of the New* (Alfred A. Knopf, 1980) that, "...the key Cubist paintings can be obscure. They seem hard to grasp; in some ways they are almost literally illegible."

Lest you think that Cubism was an invalid expression of everyday life, then Hughes is quick to point out that, "...Cubism was the first radically new proposition about the way we see that painting had made in almost five hundred years."

Like it or loathe it, Cubism offers a unique perspective on the world around us (see Fig. IV-13). In addition to his well-known paintings, Picasso was also an avid sculptor. Works like his *Mandolin and Clarinet,* made in 1914 from

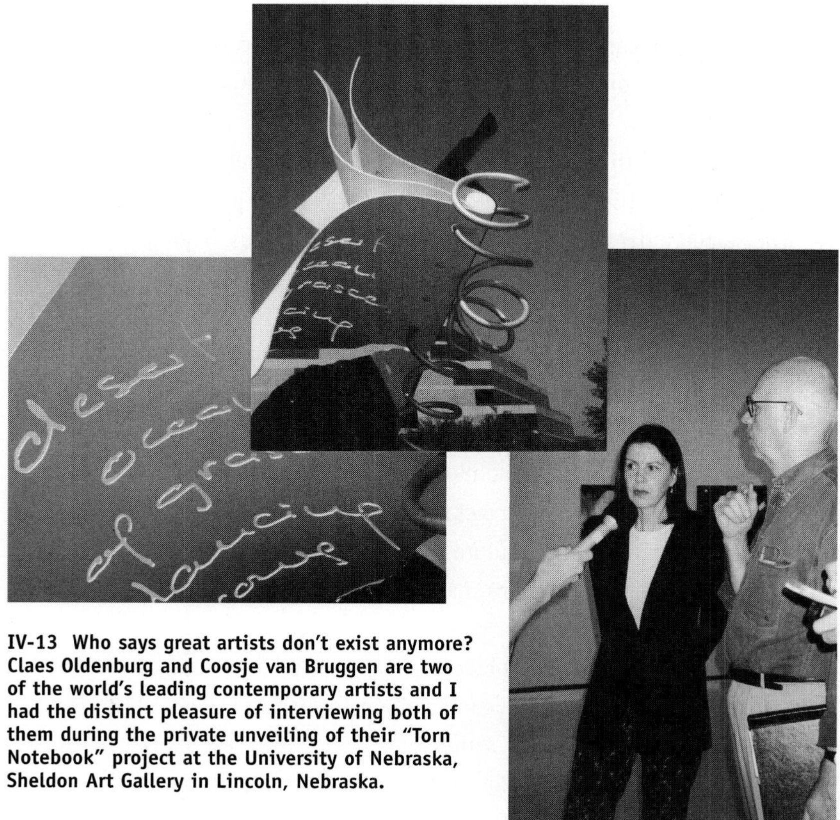

IV-13 Who says great artists don't exist anymore? Claes Oldenburg and Coosje van Bruggen are two of the world's leading contemporary artists and I had the distinct pleasure of interviewing both of them during the private unveiling of their "Torn Notebook" project at the University of Nebraska, Sheldon Art Gallery in Lincoln, Nebraska.

WHAT IS CUBISM?

Although the derivation of Cubism has been argued to be both African and Russian in its origin, one thing is certain: This art movement has a funny name. Who actually coined the term Cubism? In a remarkable book written by Picasso's art dealer, Daniel Henry Kahnweiler, titled *The Sculptures of Picasso* (Rodney Phillips, 1949) which, by the way, this book features photographs by legendary Parisian nightlife voyeur photographer Brassaï, the term "cubisme" was first used by art critic Louis Vauxcelles while writing about his initial encounter with the work of Georges Braque and Pablo Picasso.

painted wood and pencil (Musée Picasso, Paris), offer an exciting tactile feel to his view of these musical instruments.

In dealing with Cubism, Picasso had a very prosaic viewpoint about how the public should perceive his work. In Daniel Henry Kahnweiler's *The Sculptures of Picasso*, Picasso is quoted as saying, "I'd like to paint objects so that an engineer could construct them, after my pictures." This sounds like an ideal design challenge for every LEGO builder, doesn't it?

Now you might think that this story ends here, but the evolution of Cubism continues into the life and work of famed American sculptor, Alexander Calder. Yes, the same Calder who, in 1973, created the infamous "Flying Colors" for a Braniff International McDonnell Douglas DC-8-62 jetliner. [Note: In 1975, another Braniff International aircraft, a Boeing 727-291, was also painted by Calder. More airframes in the 727 fleet were also commissioned, but not completed at the time of Calder's death.]

Although an aircraft is a bold, unique canvas, Calder is better known for his floating sculptures called mobiles. Calder called them "space cages." Typically constructed from wire and metal, Calder's space cages were designed with pivot points for enabling the sculpture to move. This movement ability empowered his art to form evolutionary spatial relationships with countless perspectives—a Cubist sculpture that moves.

This final point shouldn't be taken as hyperbole. Calder actually met Picasso in Paris during a lengthy three-year development period (1929–1932) when he was exploring wire sculptures. While some critics have attempted to characterize Calder as a Surrealist sculptor, his art is certainly devoid of the psychic automatisms that define most Surrealists.

Case in point: the 1932 space cage, *Calderberry Bush*, by Calder. Fabricated from sheet metal, wire, and wood, this pivoting mobile differs markedly from the Surrealist works of Max Ernst and Joan Miró. Lacking the political substrata common to the Surrealism movement, *Calderberry Bush* drifts precariously between the Cubist influences of Picasso and Calder's own intention of art without constraints inside a continually evolving space (see Fig. IV-14).

This is "feel good" art—you can't explain it, but you'll know it when you see it.

IV-14 A plan for kinetic art sculptures.

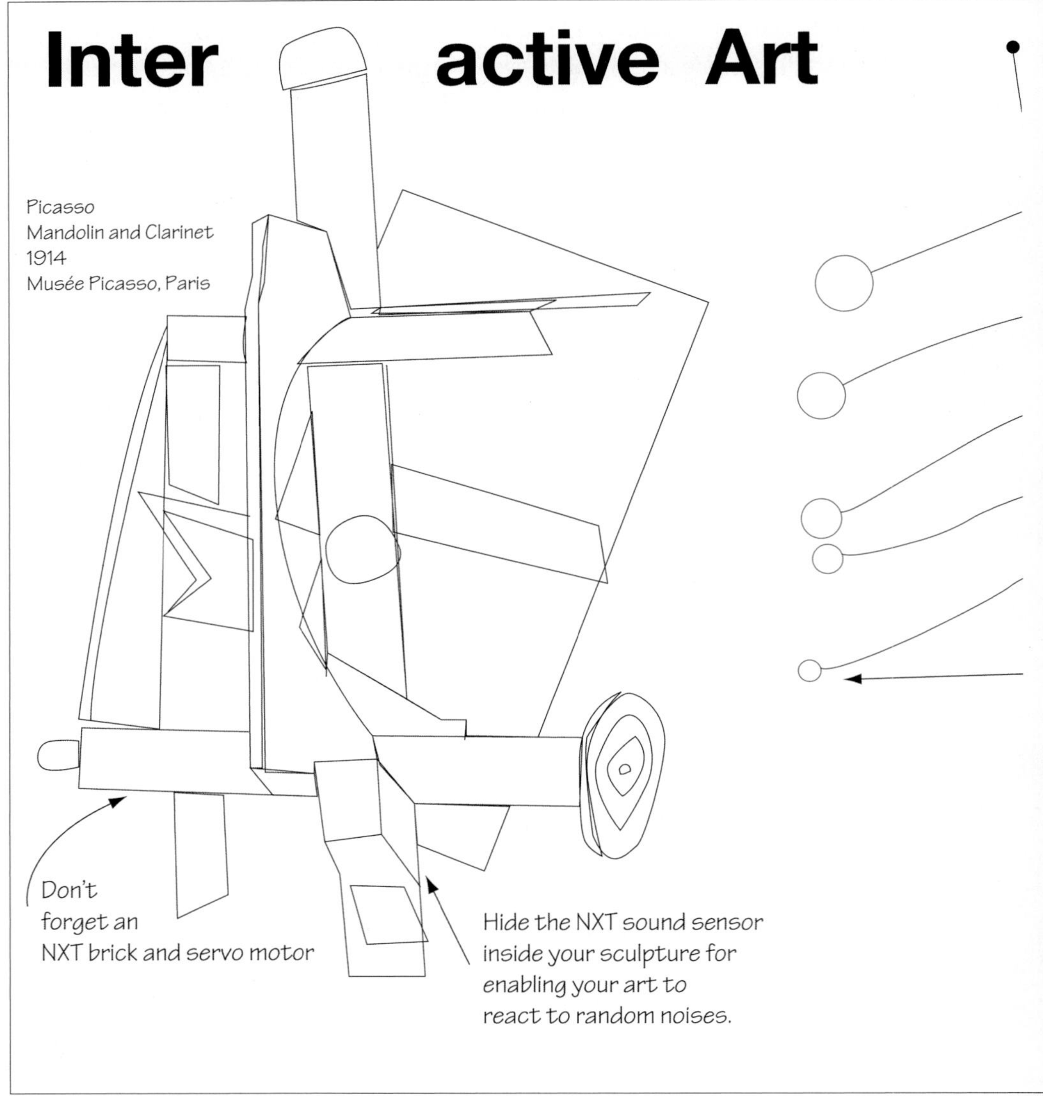

Inter active Art
Picasso
Mandolin and Clarinet
1914
Musée Picasso, Paris
Don't
forget an
NXT brick and servo motor
Hide the NXT sound sensor
inside your sculpture for
enabling your art to
react to random noises.

IV-14

with a Mind of Its Own

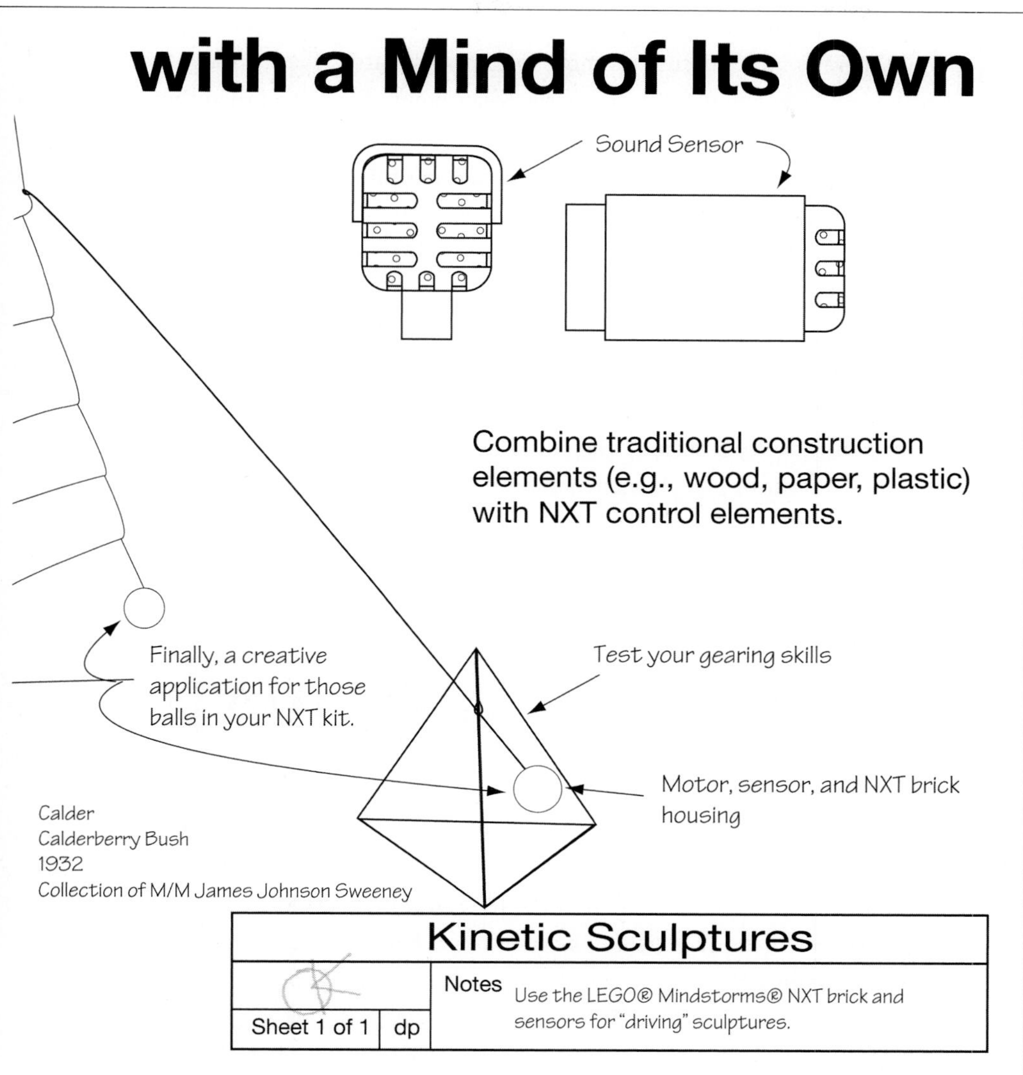

Sound Sensor
Combine traditional construction elements (e.g., wood, paper, plastic) with NXT control elements.
Finally, a creative application for those balls in your NXT kit.
Test your gearing skills
Motor, sensor, and NXT brick housing
Calder
Calderberry Bush
1932
Collection of M/M James Johnson Sweeney
Kinetic Sculptures
Sheet 1 of 1
dp
Notes
Use the LEGO® Mindstorms® NXT brick and sensors for "driving" sculptures.

IT'S LEGO TIME

Once again, we must return to France. This time it's to tell time. More specifically, telling time from a band strapped to your wrist.

Historians are uncommonly vague about the origins of today's wristwatch. While there is no clear-cut "inventor" of the wristwatch, there are two well-documented starting points. There's the Swiss point of view and then there is the French viewpoint.

Remarkably, each of these opinions share one thing in common: Each of the purported wristwatch inventors is today a well-established luxury watch purveyor. For the Swiss, it is Patek Philippe. And for the French it is Cartier.

If you just look at the dates of each discovery, then you would have to opt for Patek Philippe as the inventor of the wristwatch. The year was 1868 and the device was, in the words of the Patek Philippe Museum, a "key-winding lady's bracelet watch." Alas, this feat has been snubbed by some as nothing more than "novelty bracelet," therefore, these naysayers do not believe that it is the forebear of today's wristwatch.

Fast-forward to the beginning of the twentieth century and the French claim to this fame. In 1904, Louis Cartier was asked by budding aviator Alberto

WATCHES WIN WAR

There have been some groundless, unsupported claims that the wristwatch won the First World War for the Allies. In these fanciful tales, the Allies were able to "synchronize" their battles with better accuracy due to quick glances at wristwatches. The Huns, however, were saddled with pocket watches and unable to perform rapid, hands-free clock queries.

Sloughing through muddy trenches, you can imagine that having to reach into your pocket for consulting a delicate timepiece could be a nagging nuisance. Was this really the edge that gave us victory? This concept sounds like a scene snipped from a Hollywood cinematic drama, doesn't it? Well, I mention this claim to serve as a springboard for your own research into this topic. OK, let's synchronize our watches and it's over the top.

GIVE 'EM AN INCH AND THEY TAKE A LIGNE

If you take a wristwatch and remove the case, bezel, and crystal, you are left with the watch's movement. The movement is where *all* of the timekeeping action happens. Gears, wheels, levers, balances, springs, and escapes are the diminutive components that make up the movement.

Oddly enough, the dimensions of movements are given in a French unit of measure called the ligne. A ligne was originally conjured as 1/12th of a French inch called a pouce. Updating this measurement to our current metric system, 1 ligne is roughly equal to 2.55883 millimeters. Or, for our English system of measure, 1 ligne is roughly equal to 3/32nds of an inch.

Now, if you think that scale of measurement is odd, consider that, early English watchmakers further subdivided the pouce into twelfths or *douzièmes* and called this unit a *pied*. Conversely, in nineteenth-century United States, watchmakers used 1/30th of an inch as the unit of movement measure. They coupled this measurement with a scaling system that started at 0. In this case, 0 was roughly equivalent to 35/30ths of an inch. Yes, you read that right...35/30ths of an inch.

Finally, adding more confusion to this already confusing measuring system, some twentieth-century watch manufacturers used the jeweler's line for movement measurement. Please don't confuse this line with the watchmaker's ligne. Because a jeweler's line measures 1/40th of an inch.

Geesch, thank goodness that today almost all watchmakers have adopted the metric system for movement measurement.

Santos-Dumont to provide him with a timepiece that he could read while flying an aeroplane. Cartier enlisted the aid of master watchmaker Edmond Jaeger. The result of this collaboration was a prototype watch that could be worn on Santos-Dumont's wrist as his hands manipulated the controls of his heavier-than-air craft—a.k.a. a wristwatch (see Fig. IV-15).

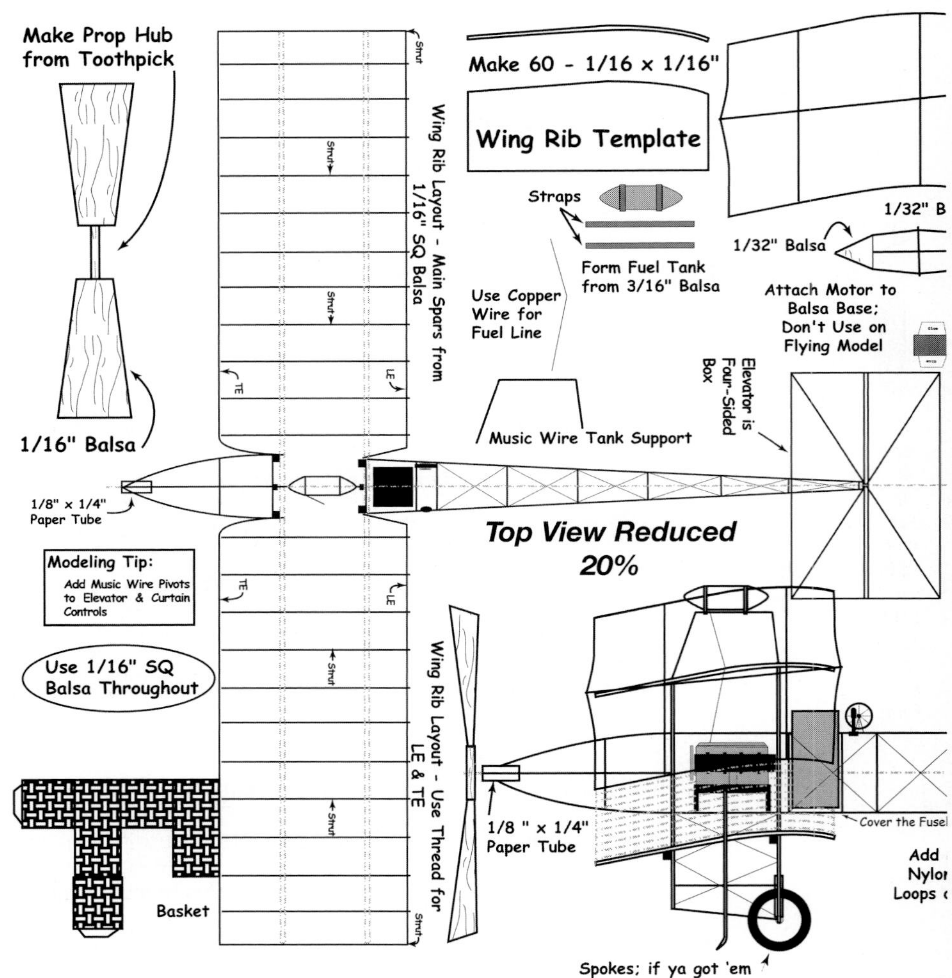

IV-15 A plan for Santos-Dumont's Number 14 aeroplane.

So take your pick; a Swiss timepiece or a French one. Whichever one you select, be careful, though. Your friends might claim that you've become a horologer. Yikes! A what? In other words, you would be a devotee of time-measuring devices. Don't be so shy, clocks and watches are fascinating pieces of engineering that enable us all to stay, well, on time.

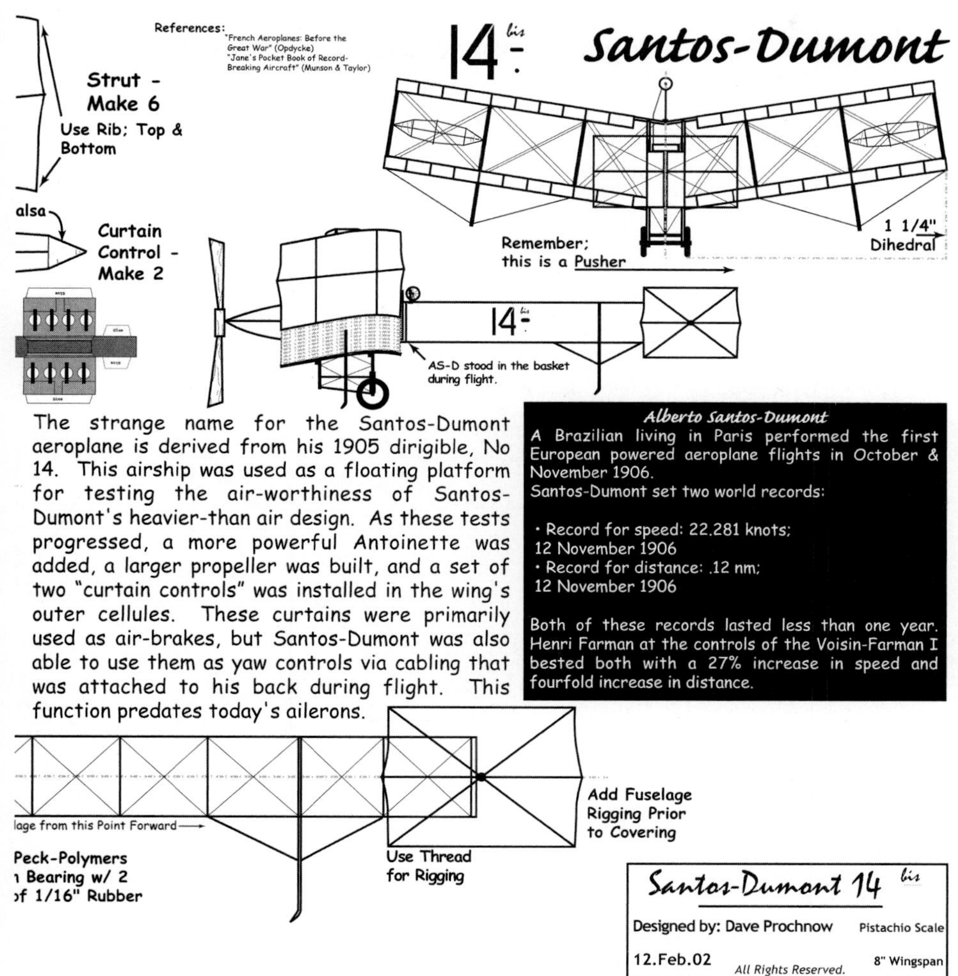

References:
"French Aeroplanes: Before the Great War" (Opdycke)
"Jane's Pocket Book of Record-Breaking Aircraft" (Munson & Taylor)
14 bis Santos-Dumont
Strut - Make 6
Use Rib; Top & Bottom
Curtain Control - Make 2
Remember; this is a Pusher
1 1/4" Dihedral
AS-D stood in the basket during flight.
The strange name for the Santos-Dumont aeroplane is derived from his 1905 dirigible, No 14. This airship was used as a floating platform for testing the air-worthiness of Santos-Dumont's heavier-than air design. As these tests progressed, a more powerful Antoinette was added, a larger propeller was built, and a set of two "curtain controls" was installed in the wing's outer cellules. These curtains were primarily used as air-brakes, but Santos-Dumont was also able to use them as yaw controls via cabling that was attached to his back during flight. This function predates today's ailerons.
Alberto Santos-Dumont
A Brazilian living in Paris performed the first European powered aeroplane flights in October & November 1906.
Santos-Dumont set two world records:
· Record for speed: 22.281 knots; 12 November 1906
· Record for distance: .12 nm; 12 November 1906
Both of these records lasted less than one year. Henri Farman at the controls of the Voisin-Farman I bested both with a 27% increase in speed and fourfold increase in distance.
Add Fuselage Rigging Prior to Covering
Use Thread for Rigging
Peck-Polymers
Santos-Dumont 14 bis
Designed by: Dave Prochnow
Pistachio Scale
12.Feb.02
All Rights Reserved.
8" Wingspan

HOW TO BUILD A LEGO MINDSTORMS NXT WRISTWATCH

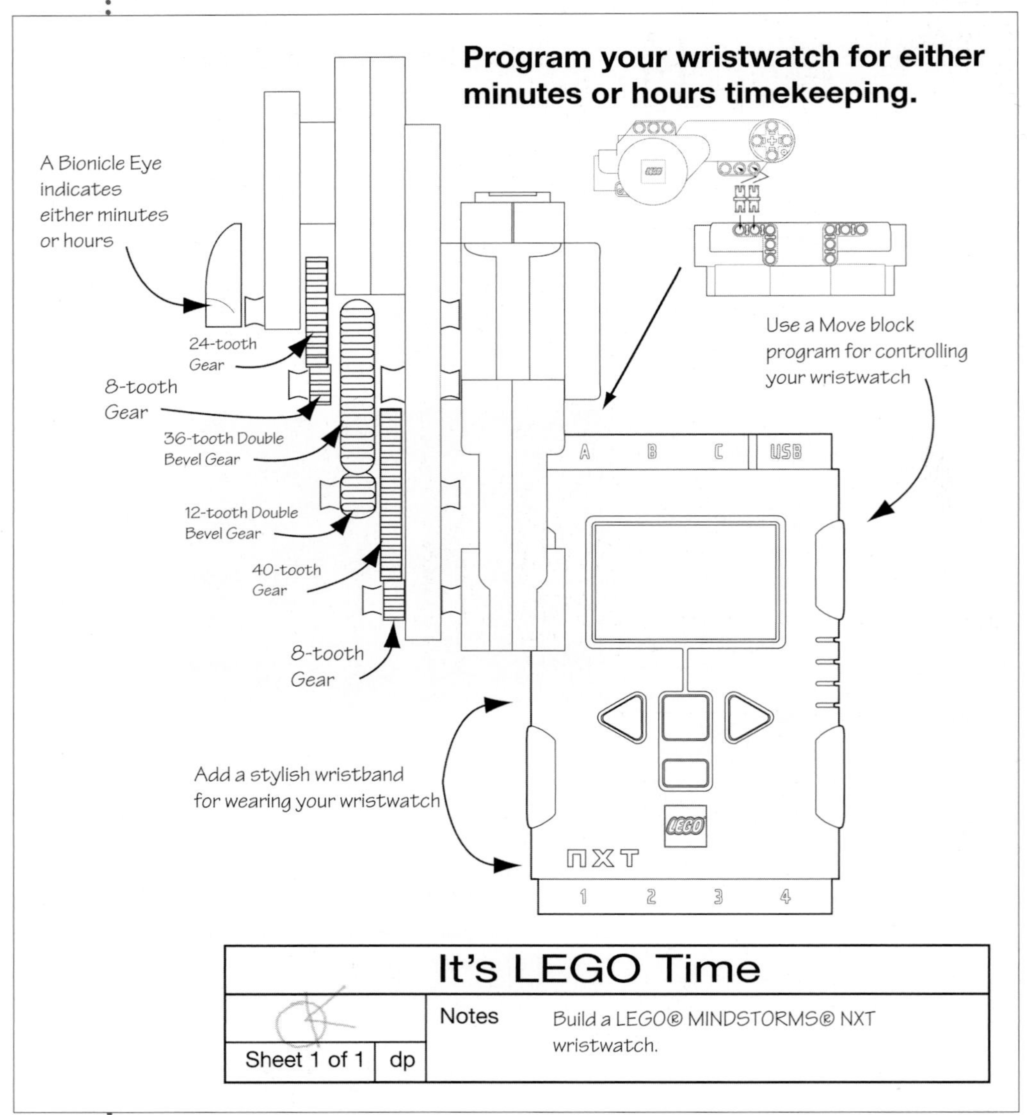

Program your wristwatch for either minutes or hours timekeeping.
A Bionicle Eye indicates either minutes or hours
24-tooth Gear
8-tooth Gear
36-tooth Double Bevel Gear
12-tooth Double Bevel Gear
40-tooth Gear
8-tooth Gear
Use a Move block program for controlling your wristwatch
Add a stylish wristband for wearing your wristwatch
A
B
C
USB
NXT
1
2
3
4
It's LEGO Time
Sheet 1 of 1
dp
Notes
Build a LEGO® MINDSTORMS® NXT wristwatch.

1

1 A plan for a LEGO NXT wristwatch.

2 Begin by adding your first gear to an 11m Beam.

3 The first gear will drive this bigger gear.

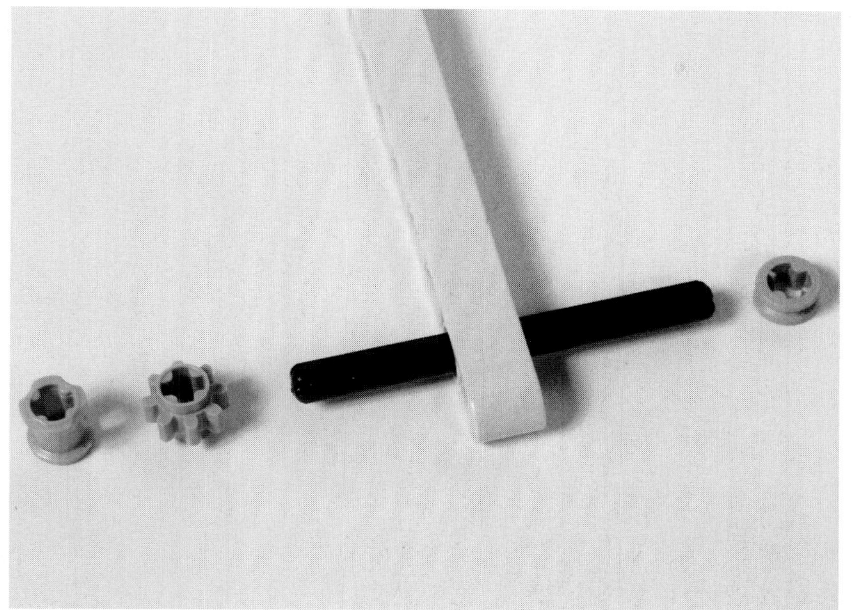

2

3

4 Fix a smaller gear to the axle of the second gear.

5 Add the axle for the fourth gear.

6 This fourth gear using a bushing as a standoff offset from the second gear.

7 Add a fifth gear to the axle of the fourth gear.

4

5

6

7

8 The first gear train is complete.

9 Begin the second gear train with another beam and a pair of three-stud connectors.

10 Attach another beam to the first beam with remaining ends of the two three-stud connectors.

11 Complete the second gear train's support with a 3m beam and two more three-stud connectors.

8

9

10

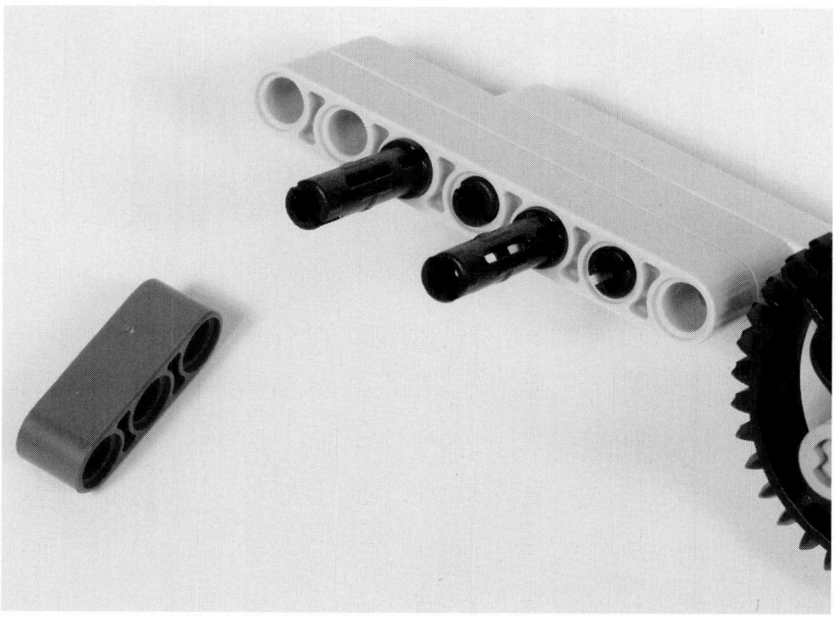

11

12 Prep the sixth gear on its own beam.

13 Invert this sixth gear and mesh it with the fifth gear. Add a Bionicle eye as the time indicator to the sixth gear's axle.

14 Attach the entire clock movement to an NXT motor.

15 Insert two connectors into the motor's underside. Mount the motor on the NXT Brick.

12

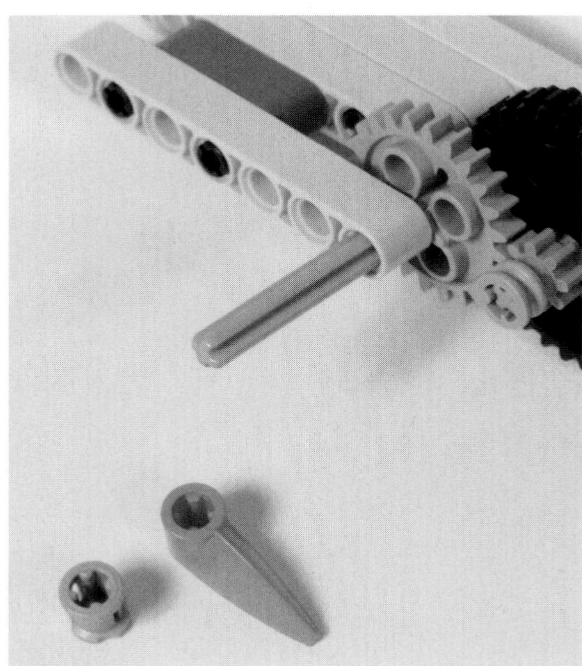

13

14

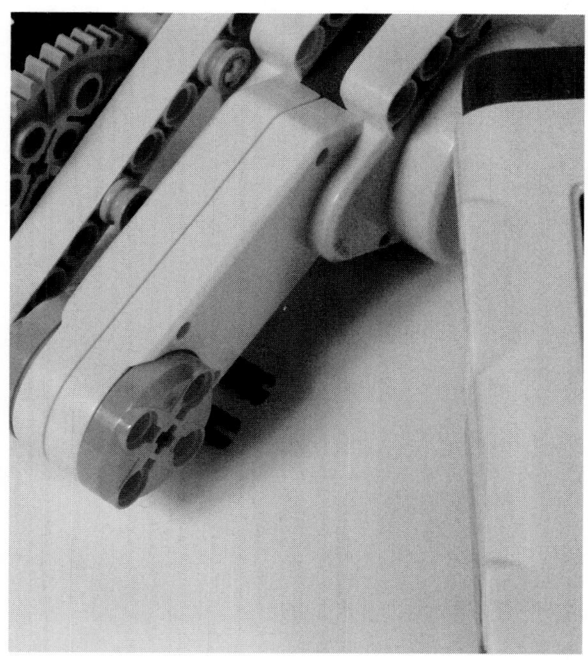

15

16 **Fashion a wrist strap from some plastic webbing. Use four-pin connectors for attaching the wrist strap to the NXT wristwatch.**

17 **Determine the spacing for your four-pin connectors.**

18 **Enlarge the holes in the plastic webbing with a reamer.**

19 **Test fit the four-pin connector in the webbing.**

16

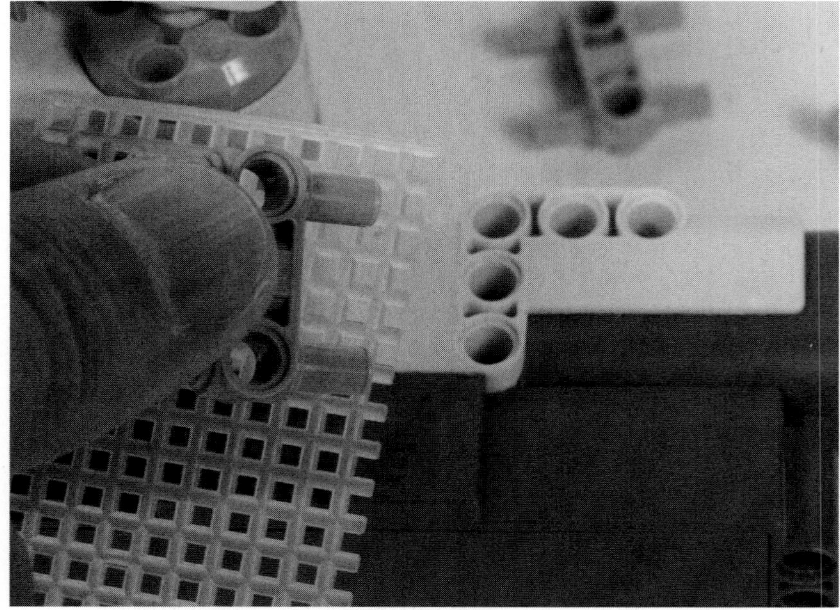

17

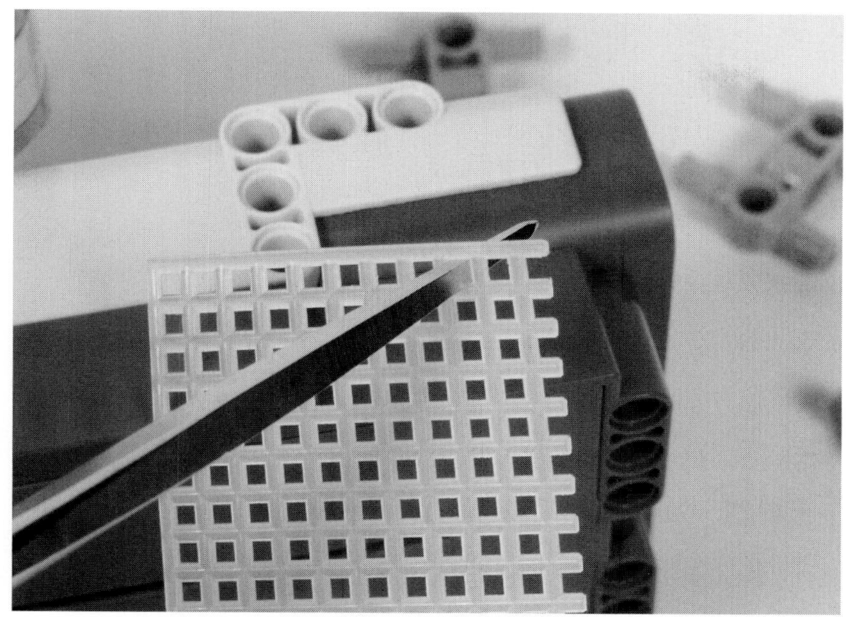

18

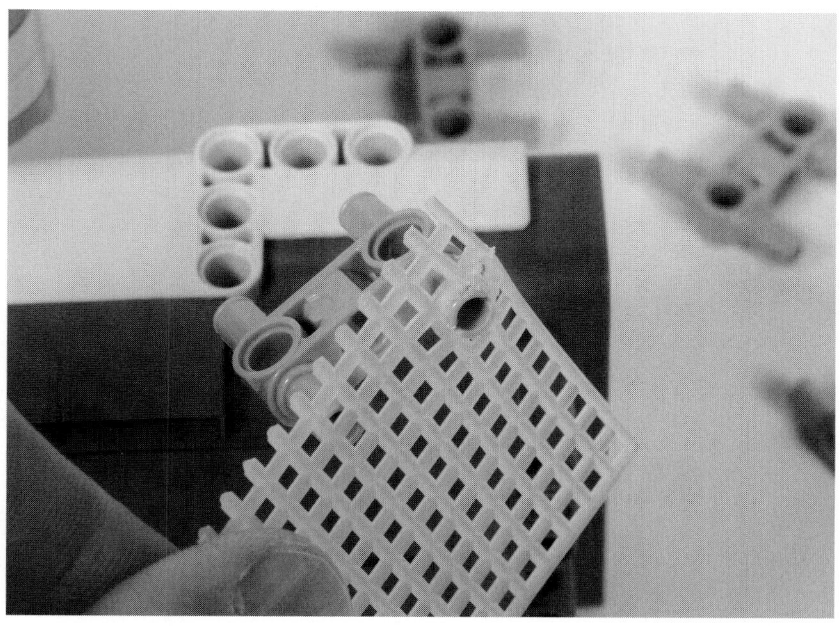

19

20 Ream another hole and insert the second pin from the four-pin connector.

21 The fit tolerance is too tight for the connector that sits closest to the NXT motor. You can ease the edge of this connector with a file and a lot of careful sanding.

22 This rounded edge should now fit against the NXT motor.

23 Attach the wrist strap to one side of the NXT Brick.

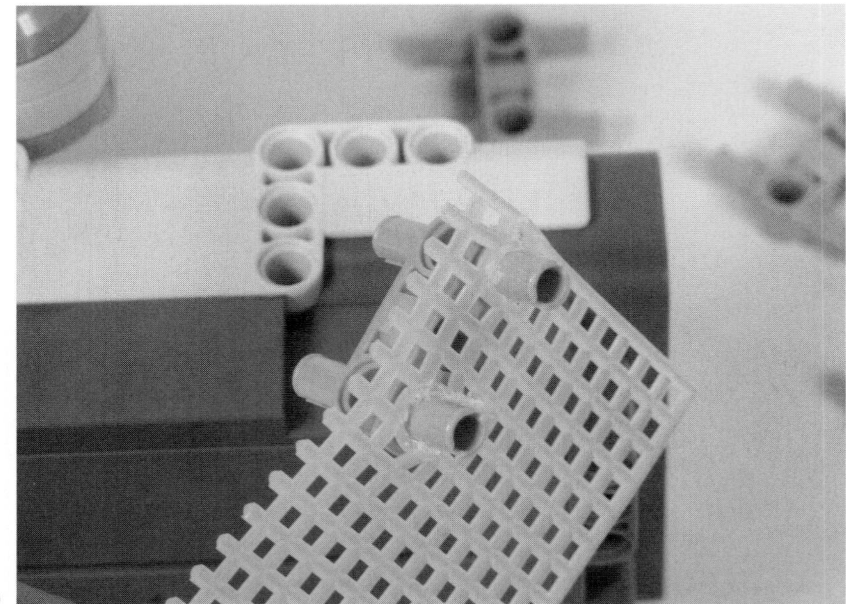

20

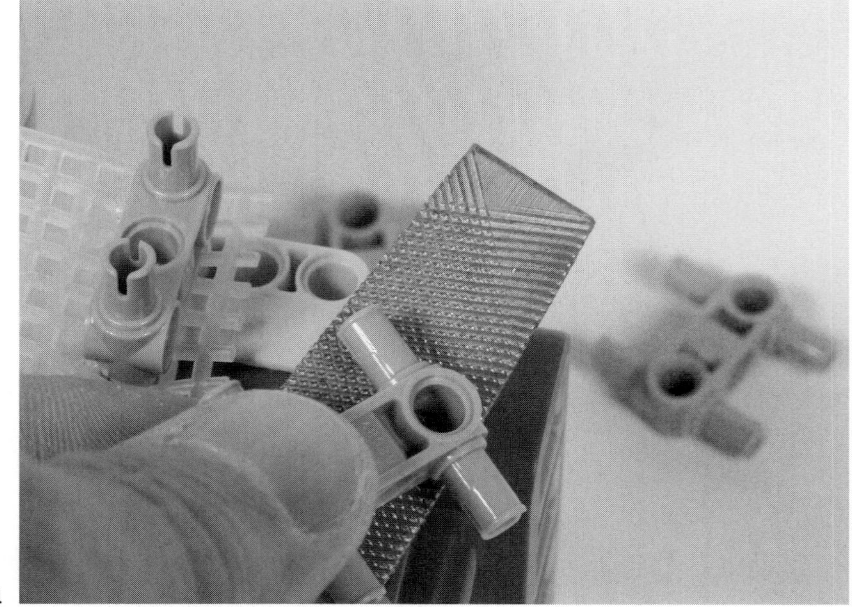

21

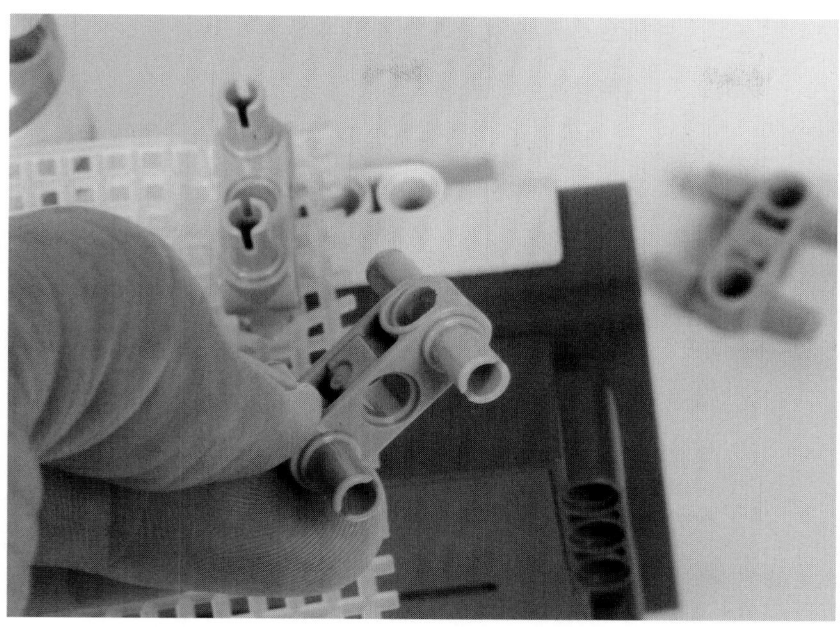

22

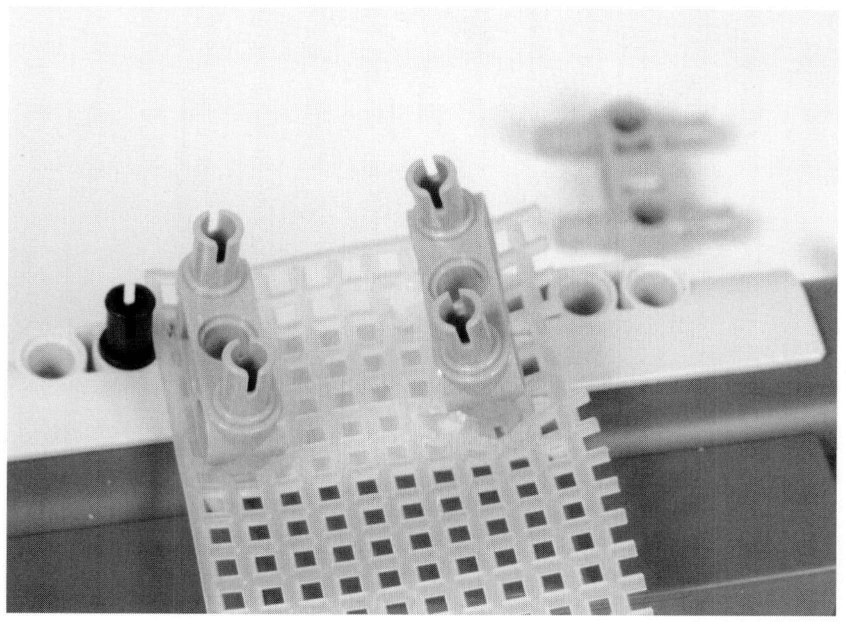

23

24 Loop the strap around the bottom of the NXT Brick and attach the other end of the strap.

25 The completed NXT wristwatch. The cabling and wrist strap have been removed for clarity.

26 A test fit.

27 Looks great and fits nice.

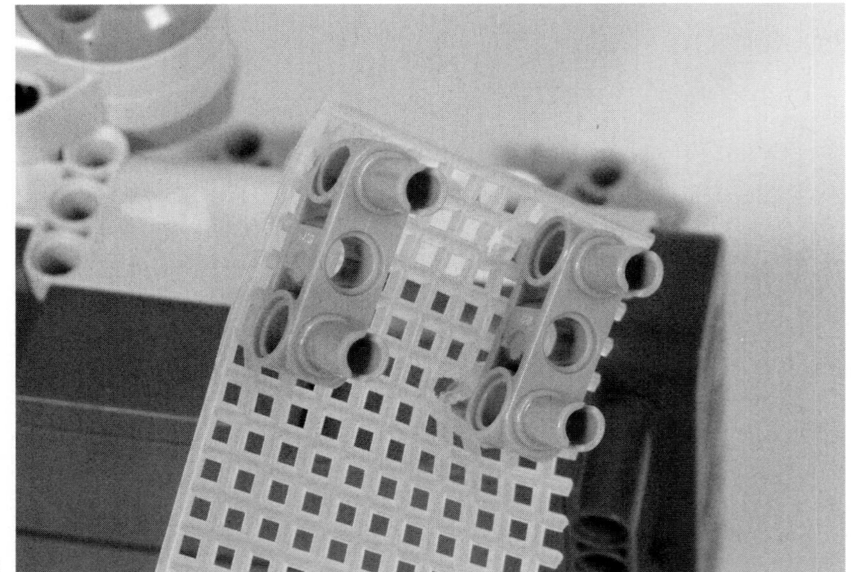

24

25

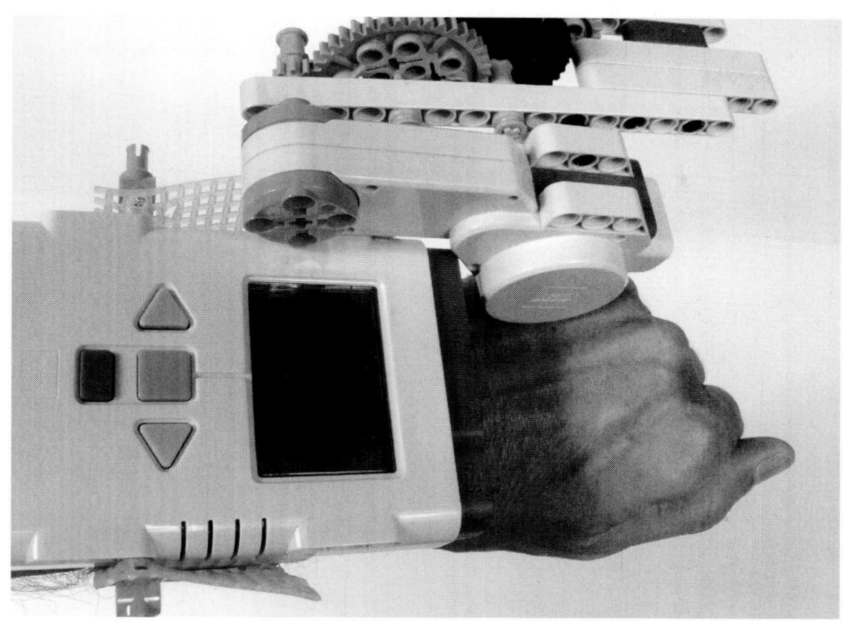

26

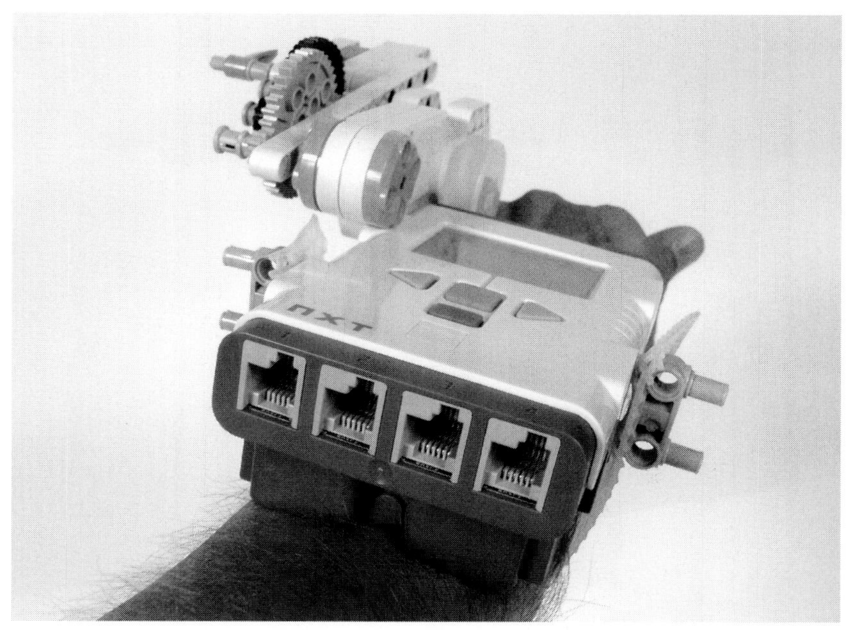

27

28 You know what time it is? Time for an NXT wristwatch. This is a prototype wrist strap that used only one four-pin connector. It did not hold very well.

29 Program your NXT wristwatch to keep meaningful time. There's no sweep second hand and no minute hand, for that matter. For the sake of your own sanity and peace of mind, you might want to print the time on the NXT Brick's display, too.

30 Oops, the Bionicle eye hand indicates that I'm late for a very important date.

31 Which would you rather wear? A Raymond Weil or an NXT wristwatch?

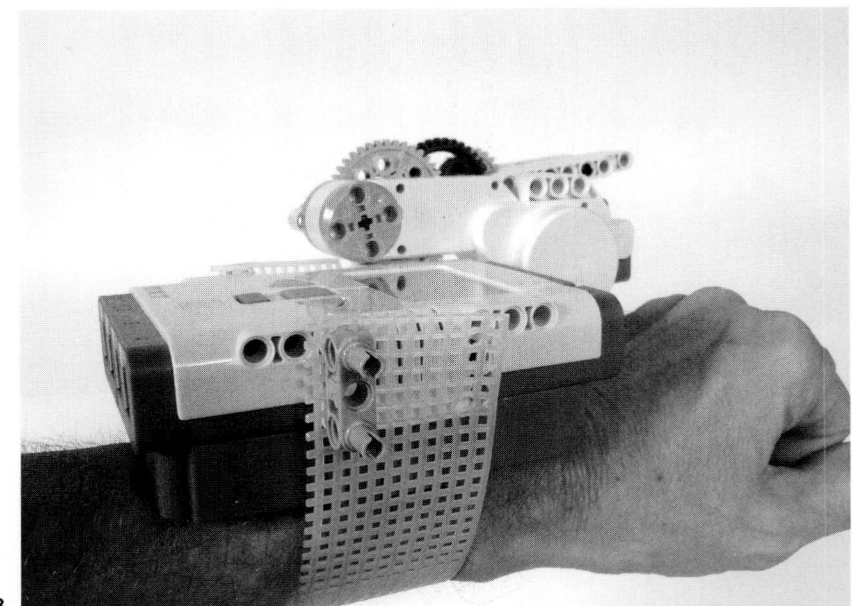

28

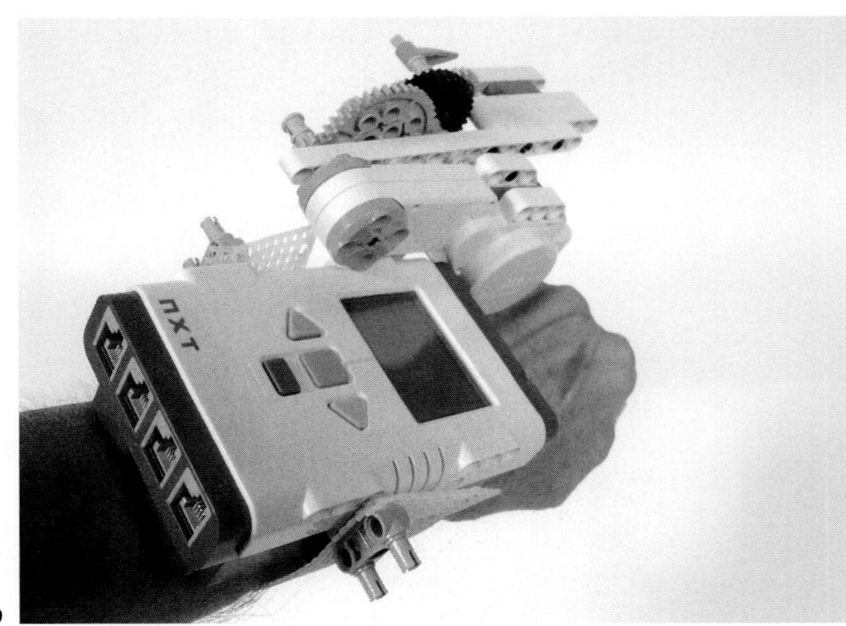

29

30

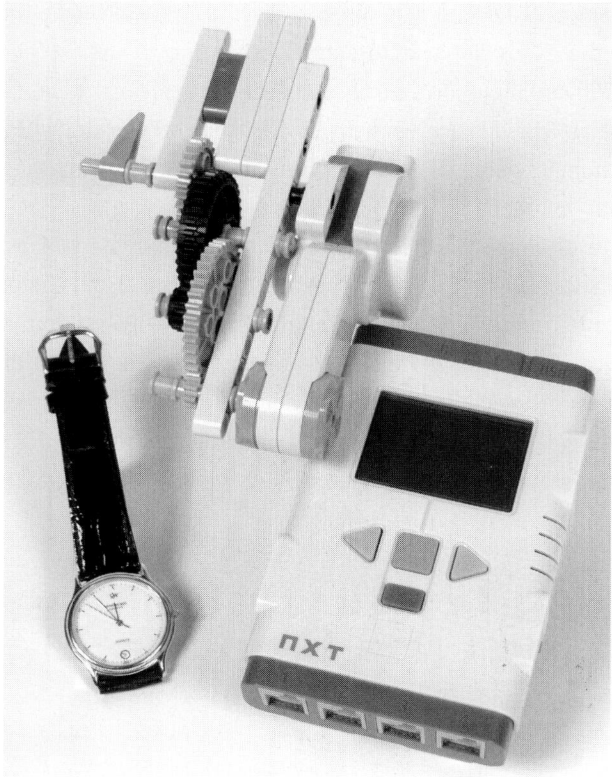

31

HEY EVERYBODY, LET'S HOST A WORLD'S FAIR

It was a veritable "who's who" of America's most creative designers. The event was the 1893 World's Columbian Exposition held in Chicago, Illinois. Architects, artists, engineers, and inventors from all over the United States were enlisted to make this exposition grander than the previous world's fair record holder, the spectacular Exposition Universelle, 1889 or Paris Exposition of 1889. This was a lofty benchmark that few people thought was achievable. I mean how can you top a show that debuts the Eiffel Tower?

Charged with organizing America's turn at hosting a grand world's fair, was architect Daniel H. Burnham. Burnham was instrumental in gathering a real cast of "characters" together for designing and building the exposition. Imagine having the likes of Henry Ives Cobb, Richard Morris Hunt, Louis Sullivan, Frederick Law Olmstead, Sophie Hayden, and George Ferris all under your command. Likewise, having all of these prima donnas under one roof must have been nerve racking.

For example, Louis Sullivan did not like the classical theme that Burnham chose for the exposition's architecture. According to the Chicago Historical Society, Sullivan is purported to have sneered at this architectural selection, claiming that "the damage wrought by the World's Fair [sic] will last for half a century from its date, if not longer." Oh, please Mr. Sullivan, don't hold back, tell us what you really think.

In spite of any petty bickering, the World's Columbian Exposition was a worldwide success. The final tally sheet for this event listed the participation of 46 nations from around the globe, attracted over 25 million visitors, carried an adult admissions price of 50¢ and cost just over $28 million to host. Yikes, the event looks like a big flop, doesn't it? Well, this exposition might have been a financial disaster if not for one small footnote.

And that story begins with George Ferris.

"My favourite building? I'd pick the London Eye."

—Lord Norman Foster, Architect of Hearst Headquarters, New York

AN EYE OVER LONDON

While you might think that Ferris Wheels are no big deal anymore, obviously you haven't been to London lately. In March 2000, the British Airways London Eye began revolving through the skies over London. Originally drawn up in 1993 on the kitchen table of architects David Marks and Julia Barfield, the London Eye has become a unique monocular vision eyeballing one of our oldest cities.

Populated with 32 passenger capsules, each of which is capable of carrying 25 people, the London Eye consists of a 30-minute ride, or, as British Airways calls it, a 30-minute flight. Therefore, during each day of operation the London Eye transports approximately 15,000 people. In fact, according to the architects over 20 million visitors have "flown" aboard this unique homage to George Ferris since its 2000 debut.

Now, for the $64,000 question: Is the London Eye a Ferris Wheel? Well, according to the architects the answer is a resounding "no." They cite enclosed capsules sitting outside of the wheel in a side-supporting A-frame as the technical differences between the two. But come on, isn't a modern jetliner still an airplane? I'm sure that Orville and Wilbur would think that there is a tangible nexus between the various forms of flight. So too then is the spirit of George Ferris revolving through the skies over London.

George Washington Gale Ferris (don't you just love people with two middle names; check it out David Jay Bird Prochnow; I wish) was an engineer who came to Burnham with an exposition saving idea. His idea sounds a little crazy today—cramming over 2,000 people into a 260-foot-tall spinning steam-driven 1,000+ ton wheel for 20 minutes. Oh, and make each person buy a ticket to sit inside this rotating death trap.

Well, it wasn't a death trap; thank goodness. Located on the Midway Plaisance, it was the world's first Ferris Wheel and exposition patrons paid 50¢ to take a ride in it. All told, George Ferris' tally sheet really didn't help the World's Columbian Exposition bottom line, but the worldwide notice that the Ferris Wheel generated helped cast this Chicago fair in a favorable light against its Parisian counterpart.

1 **The Texas Star Ferris Wheel.**

2 **The Cosmo Clock Ferris Wheel.**

3 **The London Eye.**

HOW TO BUILD A SPOKE-LESS FERRIS WHEEL

Texas Star Ferris Wheel

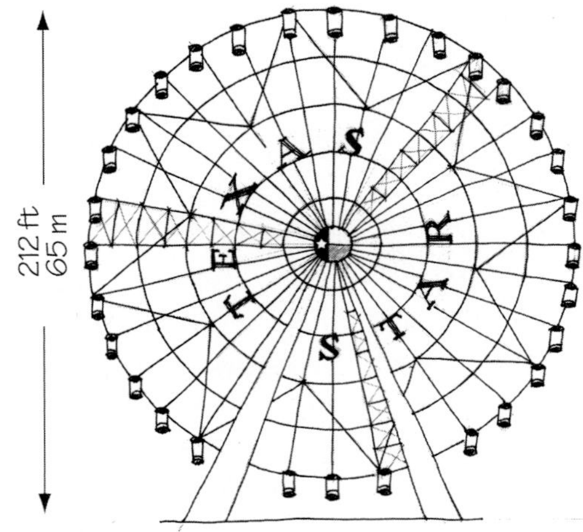

Location: Dallas, Texas, USA

1 Built: 1985

Cosmo Clock 21

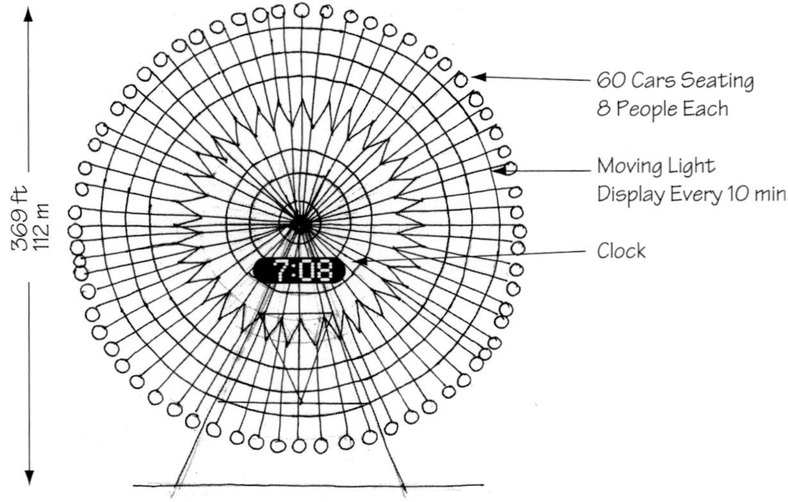

Location: Yokohama, Japan

2 Built: 1999

The London Eye

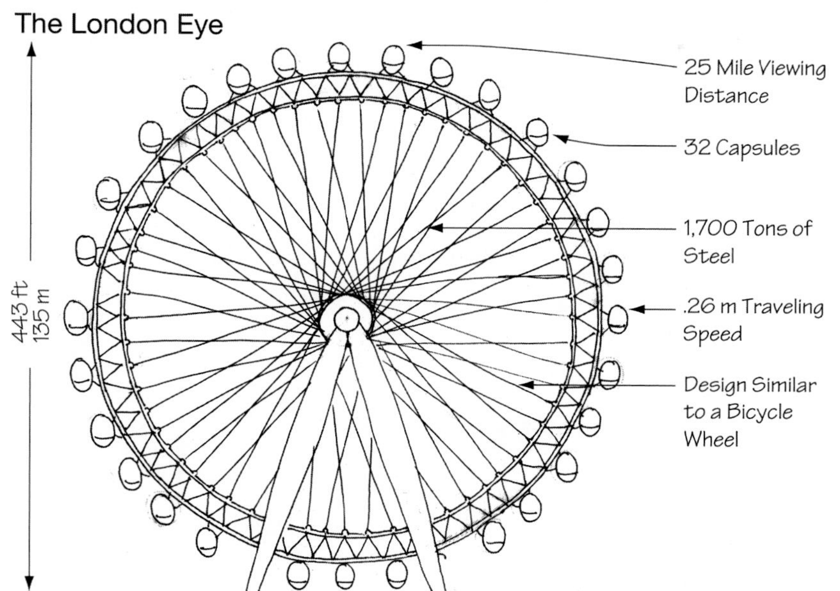

3

Location: London, United Kingdom
Built: 1999
Designer: Marks Barfield Architects
Capacity: 15,000 visitors per day
Awarded: Royal Institute of British Architects (RIBA) Award for Architecture, 2000

4 The Chicago Navy Pier Ferris Wheel.

CHICAGO'S NAVY PIER

Address - Chicago, IL USA
Description - Ferris wheel, monorail and water park
Designer - Forrec Limited, Designers of Legoland Deutschland

1

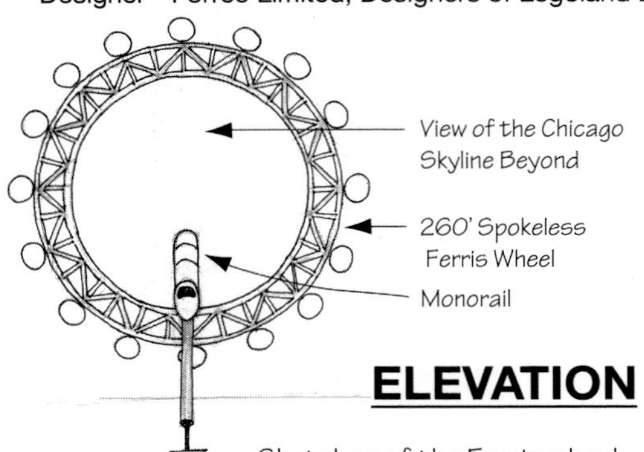

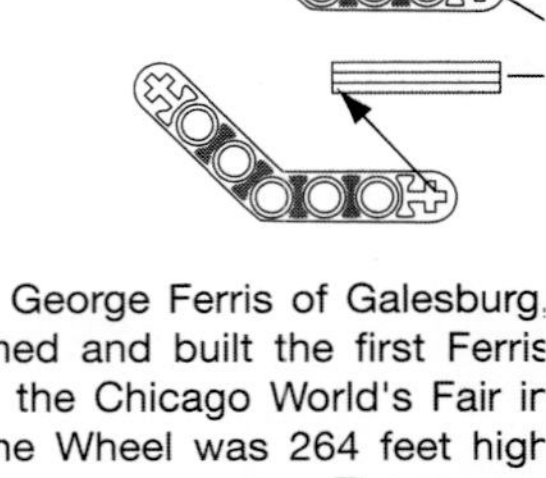

ELEVATION

Sketches of the Ferris wheel monorail and water park in Chicago.

Engineer, George Ferris of Galesburg, IL, designed and built the first Ferris wheel for the Chicago World's Fair in 1893. The Wheel was 264 feet high with 36 passenger cars. These cars could hold 60 people.

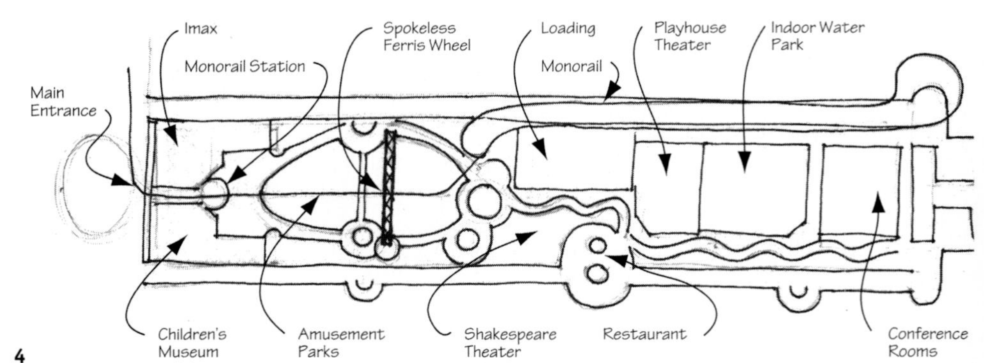

4

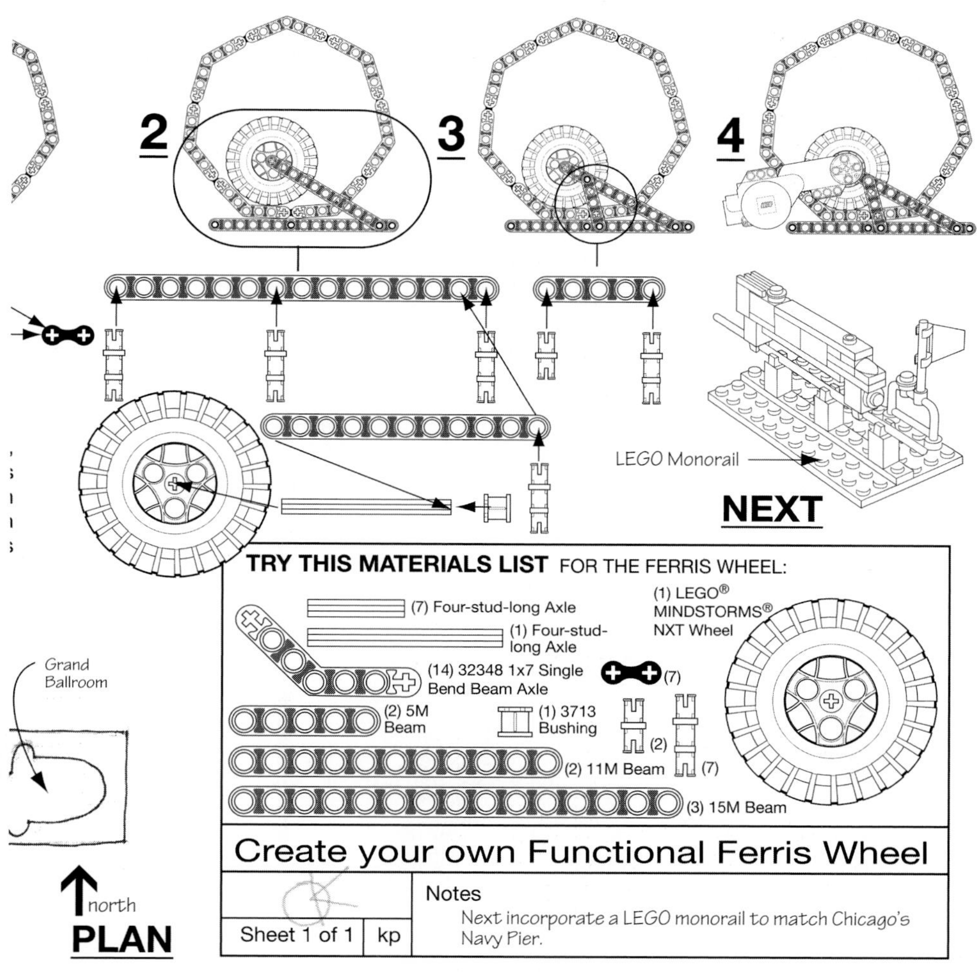
2
3
4
LEGO Monorail
NEXT
TRY THIS MATERIALS LIST FOR THE FERRIS WHEEL:
(7) Four-stud-long Axle
(1) Four-stud-long Axle
(1) LEGO® MINDSTORMS® NXT Wheel
(14) 32348 1x7 Single Bend Beam Axle
(7)
(2) 5M Beam
(1) 3713 Bushing
(2)
(2) 11M Beam
(7)
(3) 15M Beam
Create your own Functional Ferris Wheel
Notes
Next incorporate a LEGO monorail to match Chicago's Navy Pier.
Sheet 1 of 1
kp
Grand Ballroom
north
PLAN

5 A plan for an NXT spoke-less Ferris Wheel.

6 Step right up for a ride you'll never forget.

7 Build the inner wheel ring first. Connect two 1x7 angled beams together.

8 Keep adding connectors and beams.

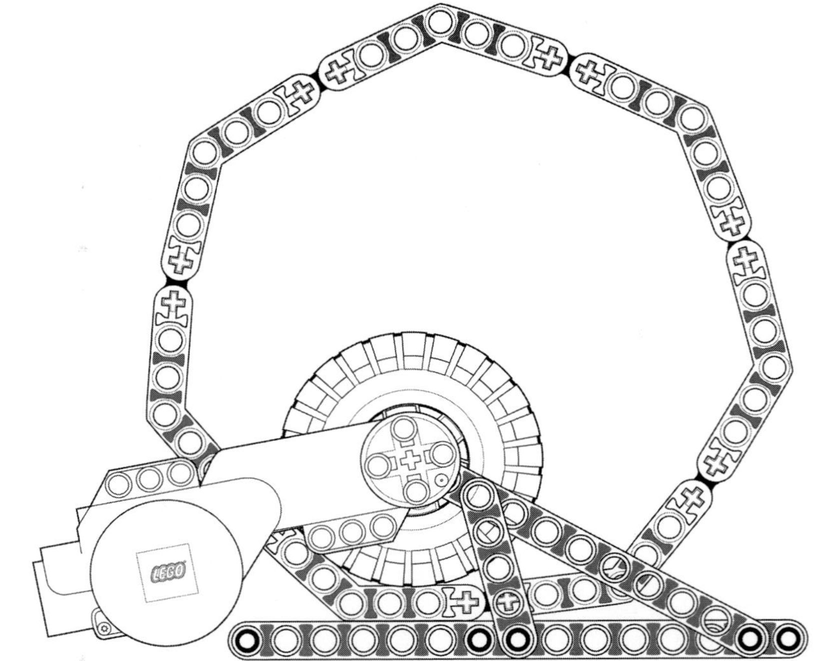

5

6

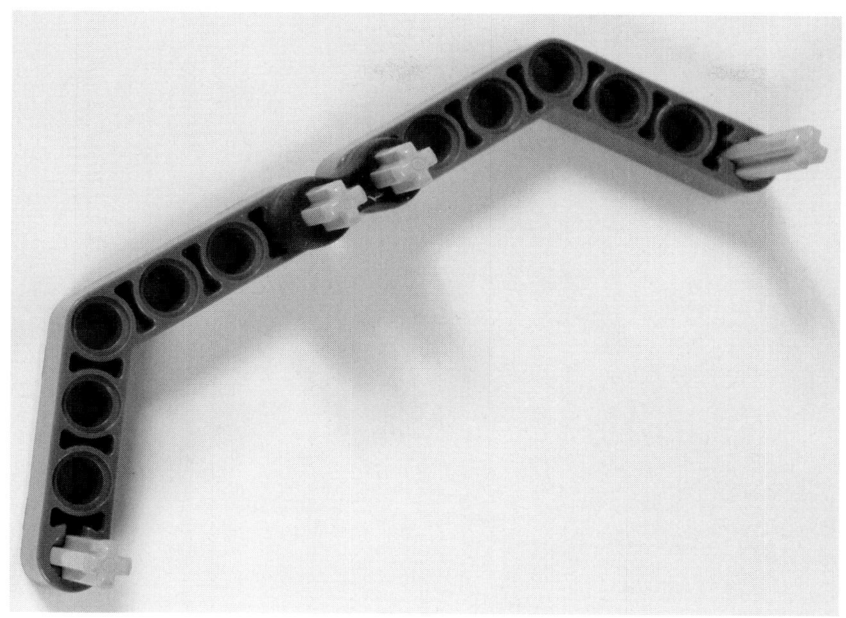

7

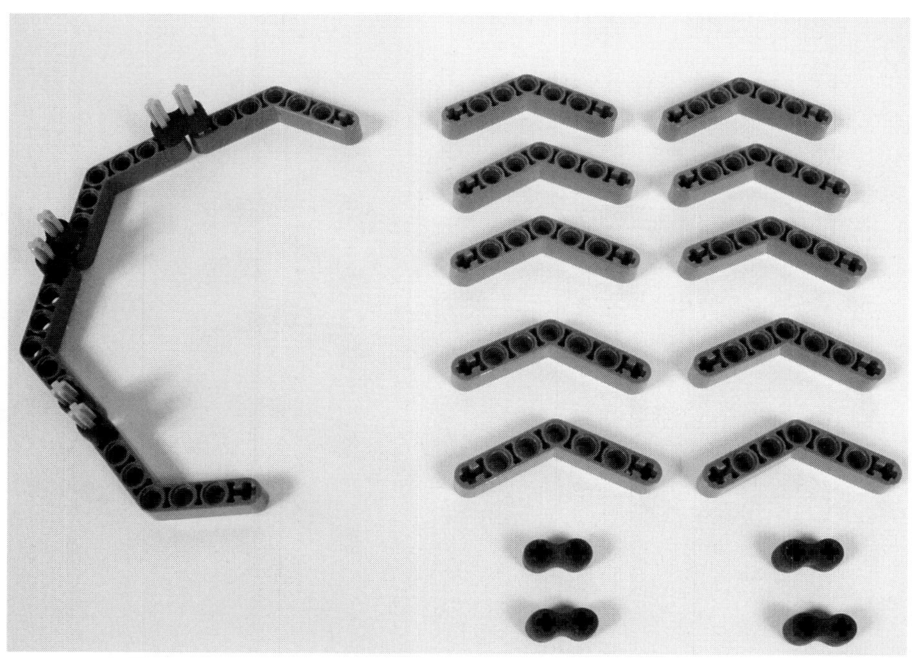

8

9 Use the rubber boot two-axle connectors for joining the angled beams.

10 The inner ring of the wheel is complete.

11 Build the outer ring on top of the inner ring.

12 The spoke-less Ferris Wheel ring is complete.

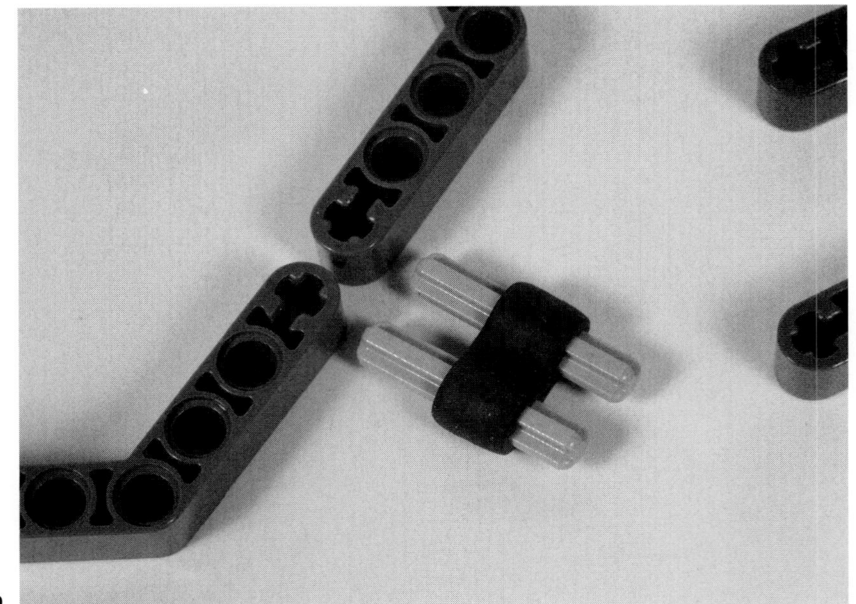

9

10

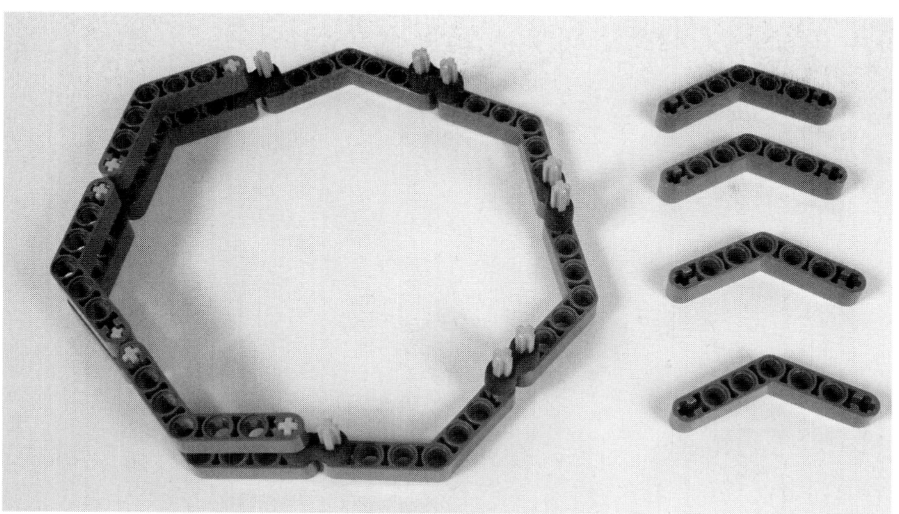

11

12

13 Assemble a support for the ring.

14 Join beams together for holding the ring's width.

15 Elevate the friction tire support. This tire drives the spoke-less Ferris Wheel.

16 Finally, something worthwhile that you can do with the big "phat" rubber wheel inside your NXT kit. Use this wheel as a friction drive for the Ferris Wheel.

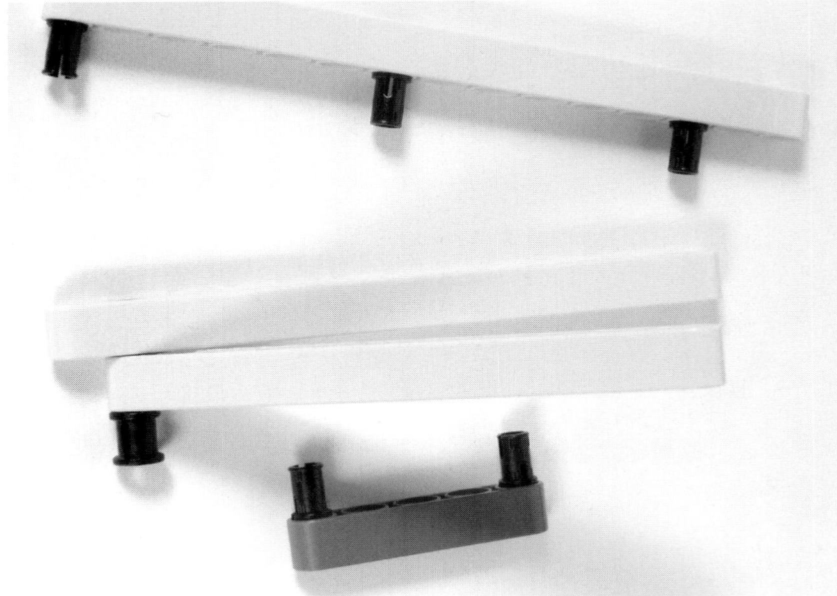

13

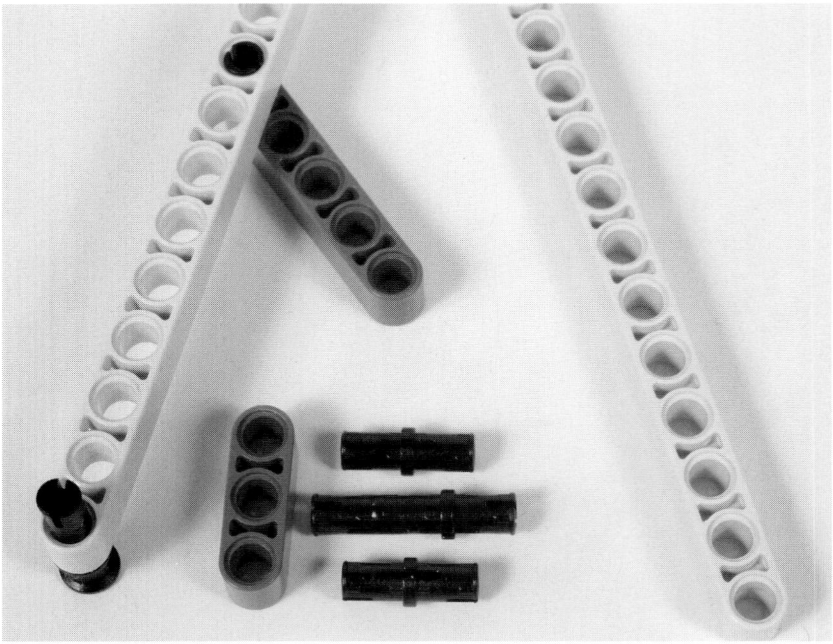

14

15

16

17 The spoke-less Ferris Wheel and friction drive mechanism are now complete.

18 Add an NXT motor for driving your Ferris Wheel. Cables and motor supports have been removed for clarity.

19 Ready for your next state fair. Cables and motor supports have been removed for clarity.

17

18

19

HOW TO BUILD A MONORAIL FOR YOUR SPOKE-LESS FERRIS WHEEL

1 The parts needed for building a small non-operational monorail. All of the parts for building this monorail are contained in my LDU LEGO Factory kit.

2 Assemble the base plate.

3 Add a rail standard and banner holder.

4 Complete the rail.

1

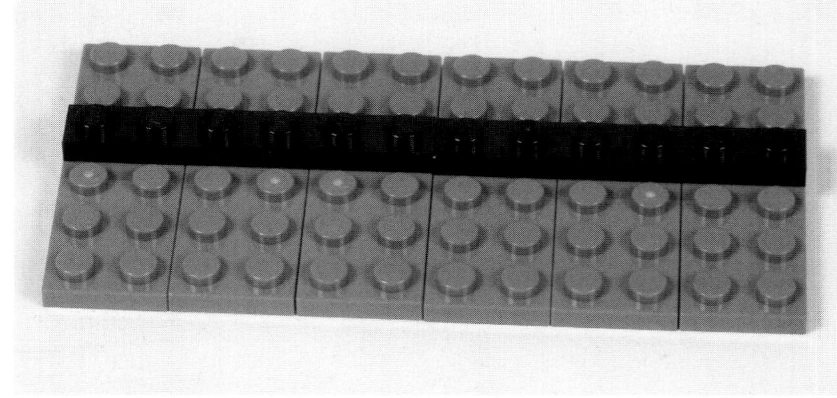

2

3

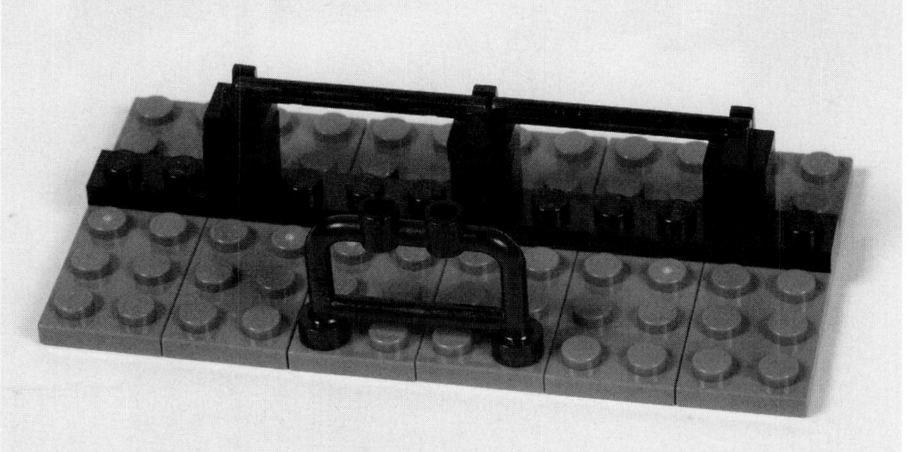

4

5 **Complete the banner and the monorail's undercarriage.**

6 **All aboard; this train's ready to depart.**

5

6

A PAIR OF PRITZKERS

Awards are a funny accolade afforded to certain members of our society. Some awards seem like a concocted public relations statement with little merit and way too much pomp and circumstance. These types of awards are usually handed out by a group that has a vested interest in promoting its own membership. In contrast to these glad-handing awards are the independent awards that serve to recognize individual accomplishment. Such an award is the Pritzker Architecture Prize (see Fig. IV-16). This award bestows a $100,000 grant on the selected individual, a bronze medallion that goes well with any type of clothing, and a formal printed citation certificate that just screams "I love me."

Seriously, the Pritzker Architecture Prize was established by The Hyatt Foundation in 1979 to annually honor a living architect whose built work demonstrates a combination of talent, vision, and commitment, which has produced consistent and significant contributions to humanity and the built environment through the art of architecture.

The prize derives its name from the Pritzker family of Chicago, Illinois. While this family is internationally known for their support of educational, social welfare, scientific, medical, and cultural activities, they are probably better known as the owners of the Hyatt Regency hotel chain.

The prize was founded by Jay and Cindy Pritzker. Today the prize is administered by the couple's eldest son, Thomas J. Pritzker, through the benevolent branch of the hotel corporation, the Hyatt Foundation.

Architects are nominated for consideration in the competition for this annual prize. Nominations are then judged by an international jury of their peers with the winner selected through a secret voting process. Hmm, this secrecy reminds me of Johnny Carson's "mayonnaise jar on Funk & Wagnall's front porch." According to the Hyatt Foundation, winners of the Pritzker Architecture Prize are called Laureates.

IV-16 The bronze medallion awarded to each Laureate of the Pritzker Architecture Prize is based on the designs of Louis H. Sullivan, famed Chicago architect generally acknowledged as the father of the skyscraper. (Photograph courtesy of the Hyatt Foundation)

RE: REM

After the Seattle Library Board of Trustees awarded the future Central Library design to Rem Koolhaas, the architect was asked what the new library would look like. "We pride ourselves in having no preconceptions, but we relish the opportunity to work on such a stable symbol of collective life," retorted Rem.

LAUREATES LAURELS

Here is a list of all of the previous Pritzker Architecture Prize winners, err, Laureates. How many of these architects do you know?

1979—Philip Johnson of the United States of America
1980—Luis Barragán of Mexico
1981—James Stirling of the United Kingdom
1982—Kevin Roche of the United States of America
1983—Ieoh Ming Pei of the United States of America
1984—Richard Meier of the United States of America
1985—Hans Hollein of Austria
1986—Gottfried Böhm of Germany
1987—Kenzo Tange of Japan
1988—Gordon Bunshaft of the United States of America and Oscar Niemeyer of Brazil
1989—Frank O. Gehry of the United States of America
1990—Aldo Rossi of Italy
1991—Robert Venturi of the United States of America
1992—Alvaro Siza of Portugal
1993—Fumihiko Maki of Japan
1994—Christian de Portzamparc of France
1995—Tadao Ando of Japan
1996—Rafael Moneo of Spain
1997—Sverre Fehn of Norway
1998—Renzo Piano of Italy
1999—Sir Norman Foster (Lord Foster) of the United Kingdom
2000—Rem Koolhaas of The Netherlands
2001—Jacques Herzog and Pierre de Meuron of Switzerland
2002—Glenn Murcutt of Australia
2003—Jørn Utzon of Denmark
2004—Zaha Hadid of the United Kingdom
2005—Thom Mayne of the United States of America
2006—Paulo Mendes da Rocha of Brazil

HOW TO BUILD A SEATTLE CENTRAL LIBRARY MODEL

1

2

1 Sometimes you have to look to other LEGO kits for the parts that you need. In this case, Police Car (7236) will be our donor.

2 Clear 1x1 roof tiles for use on the Seattle Central Library model can be removed from this kit. Six kits were used for supplying enough roof tiles.

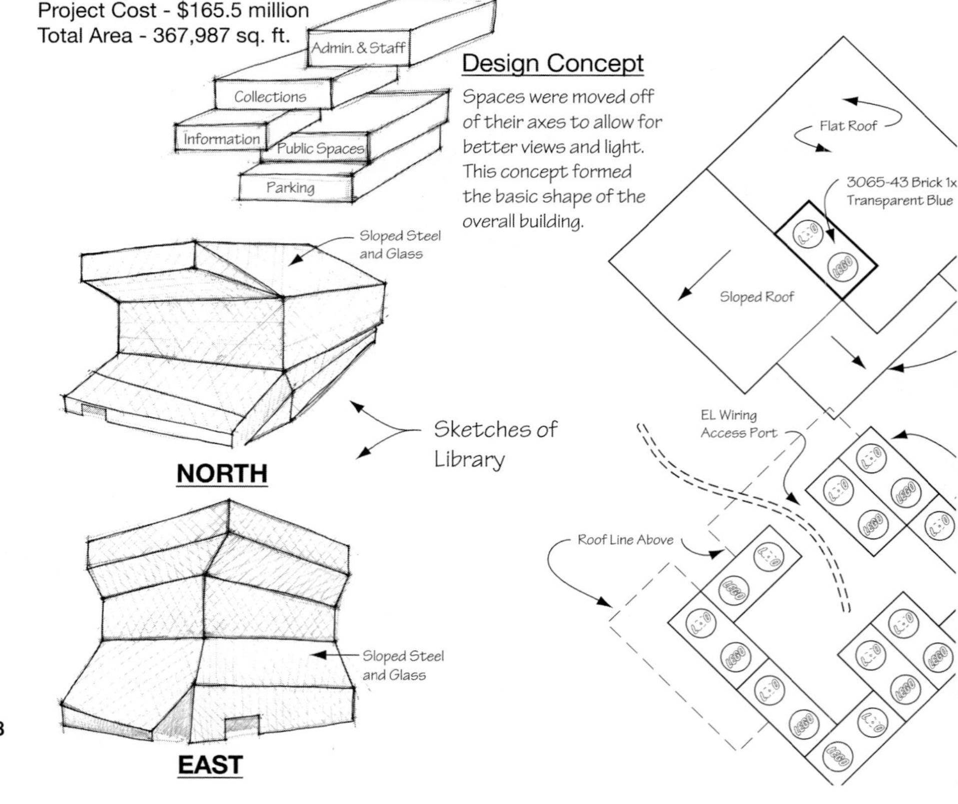
SEATTLE CENTRAL LIBR
Address - 1000 4th Ave, Seattle, WA 98104 USA
Description - 11-story library, exterior skin is of sloped steel and glass
Architect - Rem Koolhaas, winner of the 2000 Pritzker Prize
Built - 2001-2004
Project Cost - $165.5 million
Total Area - 367,987 sq. ft.
Admin. & Staff
Collections
Information
Public Spaces
Parking
Design Concept
Spaces were moved off of their axes to allow for better views and light. This concept formed the basic shape of the overall building.
Sloped Steel and Glass
Sketches of Library
NORTH
Sloped Steel and Glass
EAST
Flat Roof
3065-43 Brick 1x
Transparent Blue
Sloped Roof
EL Wiring Access Port
Roof Line Above
3

3 **A plan for Seattle Central Library.**

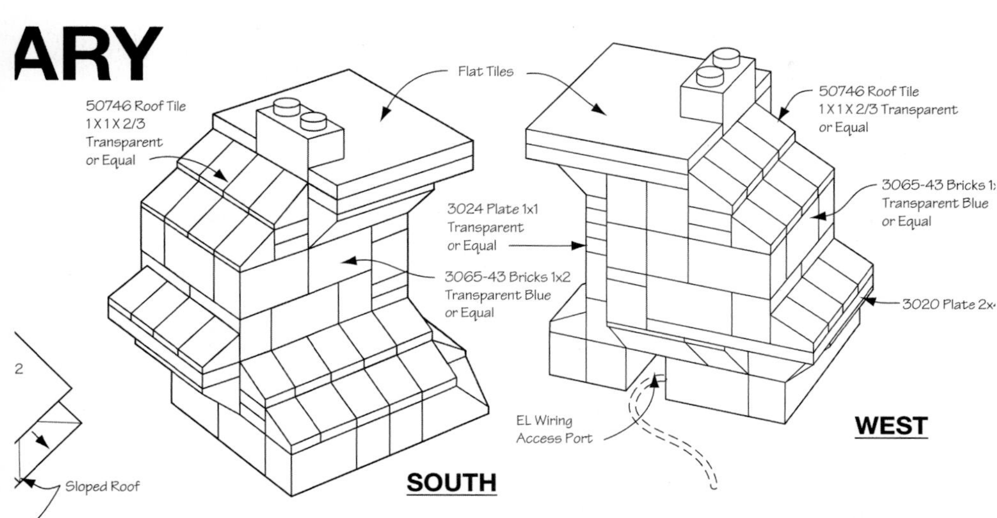

ROOF PLAN

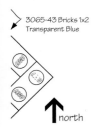

3065-43 Bricks 1x2
Transparent Blue

north

1ST BRICK COURSE

TRY THIS MATERIALS LIST

Qty	Part	Description
(26)	50746-40	Roof Tile 1 X 1 X 2/3 Transparent
(9)	3023-40	Plate 1x2 Transparent
(14)	3024-40	Plate 1x1 Transparent
(26)	3065-43	Brick 1x2 Without Tap Transparent Blue
(3)	3020	Plate 2x4
(1)	3022	Plate 2x2
(1)	3046	Corner Brick 2x2/45 Inside
(3)	3068	Flat Tile 2x2
(3)	3069	Flat Tile 1x2
(4)	3660	Roof Tile 2x2/45 Inverted
(3)	3665	Roof Tile 1x2, Inverted
(1)	3710	Plate 1x4

Number of bricks: 94

Light Up the Seattle Central Library

Sheet 1 of 1 | kp

Notes
Add EL Wiring and light up your life.

4 The parts needed for building the Seattle Central Library model. These parts are included in my LEGO Factory LDU kit.

5 Begin building the first floor. Leave a space in the center for holding some NXT-controlled EL wiring for and LED brick.

6 Insert the LED brick now.

7 Finish the first floor.

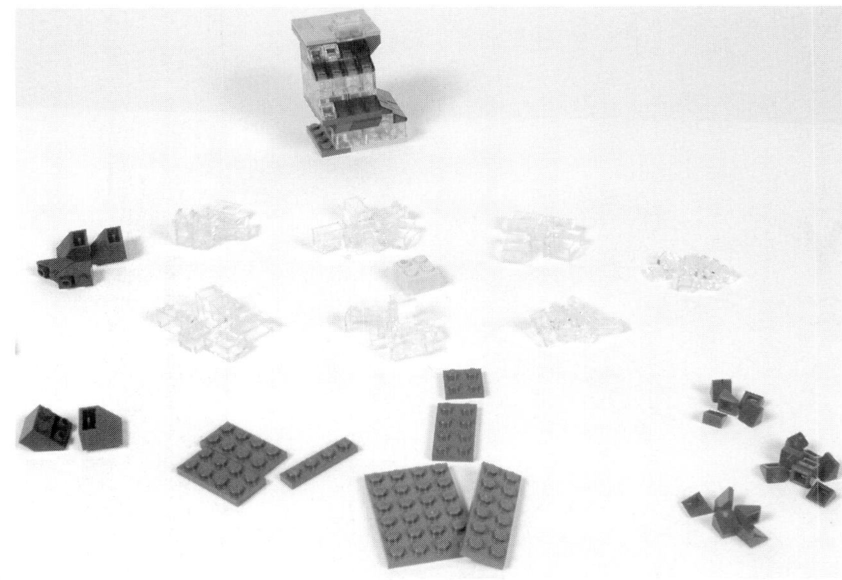

4

5

6

7

8 Begin the second floor.

9 Use the transparent 1x1 roof tiles that you gleaned from the Police Car kits shown in 2.

10 Moving on up to the third floor.

11 Finishing up the third floor.

8

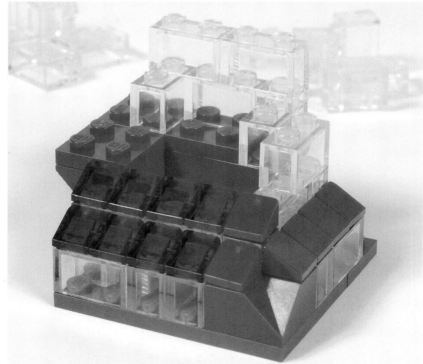

9

10

11

12 Start adding the roof elements.

13 Make sure that the LED brick has a clear path from top to bottom.

14 Add the final roof tiles.

15 Add some LEDs and light 'er up.

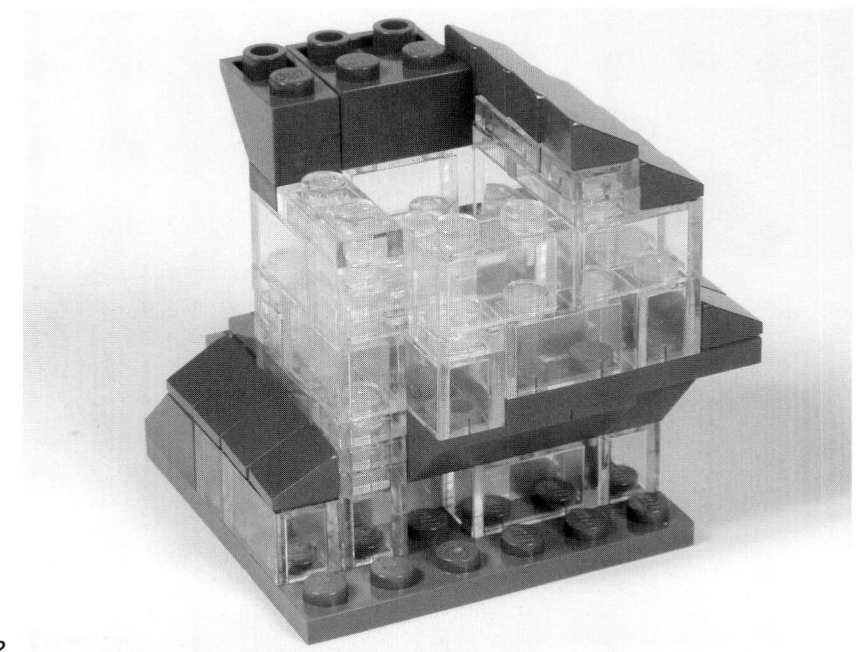

12

13

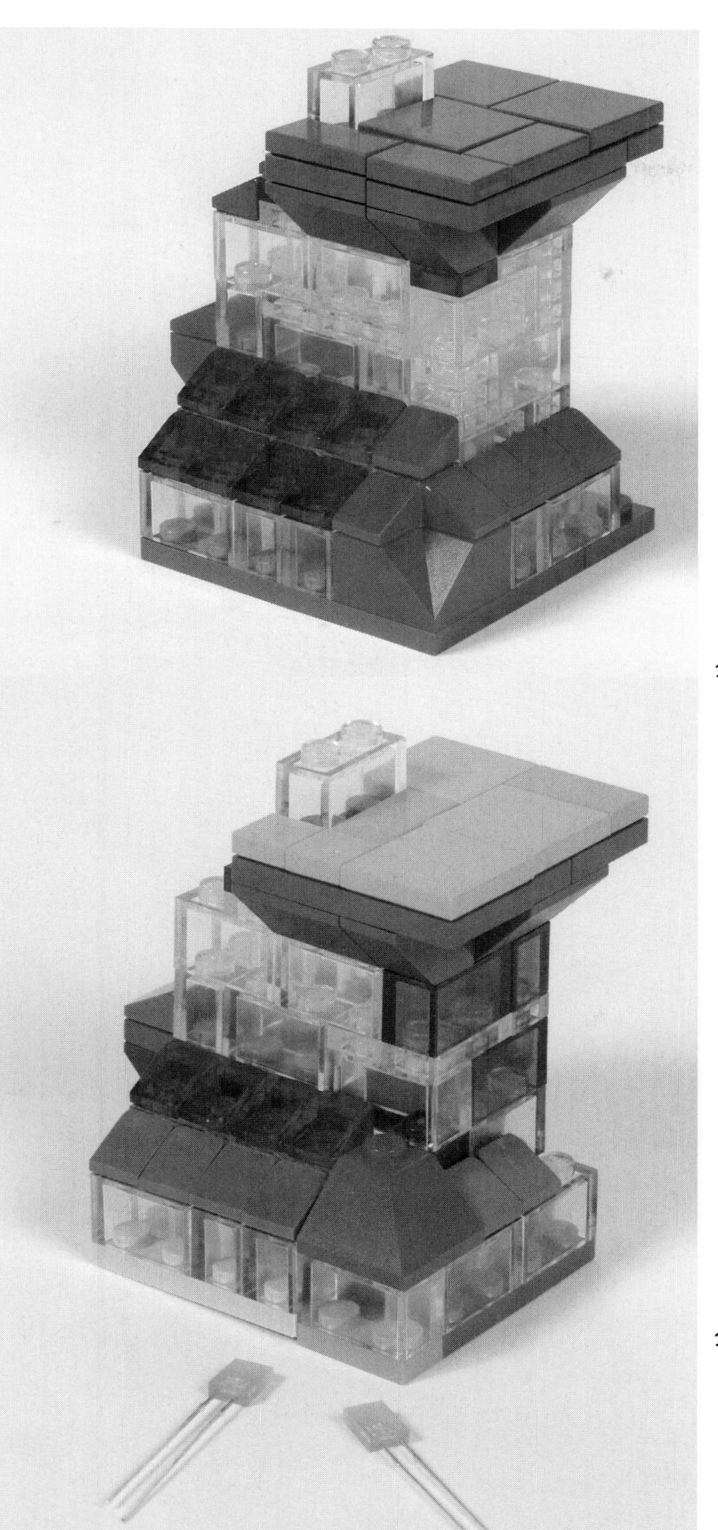

14

15

16 This is a rather small model that can be easily carried and displayed. Like, I don't know, maybe when you're reading a book at the library you could set this structure on your table. It's bound to attract an architect or two.

17 The final lighting effect with the transparent bricks and tiles is sweet.

16

17

THE LORD SAYETH

Following the completion of his addition to the Joslyn Art Museum in Omaha, Nebraska, Lord Norman Foster reflected on his design (see Fig. IV-17). "In adding the new wing," Lord Foster recalled, "we were able to regenerate the interior of the original building, redesign the outside spaces for outdoor events, and most significantly, make use of the gap between the old and the new parts of the building as another social space, in this case, a café."

During the "topping off" (a ceremony placing the highest steel beam in a building) of the 46-story Hearst Corporation's office tower in New York City, the Lord got downright misty eyed as he acknowledged the significance of this design: "The Hearst tower expresses its own time with distinction, yet respects and strengthens the existing six-storey historical structure. The tower is lifted clear of its historic base, linked to the outside only by columns and glazing, which are set back from the edges of the site. The transparent connection floods the spaces below with natural light and encourages the impression of the new floating above the old."

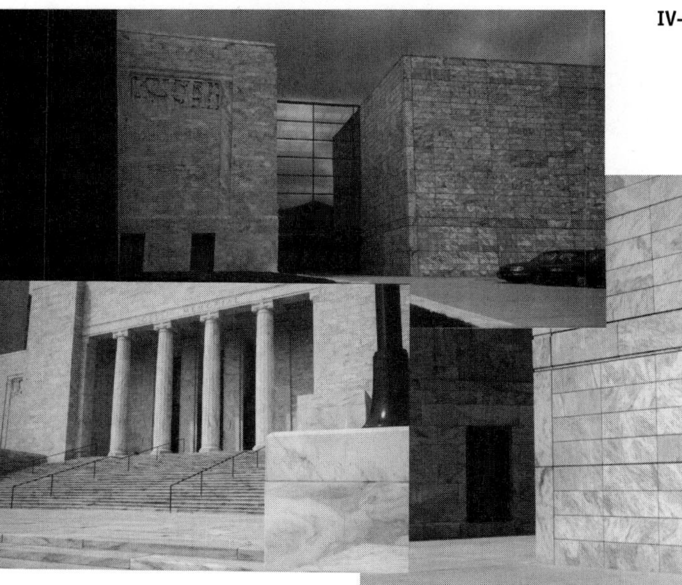

IV-17 Lord Norman Foster's addition (top) to the Joslyn Art Museum in Omaha, Nebraska sits adjacent to the original museum (left) with a glass connector atrium joining old with new (right).

HOW TO BUILD THE HEARST HEADQUARTERS TOWER AND ICE FALLS SCULPTURE

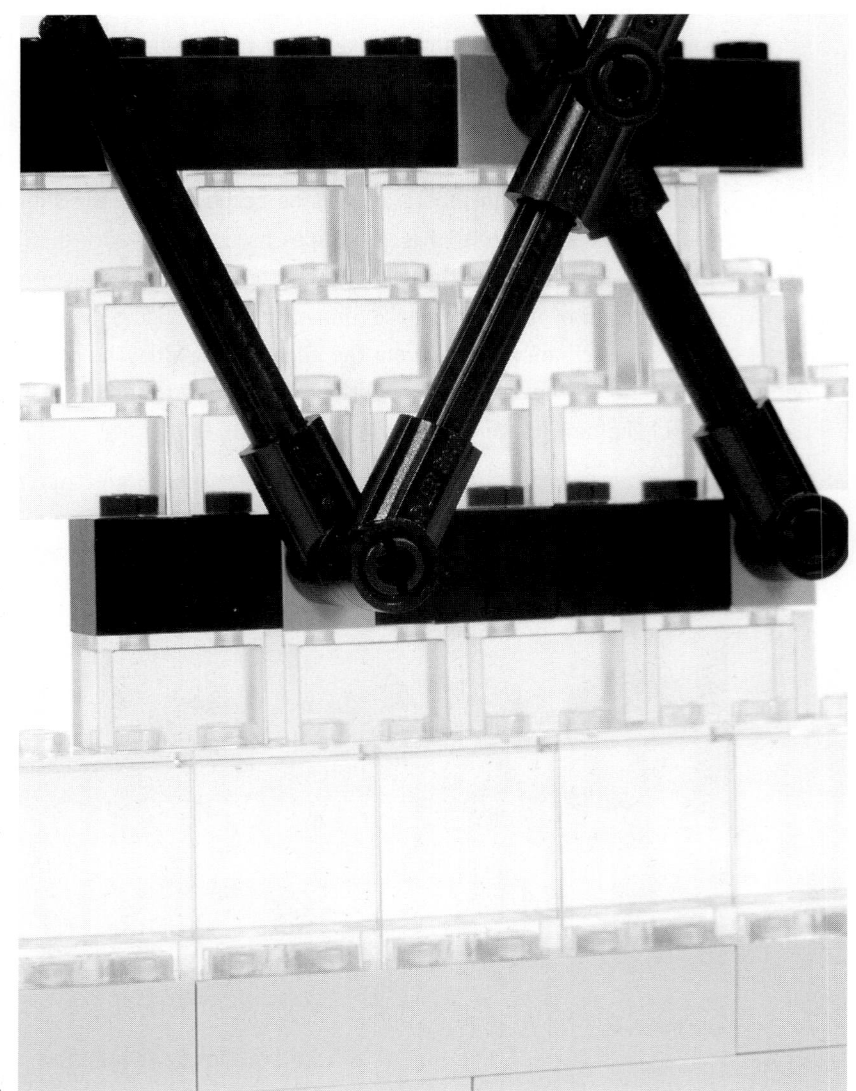

1 How would you approach the design challenges of modeling the Hearst Headquarters tower in New York City?

2 A preliminary mockup of building a steel "skin" on a glass building.

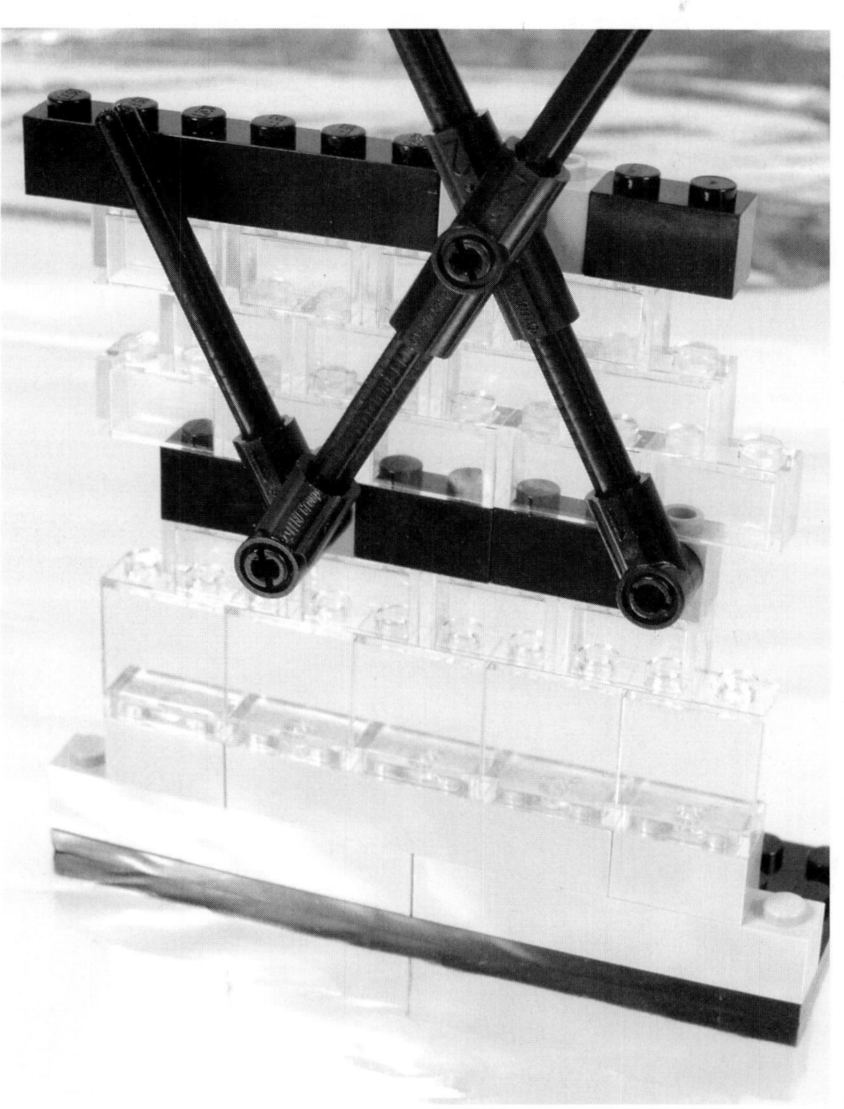

2

3 A plan for Hearst Headquarters.

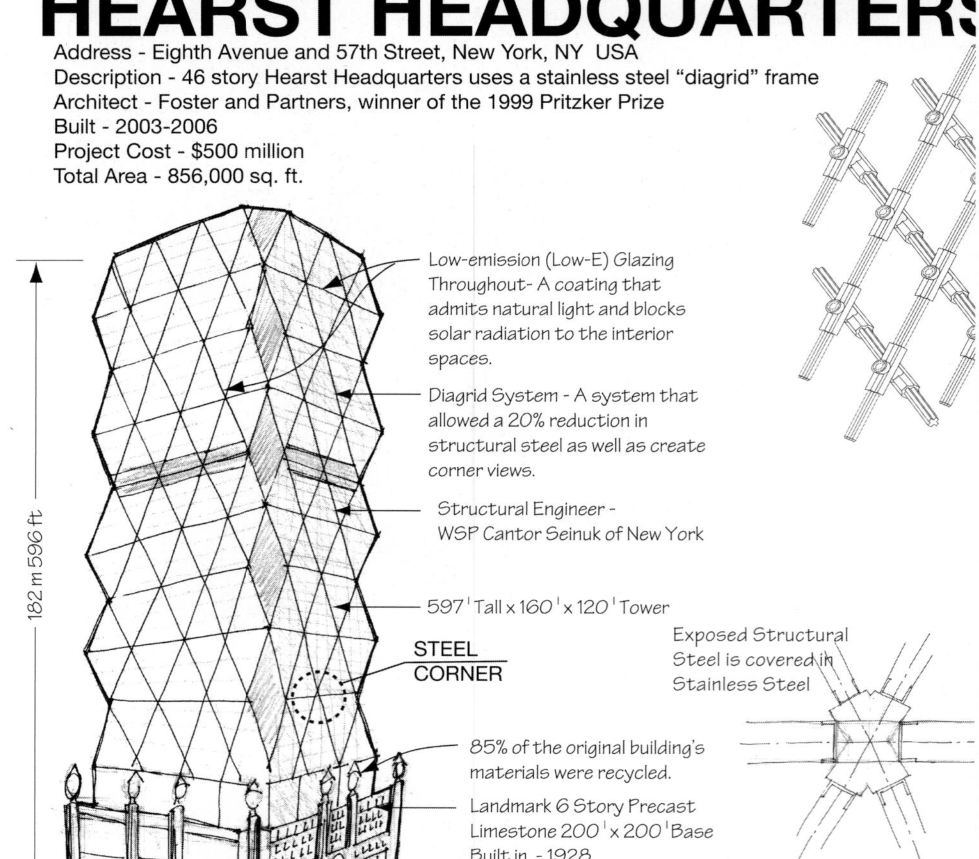

HEARST HEADQUARTERS
Address - Eighth Avenue and 57th Street, New York, NY USA
Description - 46 story Hearst Headquarters uses a stainless steel "diagrid" frame
Architect - Foster and Partners, winner of the 1999 Pritzker Prize
Built - 2003-2006
Project Cost - $500 million
Total Area - 856,000 sq. ft.
Low-emission (Low-E) Glazing Throughout- A coating that admits natural light and blocks solar radiation to the interior spaces.
Diagrid System - A system that allowed a 20% reduction in structural steel as well as create corner views.
Structural Engineer - WSP Cantor Seinuk of New York
597' Tall x 160' x 120' Tower
182 m 596 ft
STEEL CORNER
Exposed Structural Steel is covered in Stainless Steel
85% of the original building's materials were recycled.
Landmark 6 Story Precast Limestone 200' x 200' Base
Built in - 1928
Project Cost - $2 million
Total Area - 40,000 sq. ft.
STEEL CORNER
3

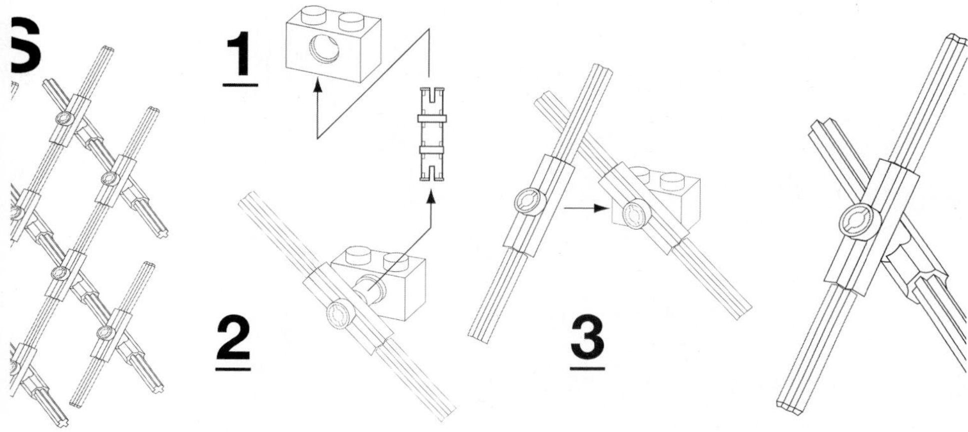

GOING GREEN

The Hearst Tower is planned to receive a gold rating from the Leadership in Energy and Environmental Design (LEED). The LEED Green Building Rating System® is the nationally accepted standard for green buildings developed by the U.S. Green Building Council (USGBC).

According to the USGBC's mission statement they are "working to promote buildings that are environmentally responsible, profitable and healthy places to live and work."

Key bricks for the LEGO model

- (842) + 3065 1x2 Without Tap Transparent Bricks
- (160) + Technic 1 x Bricks
- (128) 32034 Technic 180° Angle Element
- (190) + 32013 Technic 0° Angle Element
- (160) +Technic Pegs
- (250) + Technic Axles

Finally, add window washing equipment to the roof

Make a Skyscraper without Corners		
		Notes See the Ice Falls Plan for some interior details.
Sheet 1 of 1	kp	

R

4 TECHNIC axles and connectors were added to the glass brick building.

5 A detailed plan for the TECHNIC exoskeleton.

4

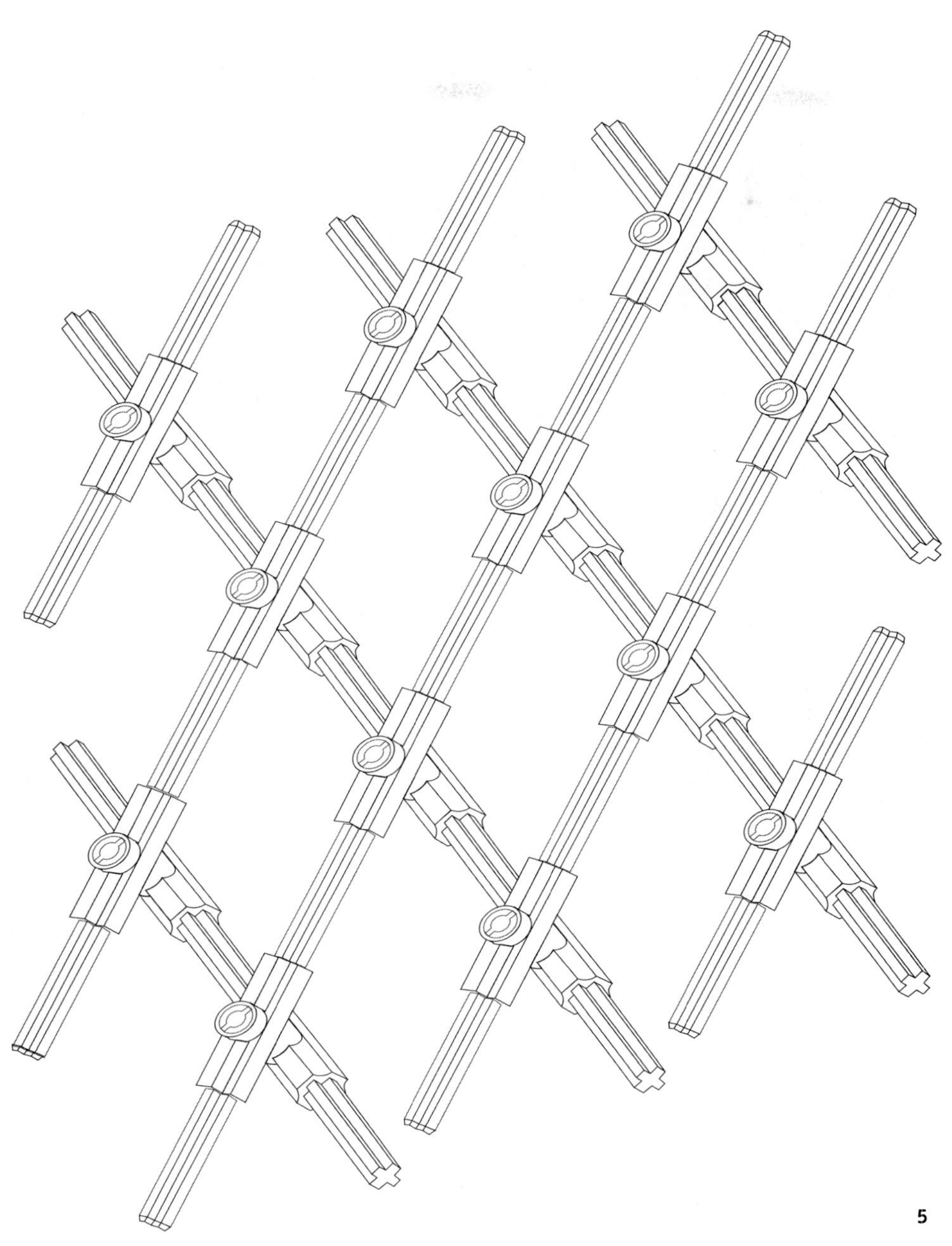

5

6 A plan for the interior Ice Falls sculpture.

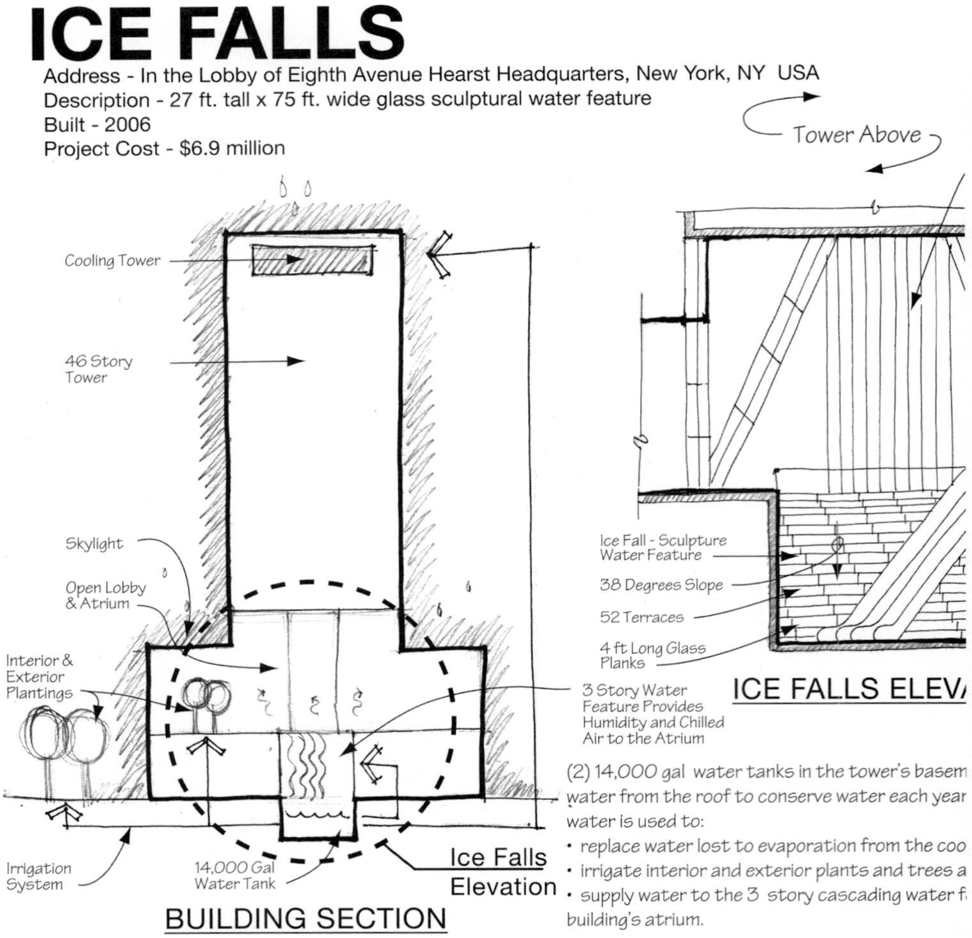

ICE FALLS

Address - In the Lobby of Eighth Avenue Hearst Headquarters, New York, NY USA
Description - 27 ft. tall x 75 ft. wide glass sculptural water feature
Built - 2006
Project Cost - $6.9 million

Tower Above

Cooling Tower

46 Story Tower

Skylight

Open Lobby & Atrium

Interior & Exterior Plantings

Irrigation System

14,000 Gal Water Tank

Ice Falls Elevation

BUILDING SECTION

Ice Fall - Sculpture Water Feature

38 Degrees Slope

52 Terraces

4 ft Long Glass Planks

3 Story Water Feature Provides Humidity and Chilled Air to the Atrium

ICE FALLS ELEV

(2) 14,000 gal water tanks in the tower's basem water from the roof to conserve water each year water is used to:

- replace water lost to evaporation from the coo
- irrigate interior and exterior plants and trees a
- supply water to the 3 story cascading water f building's atrium.

6

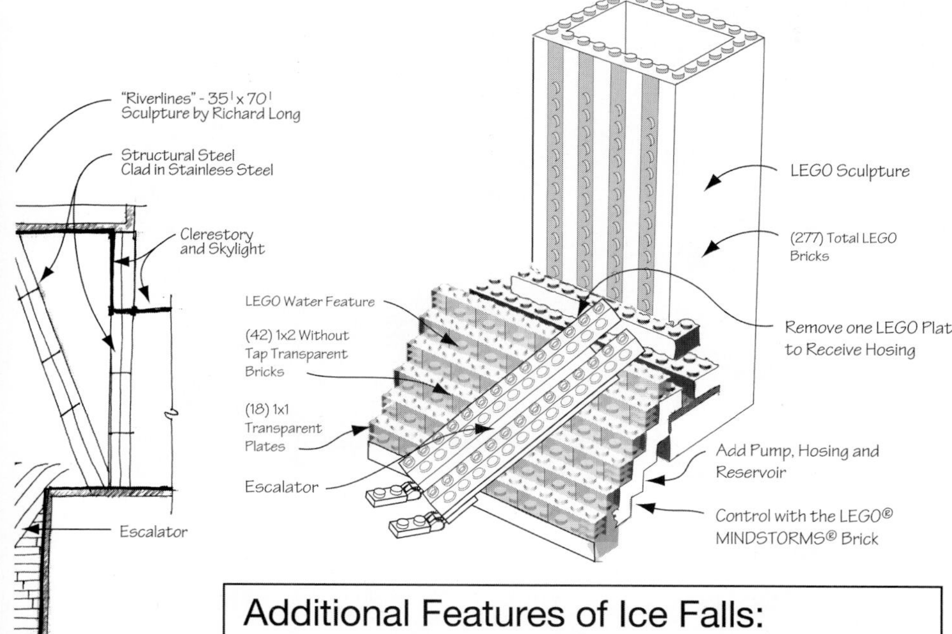

ATION

ıent collect rain-
. The collected

ling tower,
nd,
eature in the

Additional Features of Ice Falls:

- Water strength and flow rate are directed by a computer system that is divided into 6 zones, with 22 control valves.
- 50 tons of art glass were cast into (580) 4 ft long planks.
- The escalator moves people up and down through the falls.

Create your own Water Feature

Notes
Plug your water feature into your own Hearst Tower.

Sheet 1 of 1 | kp

7 LEGO Digital Designer was used for building a parts list.

8 Although it looks impressive, a successful purchase was difficult to obtain due to the unusual corners.

9 The plan required LEGO TECHNIC Connectors No. 1. These connectors were not available with LEGO Digital Designer. So LEGO TECHNIC Connectors No. 2 were substituted for each corner.

10 The base of the building was easier to design.

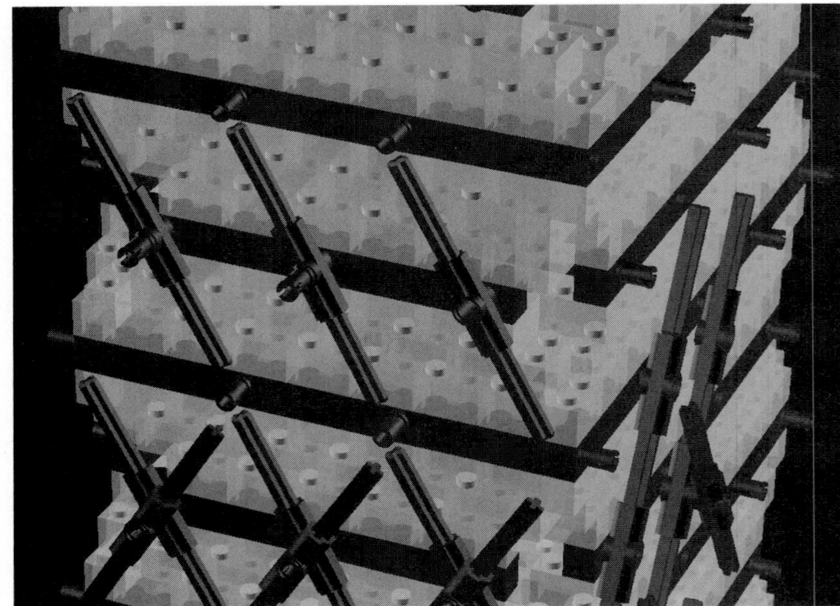

7

8

9

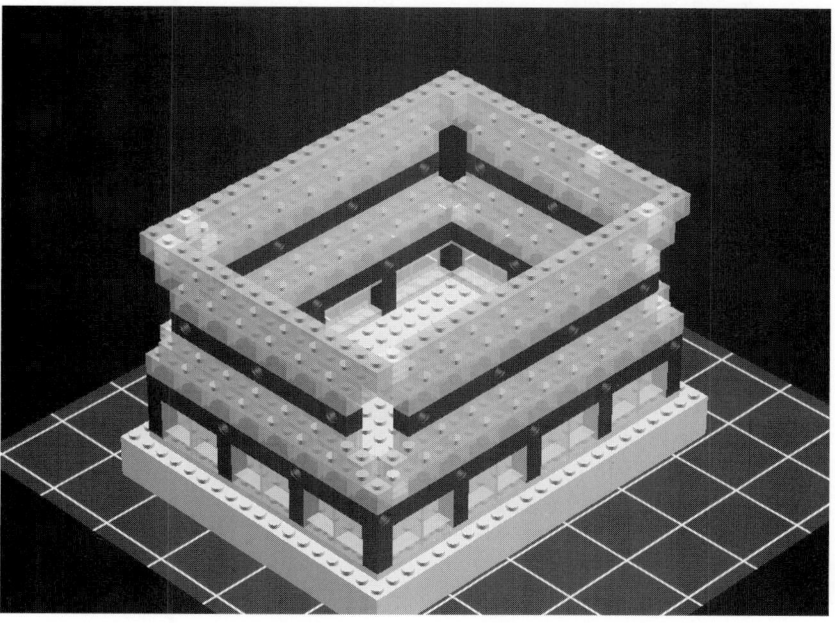

10

11 A separate model was created for the interior Ice Falls sculpture.

12 A small space was left in the model for accommodating a future water supply hose, the NXT Brick, and a submersible pump.

11

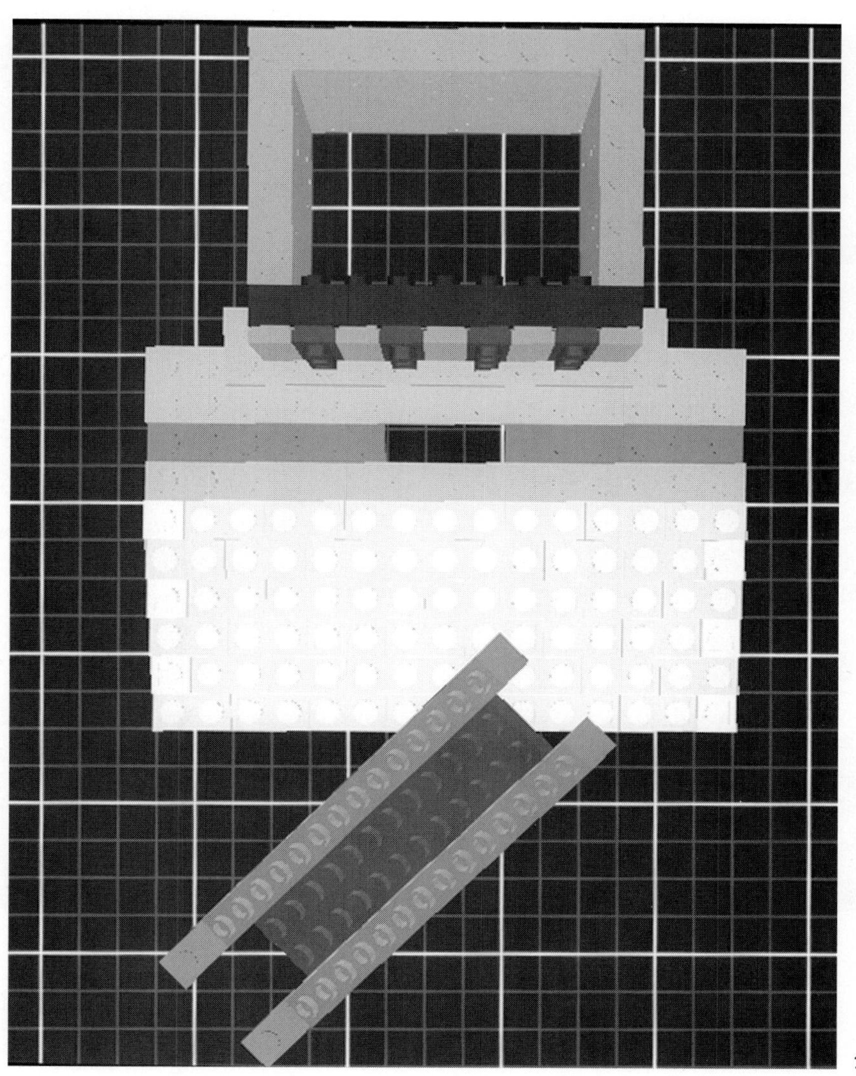

12

13 A set of escalators runs diagonally across the face of the sculpture.

14 First the inside business tower was built from black LEGO bricks.

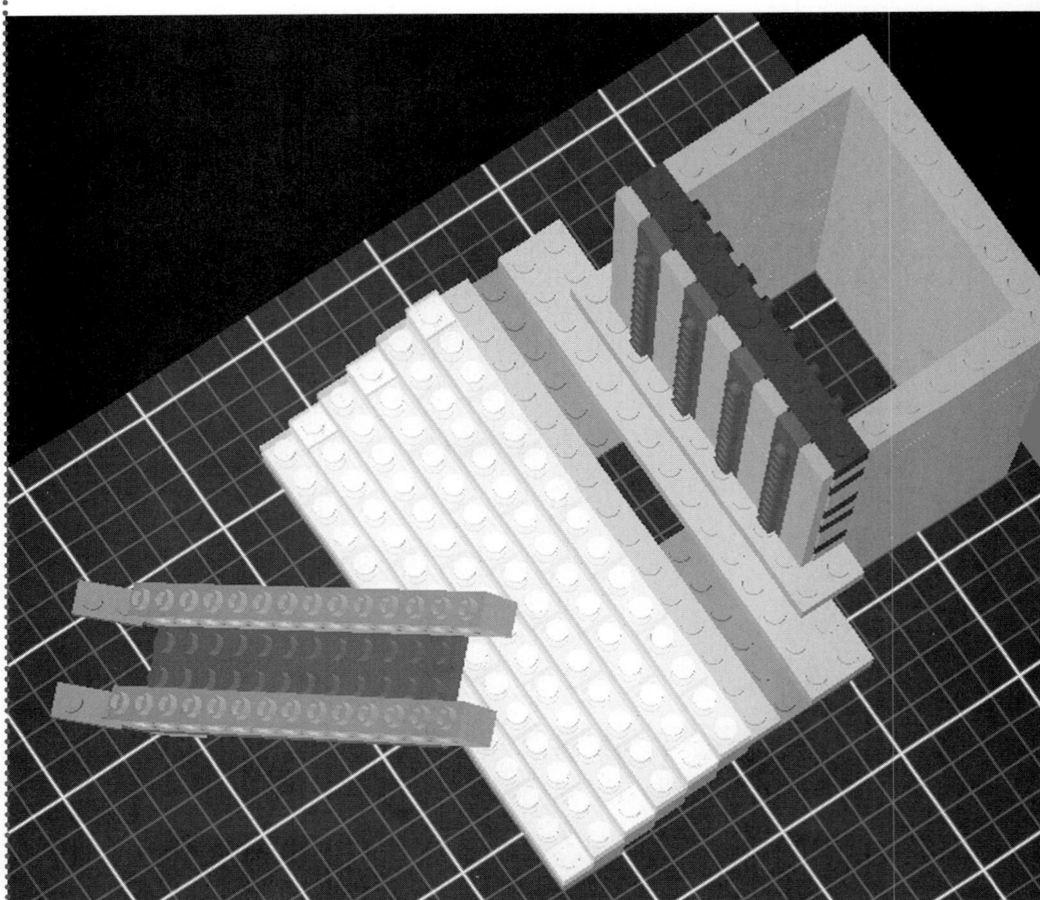

13

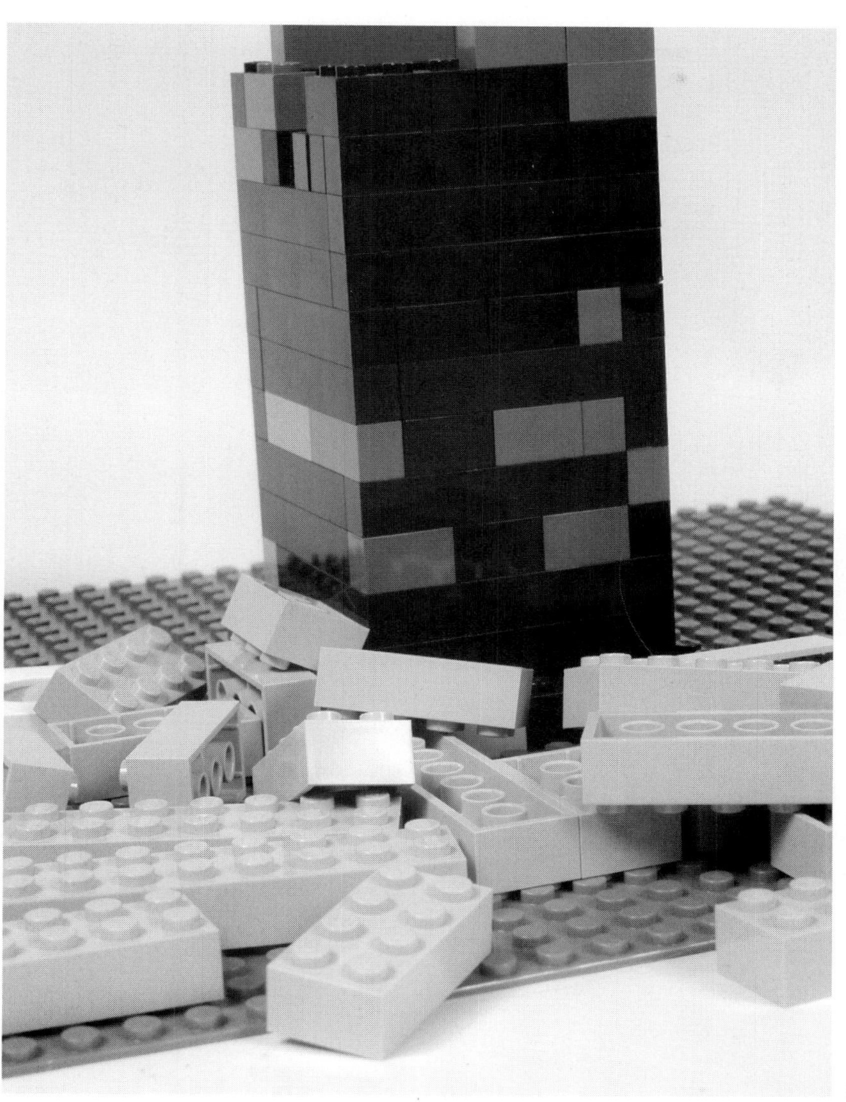

14

15 The building base was built from "brick yellow" colored LEGO bricks.

16 A skylight atrium was created from transparent, brown wall-unit pieces.

15

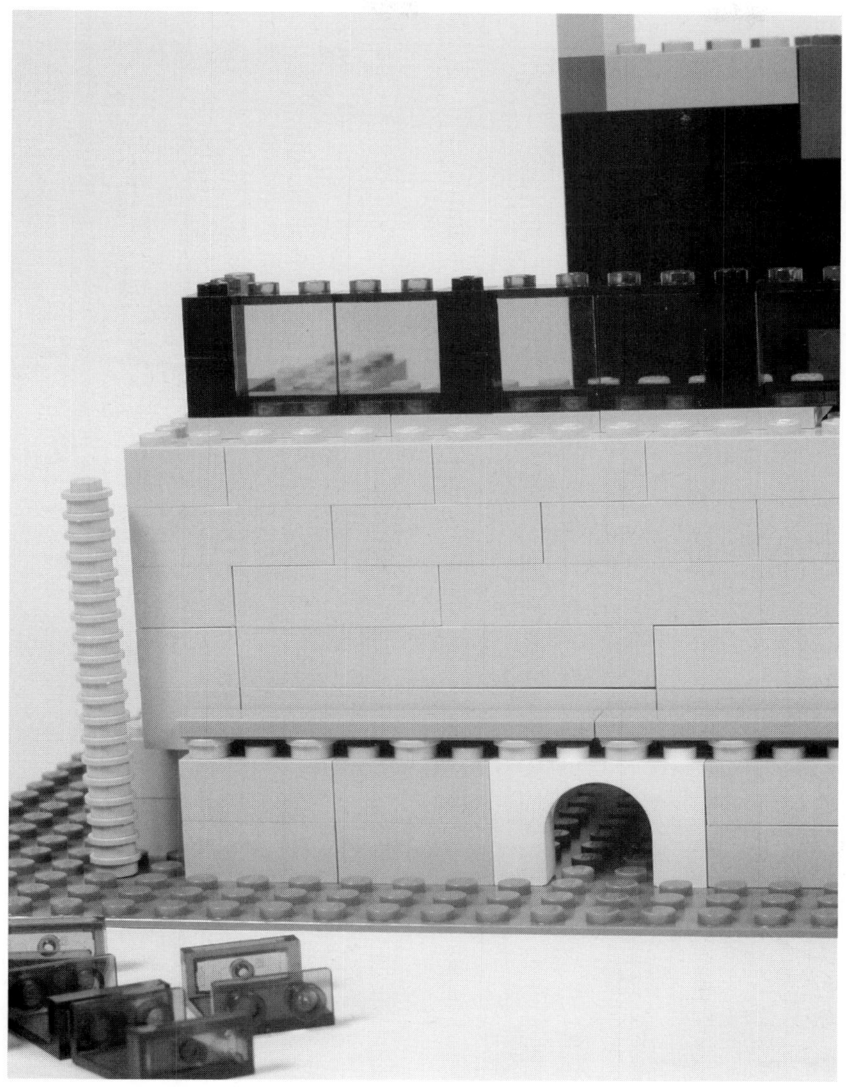

16

17 The outside column corner detail was created from a stack of 1x1 cylinder plates.

18 A close-up of the atrium level.

19 Transparent 1x2 bricks were used for the main glass walls.

17

18

19

20 **TECHNIC 1x2 bricks with one axle hole were used for joining the axis of the external support axles and connectors.**

21 **Indented corners alternate every four rows.**

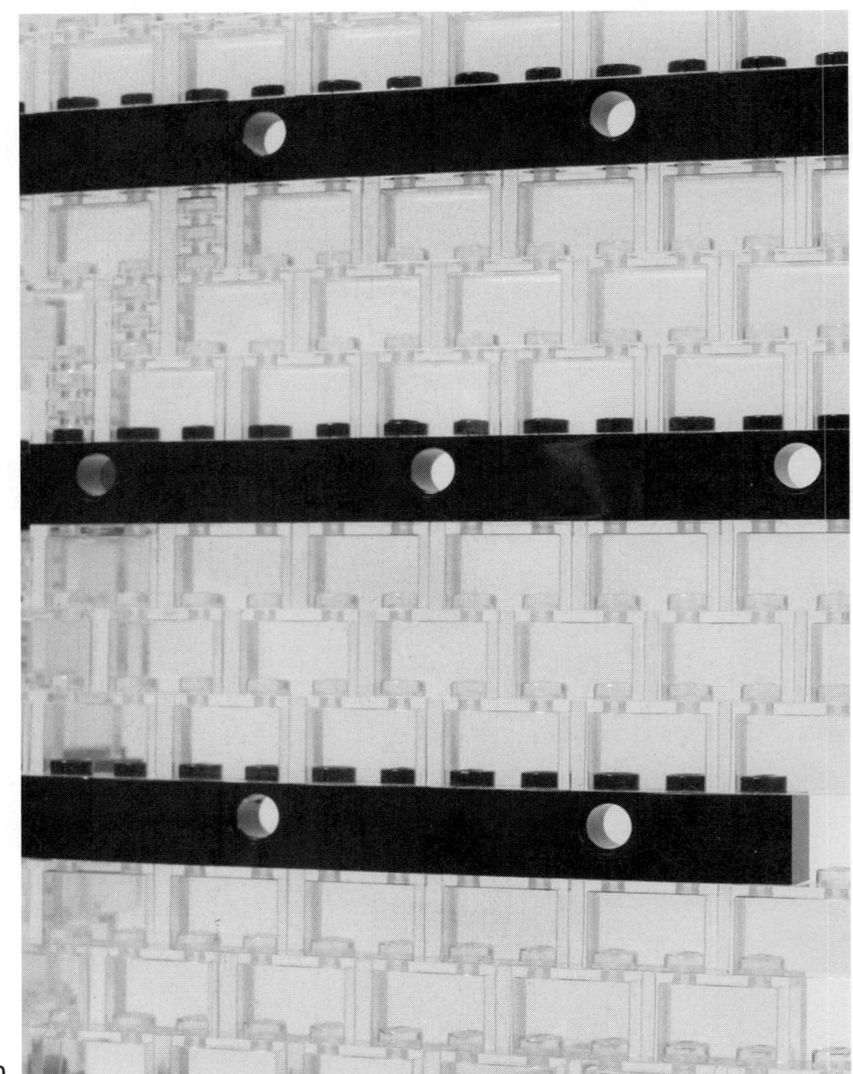

20

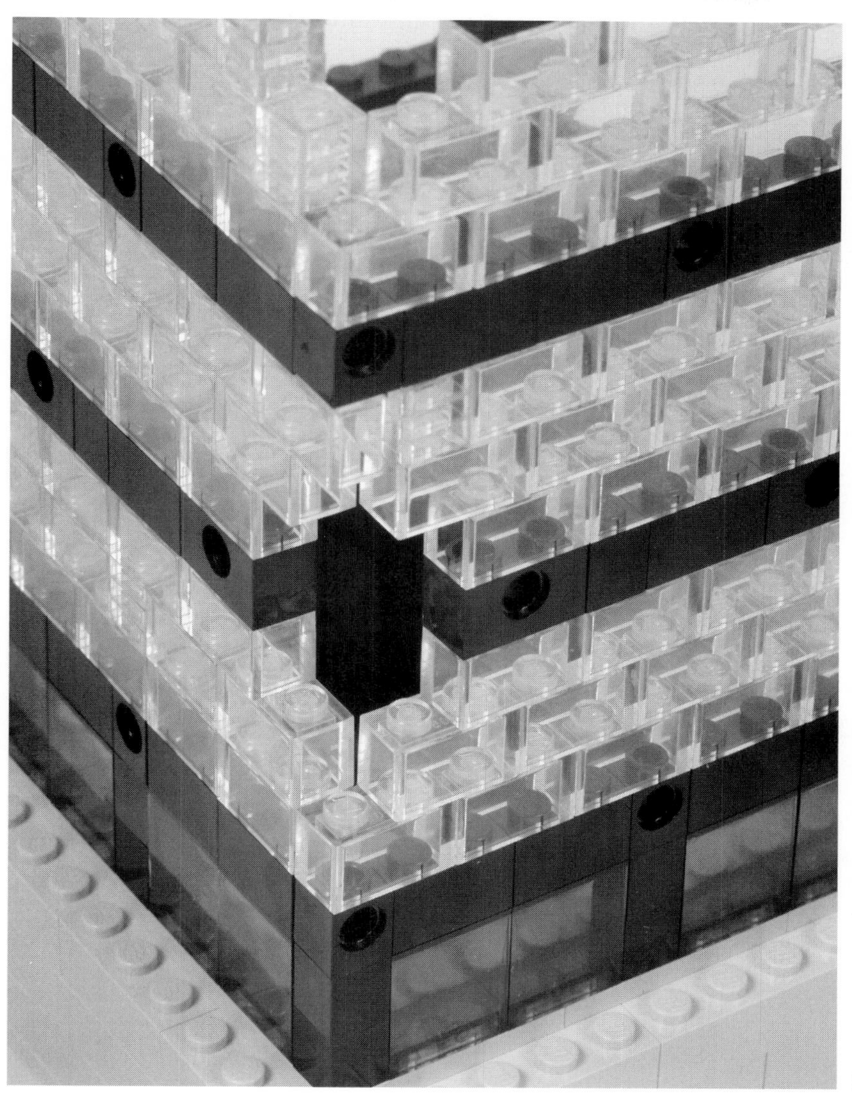

21

22 One face is almost complete.

23 Three-stud connectors were inserted into the TECHNIC Bricks.

24 LEGO TECHNIC Connectors No. 1 were used for the first floor.

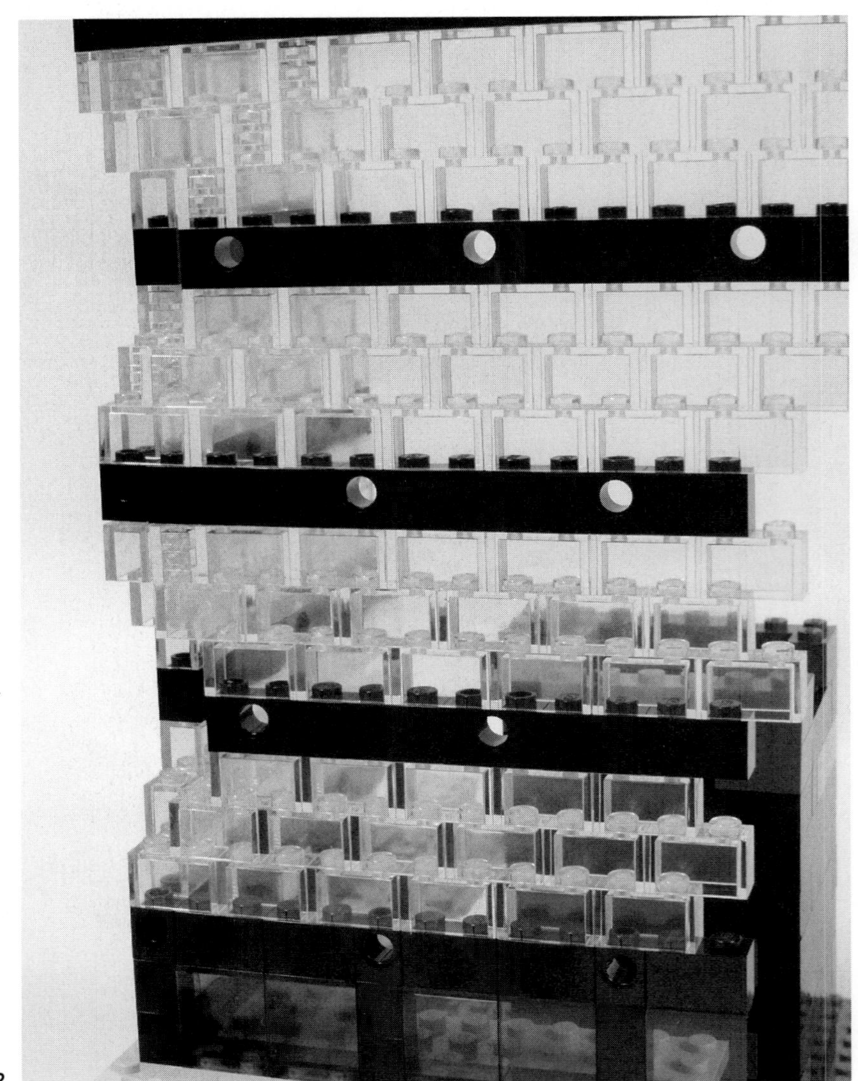

22

23

24

25 **These connectors are in perfect alignment.**

26 **The first diagonal line of external axle supports has been added.**

27 **Each external axle line should be parallel with the other.**

25

26

27

28 The first layer of diagonal axles are in place on this face.

29 Crisscross applesauce—both directions of diagonal axles are in place on this face.

28

29

30 Looking down from the top floor.

31 Prepping the next face for receiving its diagonal bracing face.

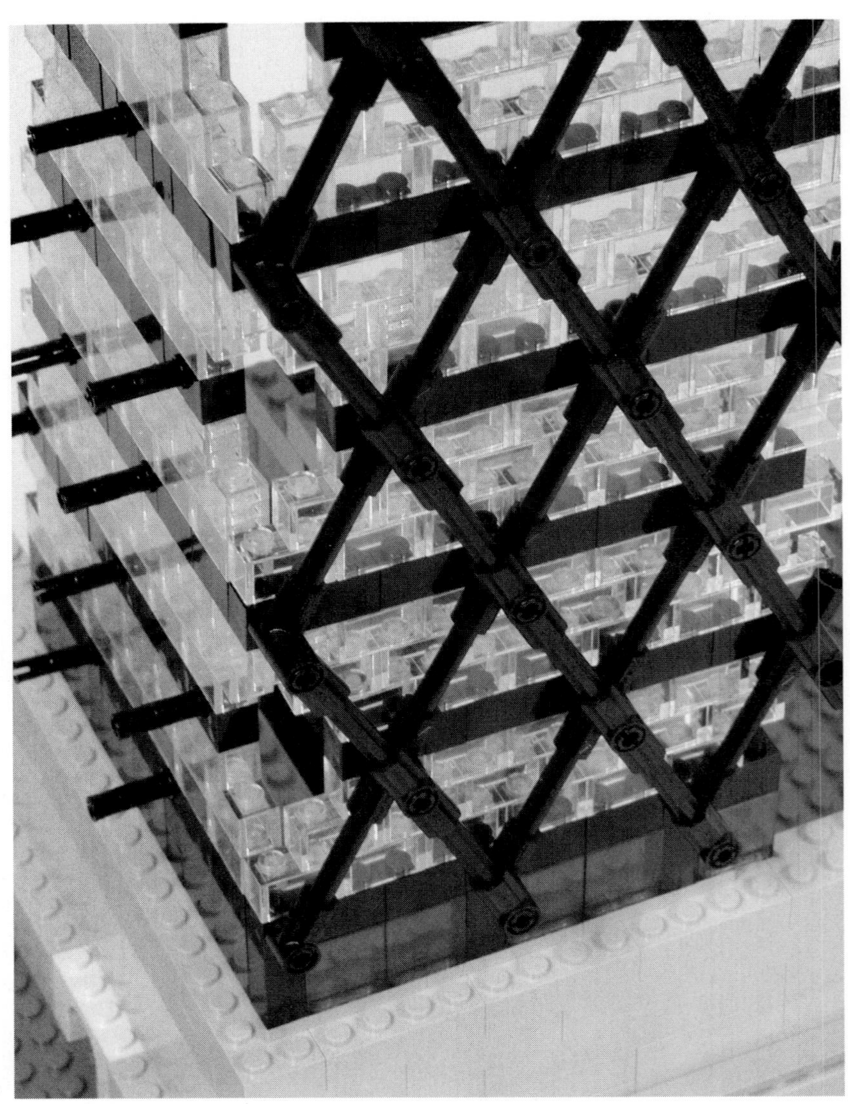

30

31

32 A corner detail.

33 A view from ground level.

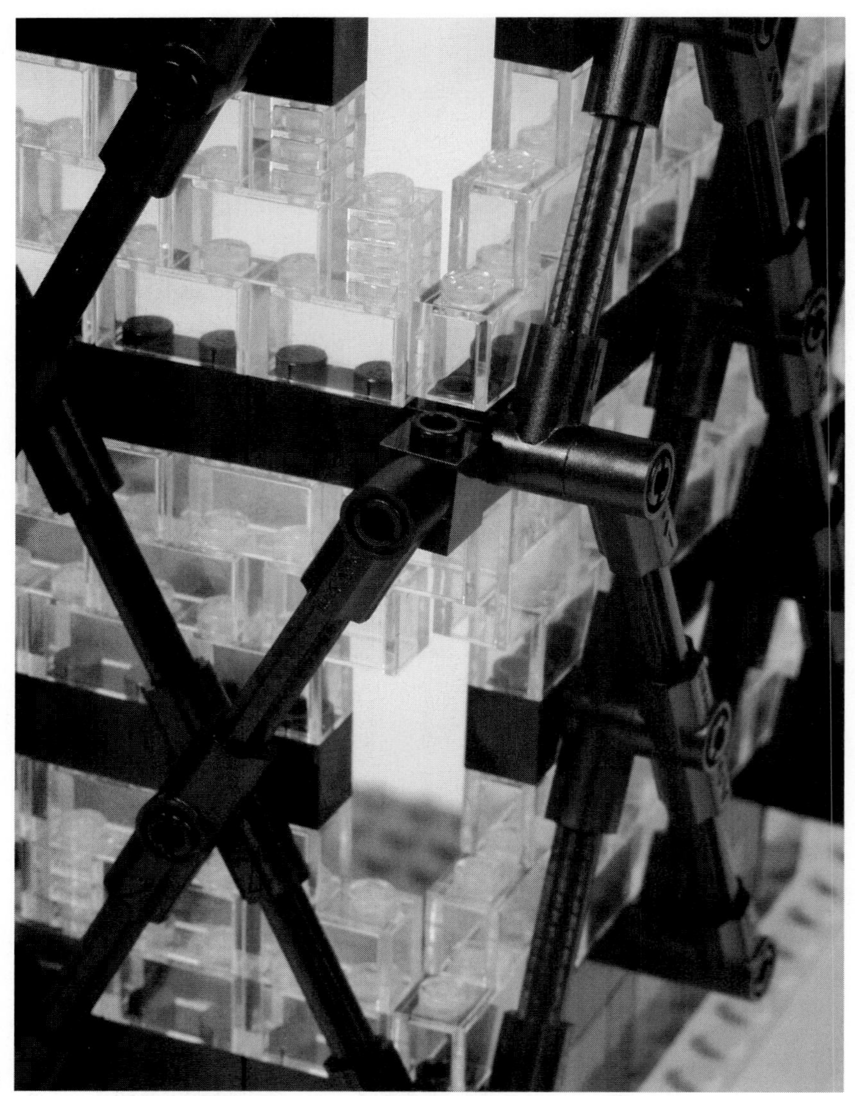

32

33

34 An atrium level detail.

35 A diagonal bracing detail. Just like its progenitor, this LEGO model is an architectural gem.

34

35

36 Now, before we finish all four sides, we've got to hoist the NXT Brick into place.

37 The foundation for the Ice Falls.

38 Let the water flow over Ice Falls.

36

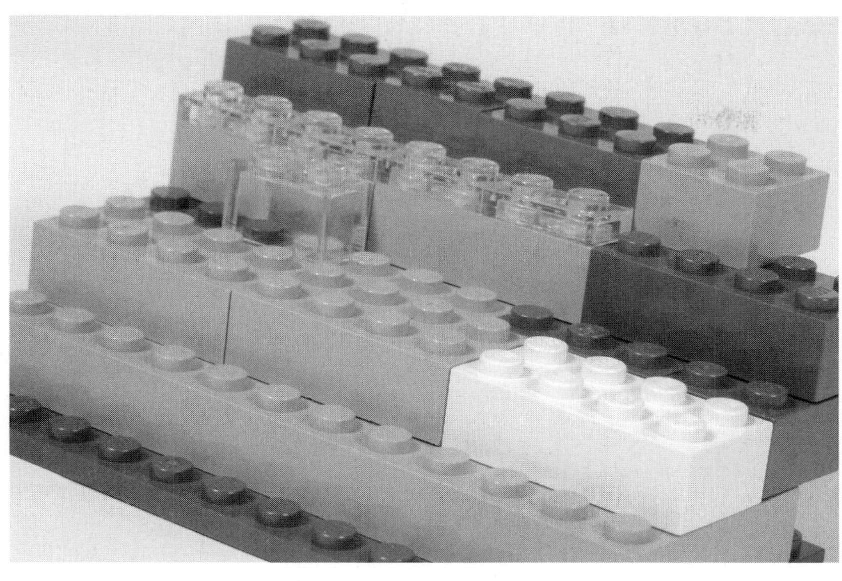

37

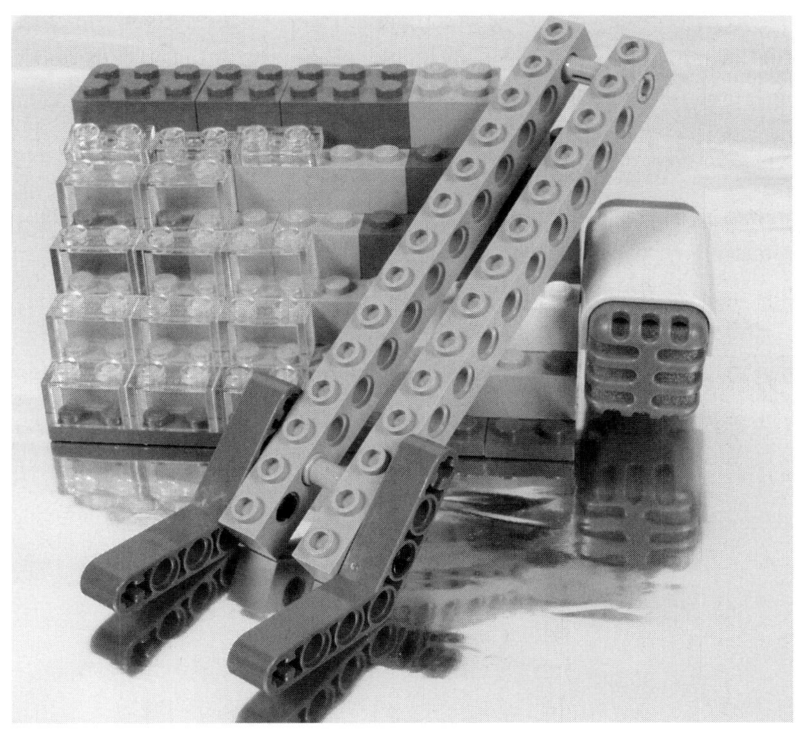

38

THERE'S DOLLARS A BLOWIN' IN THE WIND

Did you know that on a windy day dollars are invisibly blowing right through your fingers? Yup, the power of the wind can actually be harnessed for making electricity...for free. That is, if you have a wind turbine.

The basic principle for operation of a wind turbine power generator is very similar to the alternator found in most automobiles. An alternator consists of a rotating magnetic field coil, or rotor, which generates magnetic lines of force. These lines of force are intercepted by the outside concentric stationary windings, or stator coil, which are fixed in the alternator's frame. Now the rotor turns and the magnetic poles of N and S poles flip positions.

As current flows through the field coil (this current is typically supplied by a battery) an electromagnetic field is produced. The waves of electromagnetic force traveling over the stator coil make the electricity which recharges the battery. Since the poles of the rotor's field coil flip their polarity, an alternating current (AC) is produced.

This AC is rectified through a diode bridge which results in direct current (DC). The DC is then used for recharging the battery. Since the current flowing through the field coil from the battery is DC, the alternator can supply its own field coil current after it has recharged the battery.

A wind turbine system consists of a horizontal-axis propeller, generator, frame, and tail. A special generator with a permanent magnet and a low-speed direct-drive gear generator eliminates both the need for alternator-like coils and expensive and complex gearboxes.

Additionally, the frame acts as a skeleton for holding all of the pieces and parts together, while the tail helps orient the wind turbine into the wind.

YOU BIG BLOW HARD

According to the American Wind Energy Association, the top 20 states with wind energy potential are:

North Dakota	Texas	Kansas	South Dakota
Montana	Nebraska	Wyoming	Oklahoma
Minnesota	Iowa	Colorado	New Mexico
Idaho	Michigan	New York	Illinois
California	Wisconsin	Maine	Missouri

Collectively, the wind turbine is then hoisted up off the ground and installed on a tower. The purpose of this tower is to elevate the wind turbine above the wind power–sapping turbulence that increases closer to the ground. This loss of power due to turbulence can be significant.

According to the Bergey Windpower Company, wind turbines should be mounted at "least 30 feet above and at least 300 feet away from any obstacles." Likewise, taller towers are used for bigger turbines. For example, a 250-watt wind turbine should be installed on a 30- to 50-foot-tall tower, while a larger 10-kilowatt turbine (an average home-based model) would need an 80- to 120-foot tower.

So how much does a wind turbine system cost? Well, first of all make sure that your home can handle a wind turbine. A home that is suited for wind power must have:

- an average wind speed of 10 mph or more
- at least one acre of land
- commercial electricity costs of at least 12¢ per kilowatt-hour

In this case, a small home-based 10-kilowatt wind turbine system will cost $28,000 to $35,000 installed. Once you are up and running, you could expect your wind turbine system to pay for itself within 8 to 16 years of service. Never fear, most quality wind turbines have a life expectancy of 30 years. Like a modern-day Don Quixote you'll be tilting at power companies rather than windmills (see Fig. IV-18).

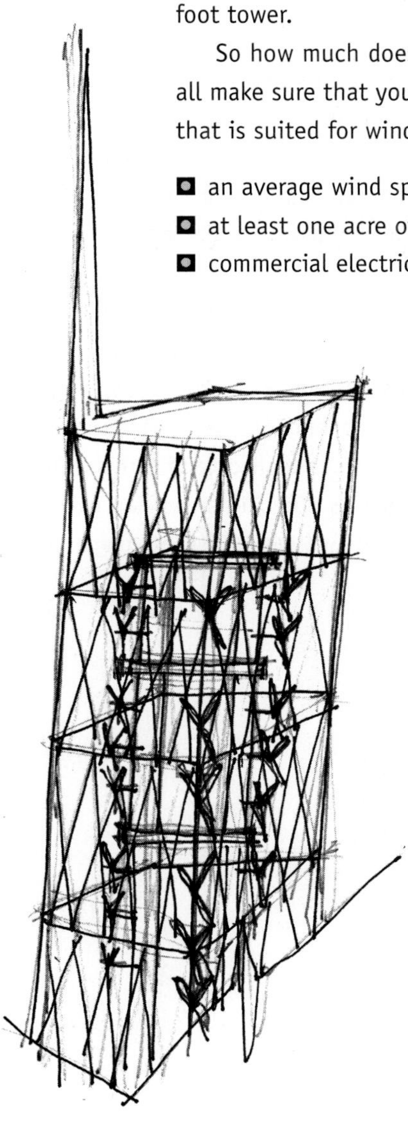

IV-18 Freedom tower with wind turbine generators. Katherine's sketch of the Freedom Tower, New York City as originally designed by Skidmore, Owings, Merrill LLP (SOM). I had planned to reprint a glamorous artist rendering of this historic building, but SOM flatly refused to provide one. According to SOM spokesperson, Elizabeth Kubany, "It isn't clear to me what Freedom Tower has to do with Lego [sic] so I'm afraid I have to say no at this time." Duh, hello, LEGO bricks are the premiere building design modeling system. And you thought that book writing was easy.

DRIVING MISS CRAZY

Driving an automobile in the United States has become a nuisance that presents a new major problem every decade. The air pollution of the seventies gave way to the gasoline shortage of the eighties and the SUV "size" craze of the nineties evolved into the skyrocketing price of gasoline in the naughts. Regardless of your driving generation, though, there has always been one nagging problem that has persisted throughout the lifespan of the automobile—where can I park?

A quick jaunt into any urban area quickly disintegrates into a "circle and wait" search for a parking place. This dilemma is so pervasive that we have even concocted temporary, albeit illegal stopgap, fixes for finding a space to park our car. Double-parking, triple-parking, forged delivery stickers, and fake handicapped licenses are regularly seen in every major urban area of the United States on a daily basis. Is there a viable legal option to urban parking woes? I doubt it, but clearly we must develop parking solutions that aren't rooted in the design of the original parking garage.

Where was the first parking garage built? According to Mary Beth Klatt in her article "Car Culture: Some Cities Convert Historic Parking Garages into Lofts or Lots" for *Preservation Magazine*, the parking garage for the 1918 Hotel LaSalle in Chicago was the nation's first vertical parking lot. Although designed by Holabird & Roche, this noteworthy structure was denied Landmark status by the city's Commission on Chicago Landmarks in 2004. This decision paves the way for the building to be demolished. What will be built in its place? A modern parking garage.

HOW TO BUILD A PNEUMATIC LIFT SYSTEM

1

1 The LEGO Mobile Crane kit contains two pneumatic pistons.

2 **A plan for Carloft.**

CARLOFT®

Address - Berlin, Germany
Description - Residential building that contains a "CarLift"
Architects - Manfred Dick and Johannes Kauka
Built - 2007
Total Area per Loft- 2,411 - 5,802 sq. ft.

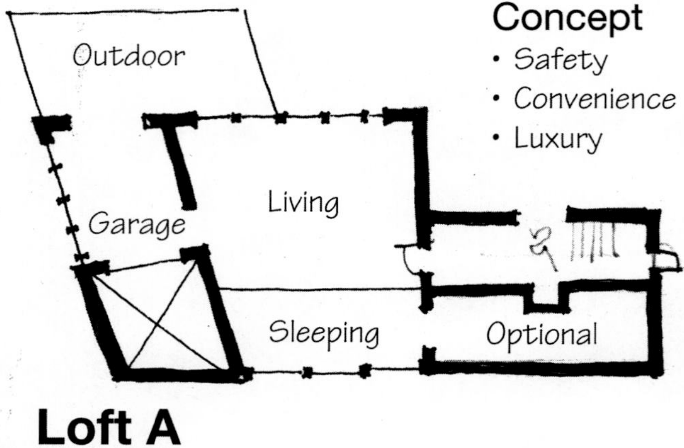

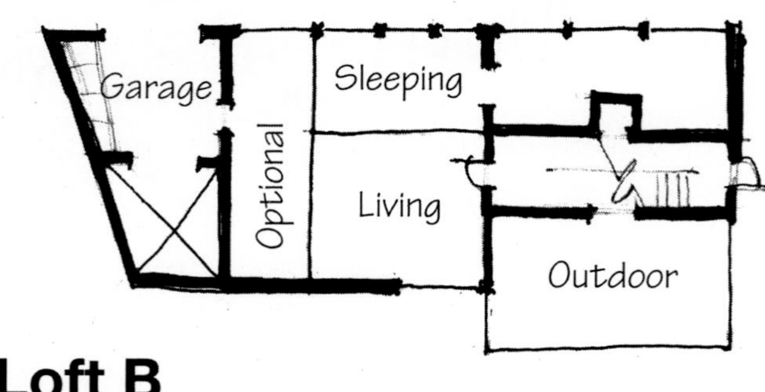

2

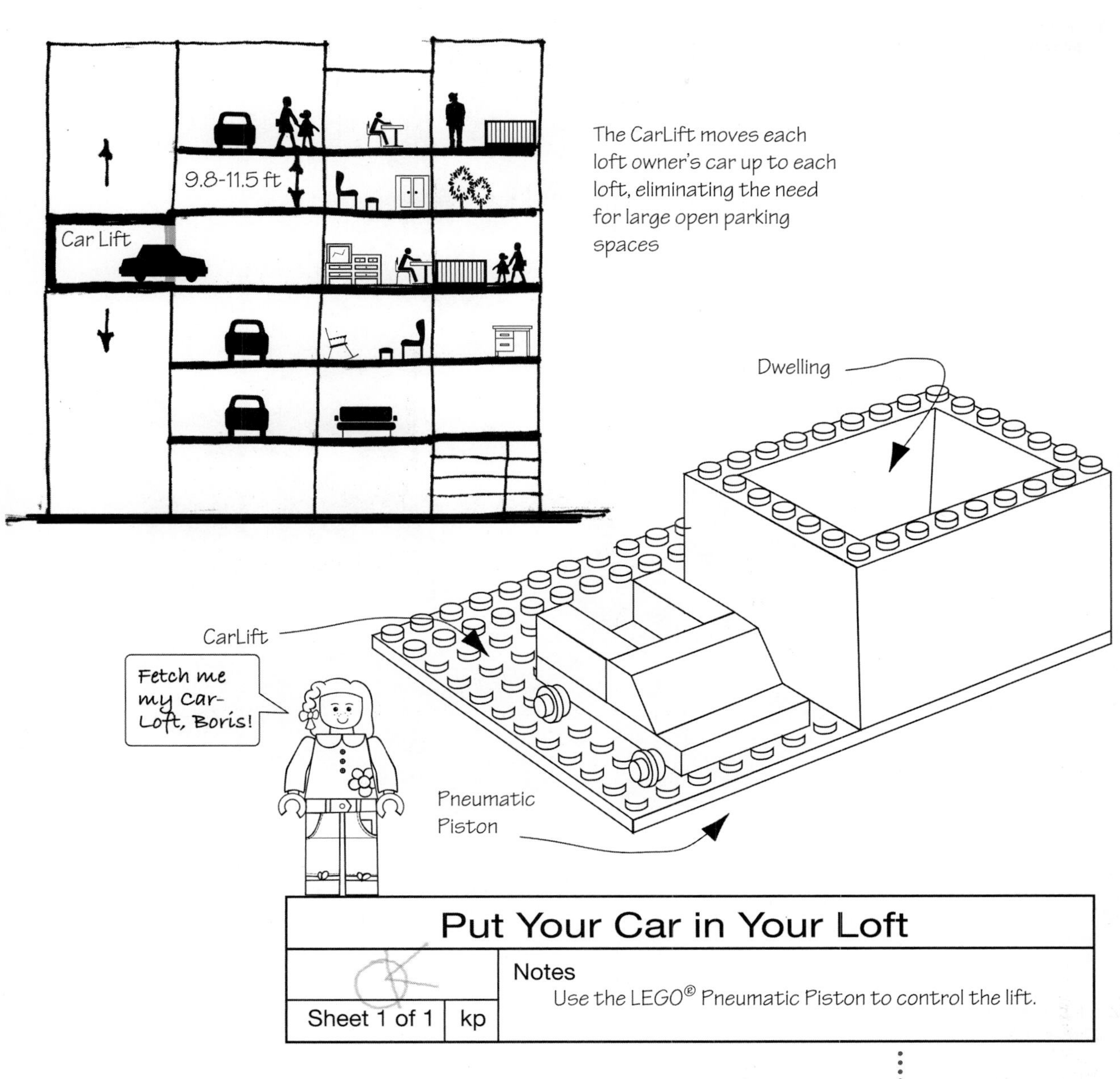

9.8-11.5 ft
Car Lift
The CarLift moves each loft owner's car up to each loft, eliminating the need for large open parking spaces
Dwelling
CarLift
Fetch me my Car-Loft, Boris!
Pneumatic Piston

Put Your Car in Your Loft

	Notes
Sheet 1 of 1 | kp	Use the LEGO® Pneumatic Piston to control the lift.

3 Extract the pistons and pump from the Mobile Crane for use in your own Carloft.

4 Watch out, these pistons don't hold the air too long.

3

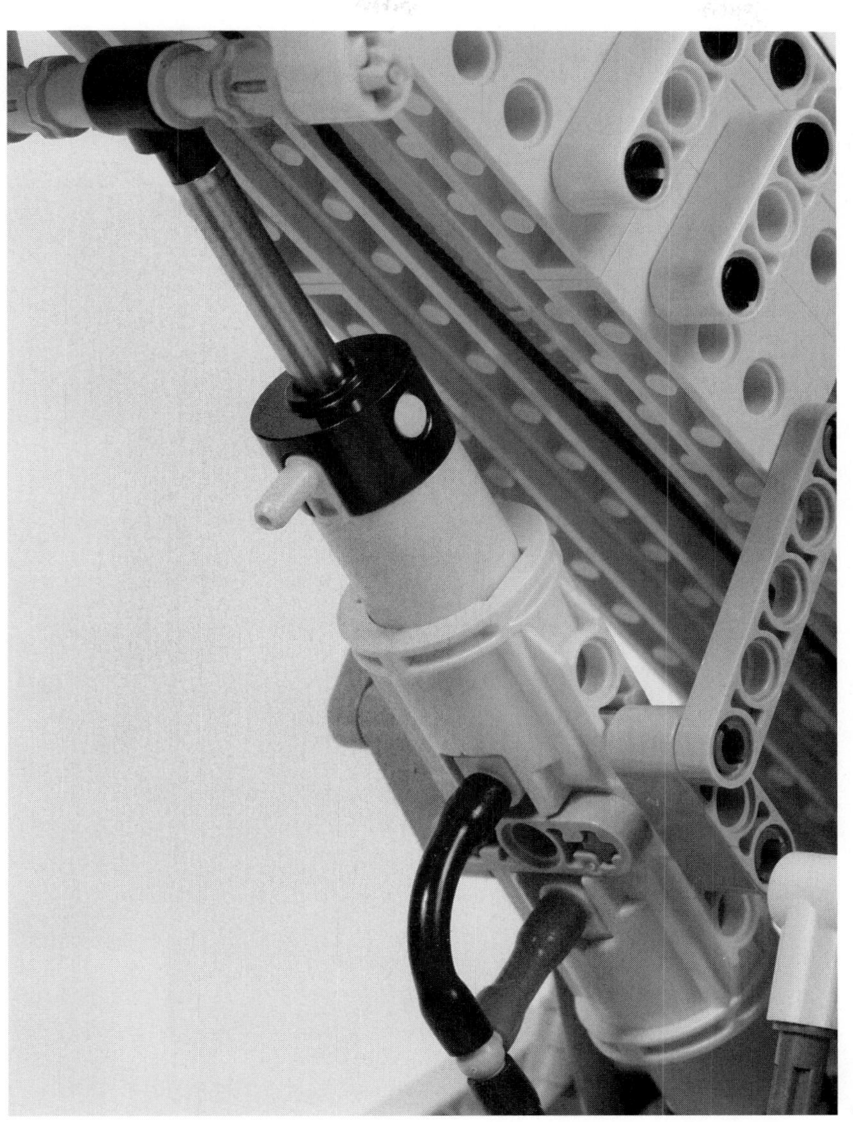

4

5 **Additional tubing can be obtained from your local aquarium supply store.**

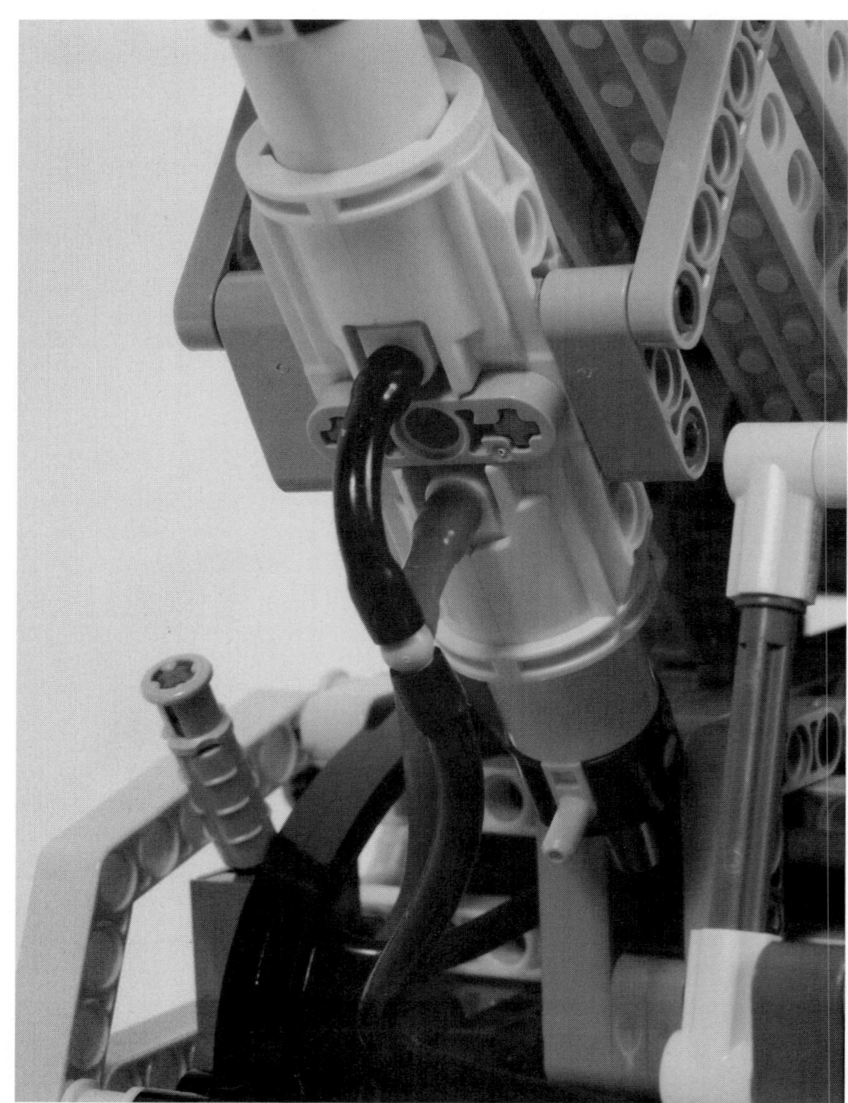

5

IF DRUIDS DROVE CARS

Somewhere on the wind-swept plains of Alliance, Nebraska there is a ring of half-buried automobiles. Mimicking or mocking the stone ring of Stonehenge, 38 different makes and models of cars were arranged in a 96-foot ring and sprayed with gray paint (see Fig. IV-19). Originally designed in 1987 as a tribute to sculptor Jim Reinders' father, Carhenge sits on a 10-acre site where a regular trickle of visitors come and sacrifice a few minutes in quiet reflection on, well, I don't know what.

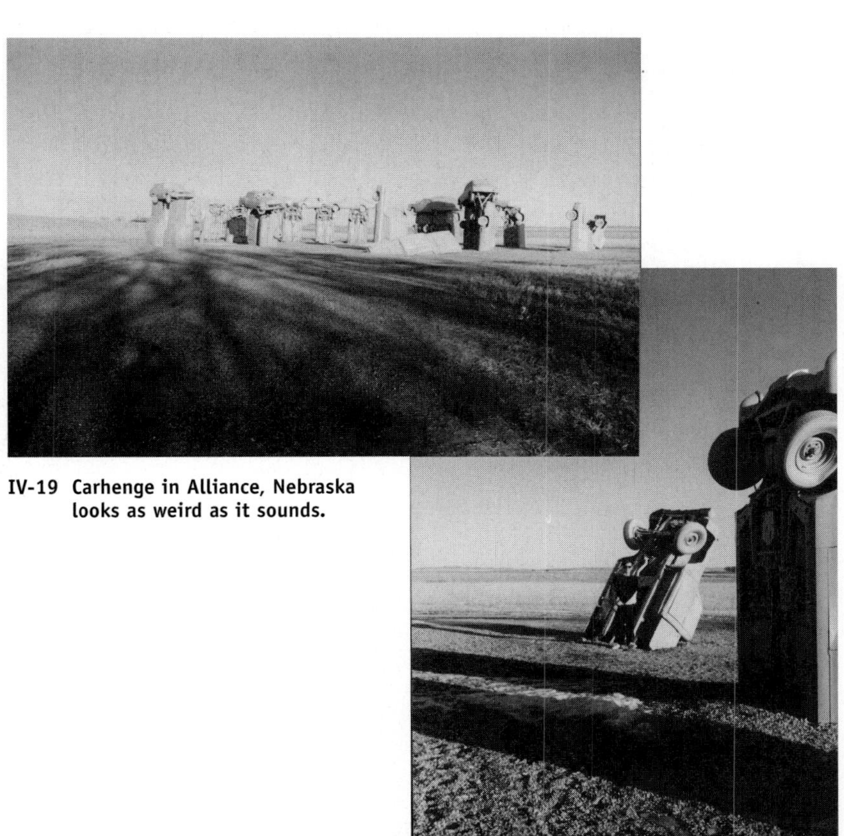

IV-19 Carhenge in Alliance, Nebraska looks as weird as it sounds.

THE GAME'S AFOOT

Every so often, the LEGO Group attempts to make brick building more fun by introducing a LEGO game. For example, in 2004, *X-Pod Play Off Game Pack* (65535) tried to instill some excitement into the tiny egg-shaped X-Pods. According to the LEGO Group, this game was going to enable you to "pick your plans, build your units, and unleash them on your opponent's team in an exciting head-to-head game of strategy, construction, and skill."

Ironically, while the X-Pod remains a viable product line (inside the Creator series), the game is no longer being commercially sold (you can play the game online, however). Luckily, there is a game hidden inside your LEGO MINDSTORMS NXT kit. *Over There* is hidden deep inside your robot design kit. Designed for two players, *Over There* can be set up in a snap and has endless variations for victory. Best of all, no batteries are required.

OVER THERE

What You Need

GAME BOARD:

Almost all of your kit's 1x beams;
  you won't need 7 15m, 3 11m, 3 9m, and 2 7m beams
16 3m beams
32 TECHNIC Friction Pegs—Black (2780)
4 TECHNIC 4-Peg Connectors

GAME PIECES:

All 1x7 Single Bend
All 1x9 Single Bend
All 2x4 Single Bend
All 3x5 Single Bend

GAME PEGS:

All TECHNIC Pin/Axle Pegs—Blue (43093)
All (Remaining) TECHNIC Friction Pegs—Black (2780)
All TECHNIC Three-Stud-Long Pegs—Black
All TECHNIC Two-Stud-Long Axles (32062)

Setting Up

1. Use all of the 1x beams to build a four-sided square. This is your game board.
2. Place all of the remaining TECHNIC game pegs in a cup or container.

CHOOSE SIDES:

3. One player secretly holds a 1x Single Bend piece in one closed hand and a 3x5 Single Bend piece in the other closed hand.
4. The second player picks one of the closed hands. This player receives all similar game pieces (i.e., if the 3x5 Single Bend piece is picked, then this player receives all 2x4 and all 3x5 Single Bend pieces). This player will also move first.
5. The other player receives all of the remaining game pieces.

How to Play

1. Players move alternately by blindly selecting a game peg, inserting the selected peg anywhere on the game board where it will fit or in one of their game pieces, and connecting one of their game pieces to this newly placed peg.
2. Whenever a TECHNIC Three-Stud-Long Peg is selected, up to two game pieces can be connected to this peg during the same turn.
3. Game pieces may be freely rotated at any time during a player's turn.
4. No player may connect to another player's game pieces.
5. Game pieces may go under, around, or over another player's pieces. Players may not connect game pieces to an opposing player's pieces. Game pieces may touch the ground.

Winning the Game

The winner of *Over There* is the first player to connect two opposing game board sides with an uninterrupted chain of connected game pieces. If neither player can build such a chain, then the game is declared a draw.

HOW TO BUILD AND PLAY THE HIDDEN NXT GAME—OVER THERE

1 **Build your game board first.**

2 **Divide the game board's beams into separate piles.**

3 **Discard the beams that you won't need. No, don't really throw them away, just put them back into your storage container.**

4 **Use the 3m beams for attaching the game board's beams.**

1

2

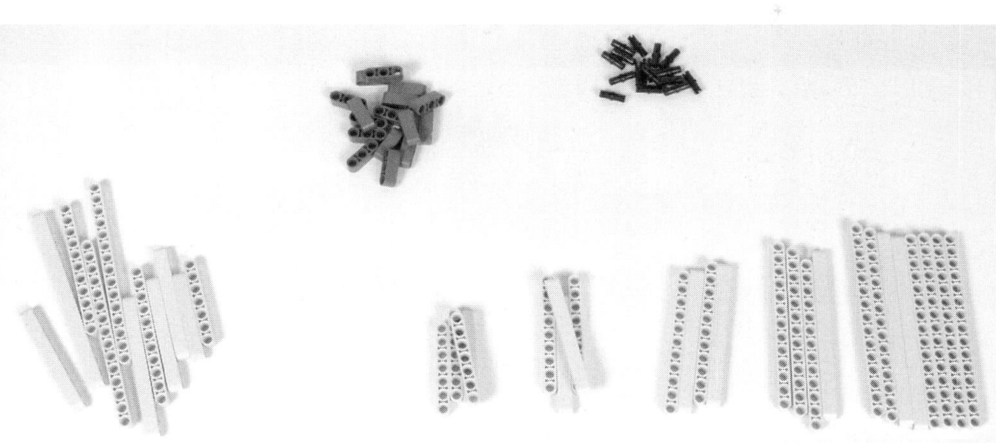

3

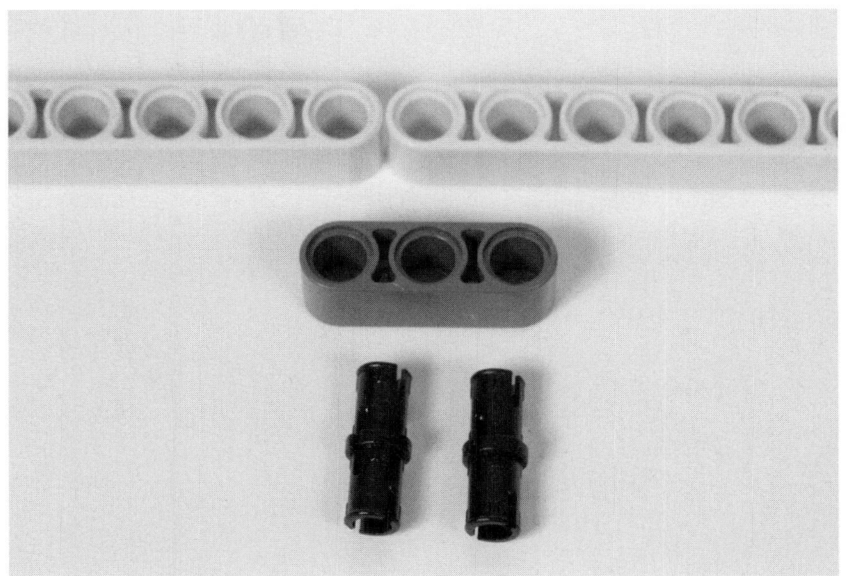

4

5 Insert connectors into the 3m beams.

6 Connect the beams with the 3m connector beams. The first game board side is complete.

7 Make four sides.

8 Use four-pin connectors for joining the sides together.

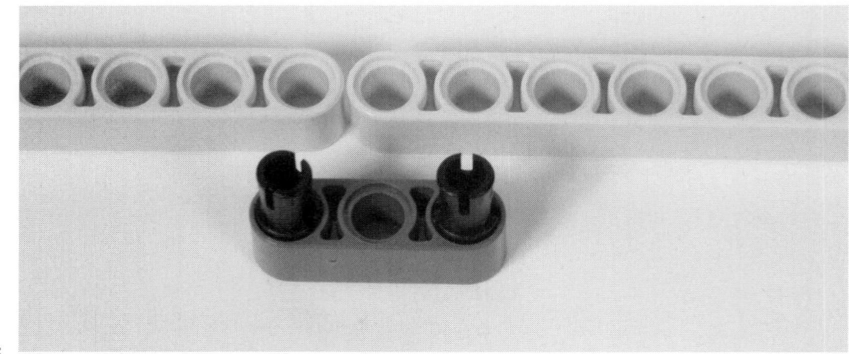

5

6

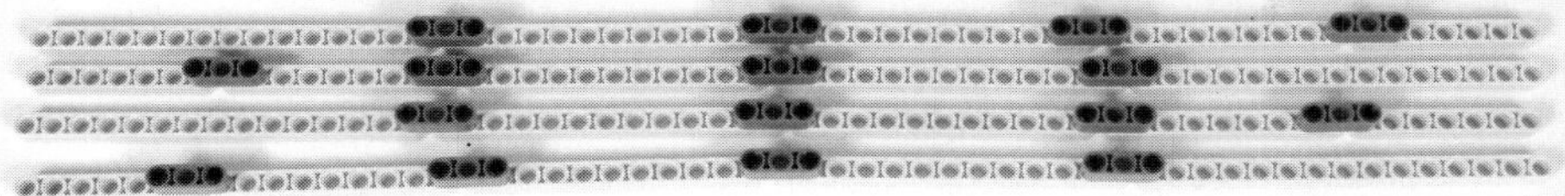

7

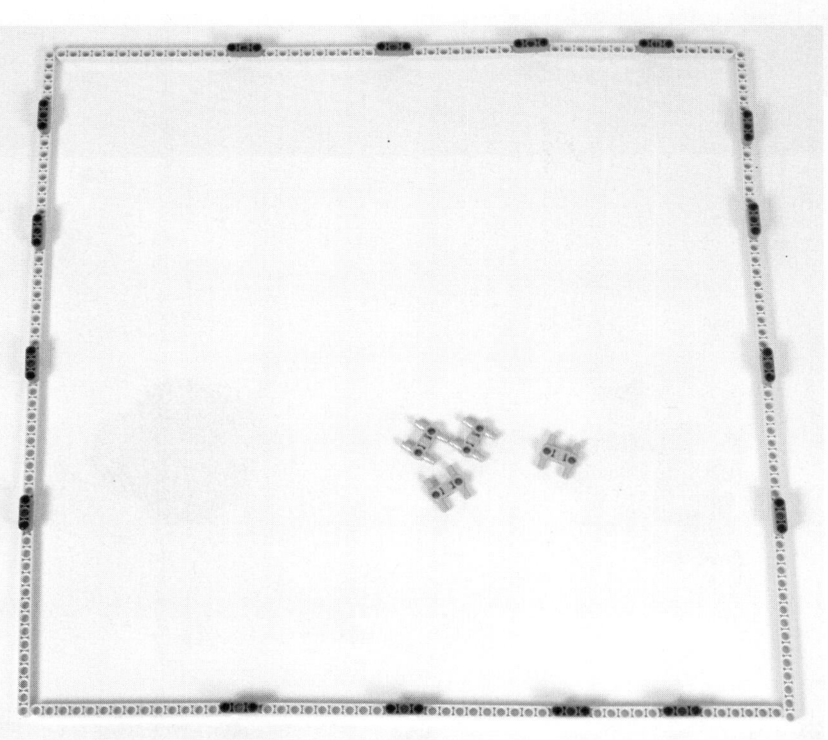

8

9 Assemble the sides into a four-sided square.

10 Make two piles of game pieces.

11 One player will use 1x Single Bend pieces, while the other player will use the 2x and 3x Single Bend pieces.

12 Only use these connectors for game play. Discard all other NXT kit connectors. Once again, you don't really throw them away, just tuck them away someplace safe.

9

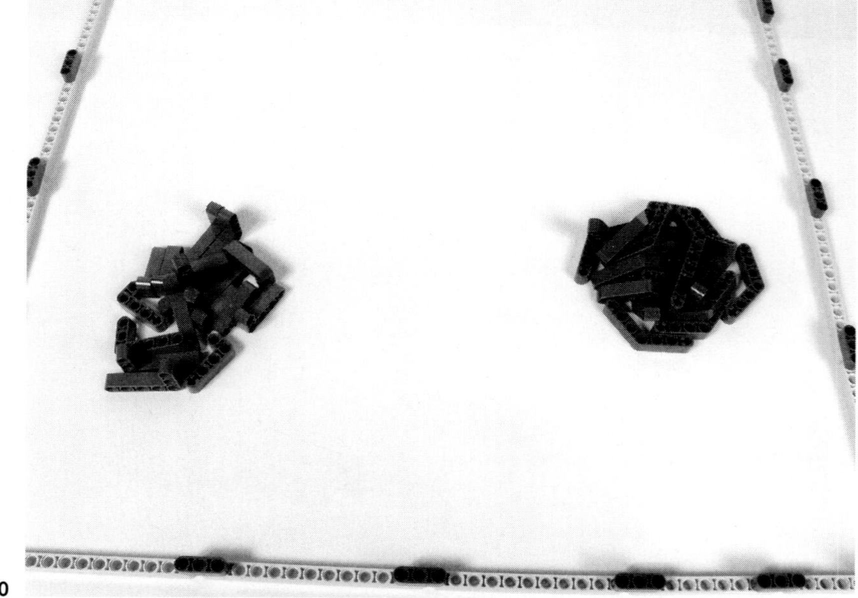

10

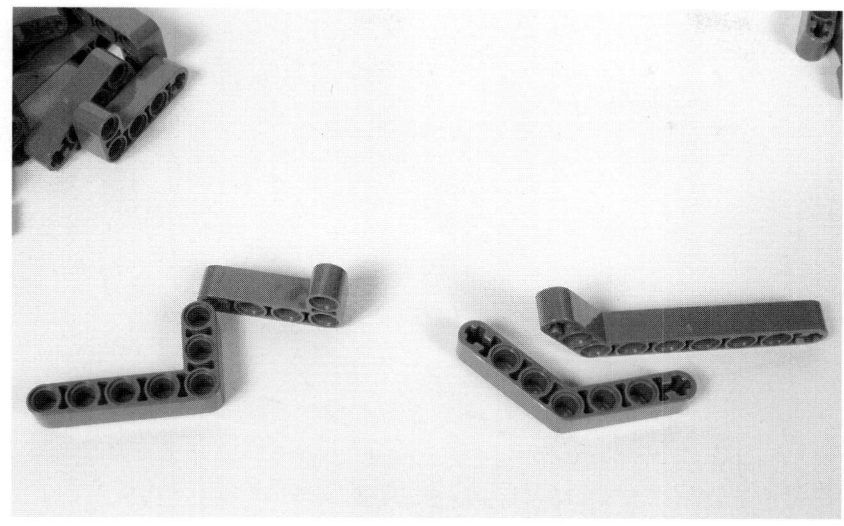

11

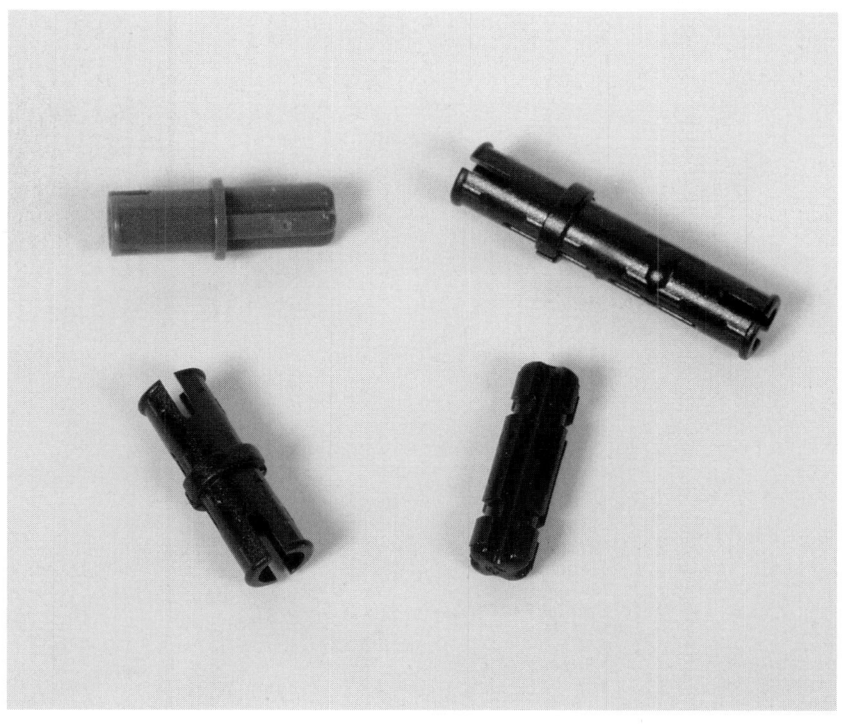

12

13 You won't be using these types of connectors and bushings for game play.

14 Here are all of your game piece connectors.

15 Locate a cup that is suitable for holding all of the game's connectors. Now this cup is perfect—a fusion of technology and personality.

16 One player holds one of each type of game piece in each of her hands and hides them behind her back. The other player chooses one of the hidden hands. The chosen piece becomes that player's game piece.

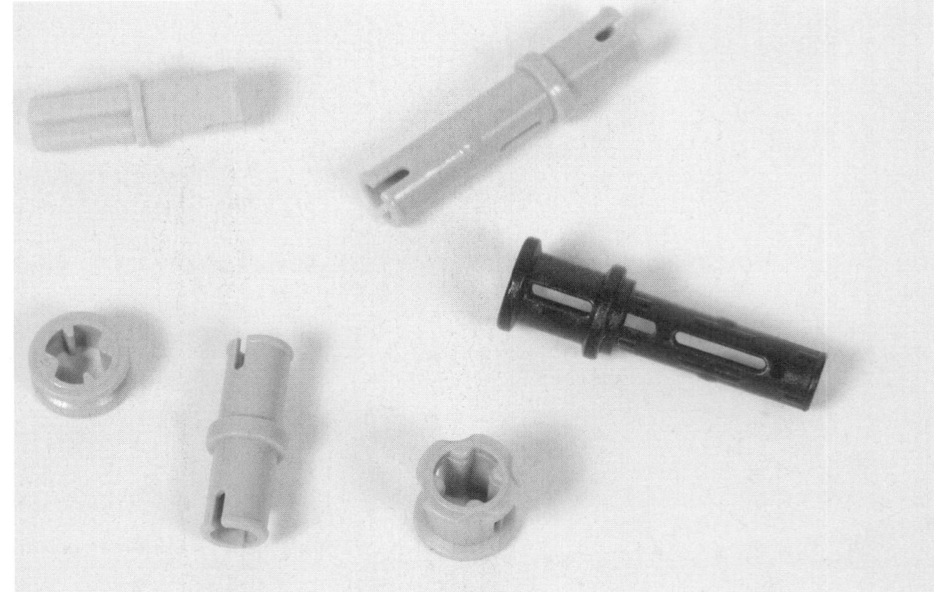

13

14

15

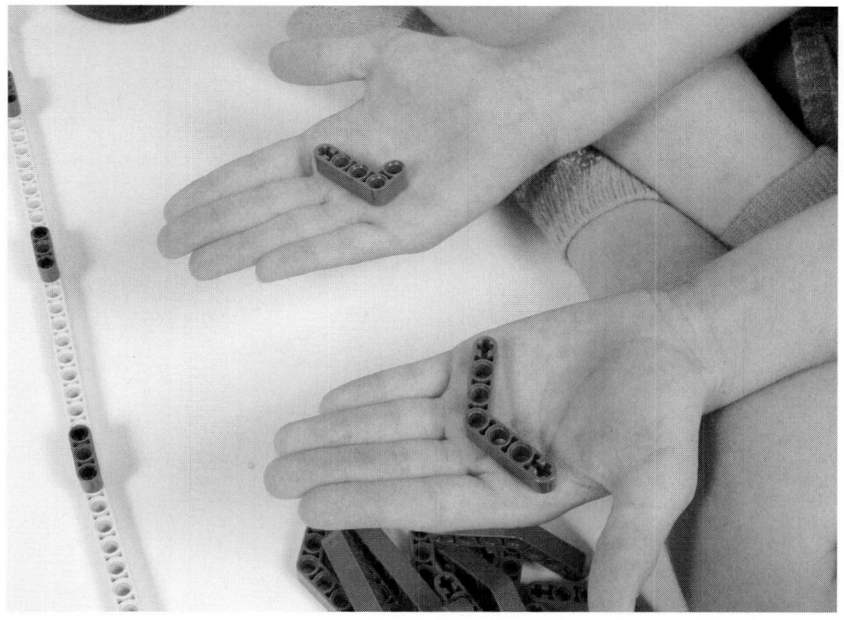

16

17 One connector is blindly drawn from the cup.

18 This selected connector is attached to a game piece.

19 If the selected connector is a three-stud connector, then two game pieces are attached to this connector.

20 The game piece is put into play.

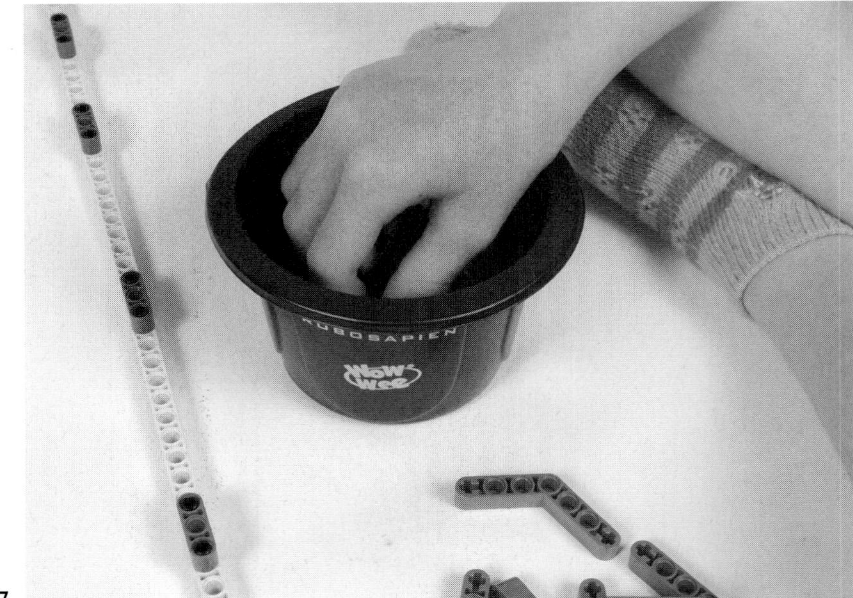

17

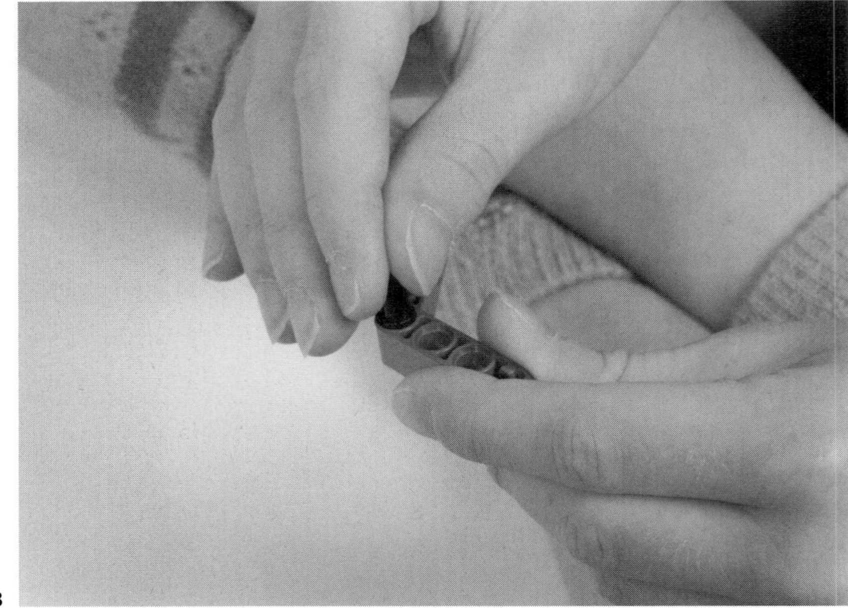

18

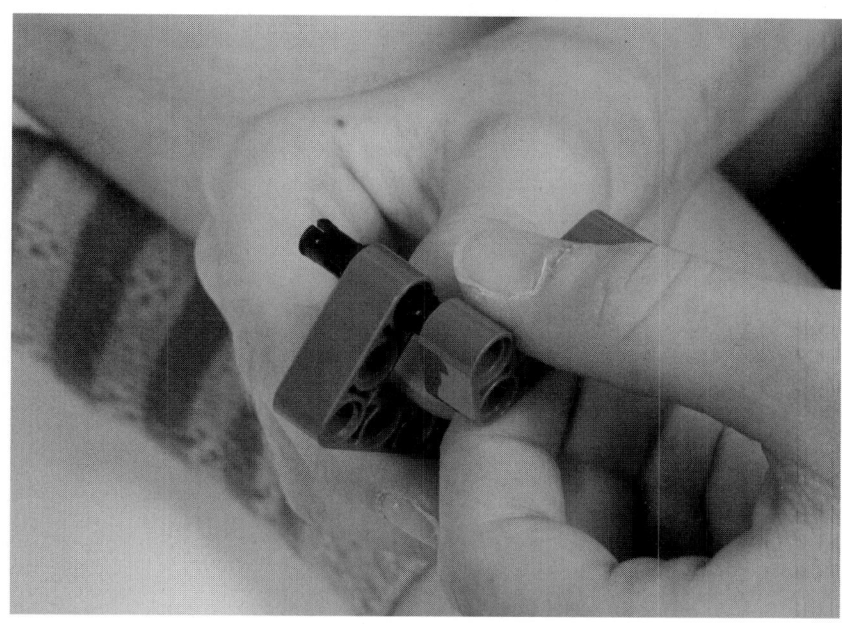

19

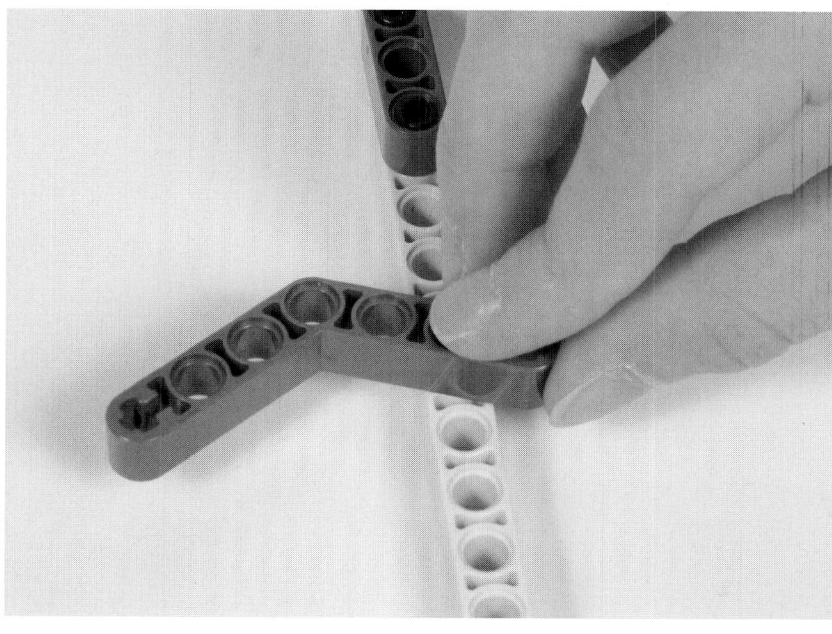

20

21 Game pieces can be pivoted by the player at any time.

22 Watch those axle holes, only one type of connector will work with them.

23 Start building your attempt to get Over There.

24 Ooo, this is getting exciting.

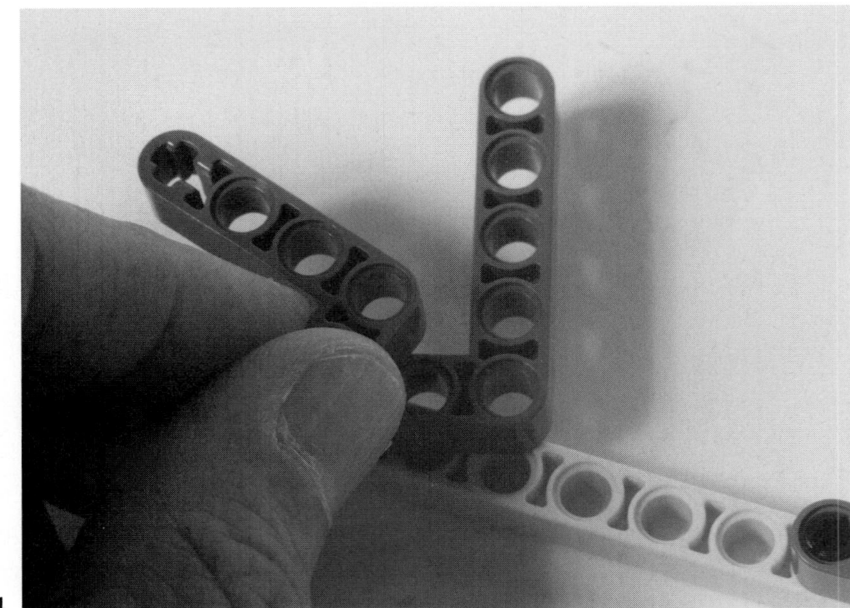

21

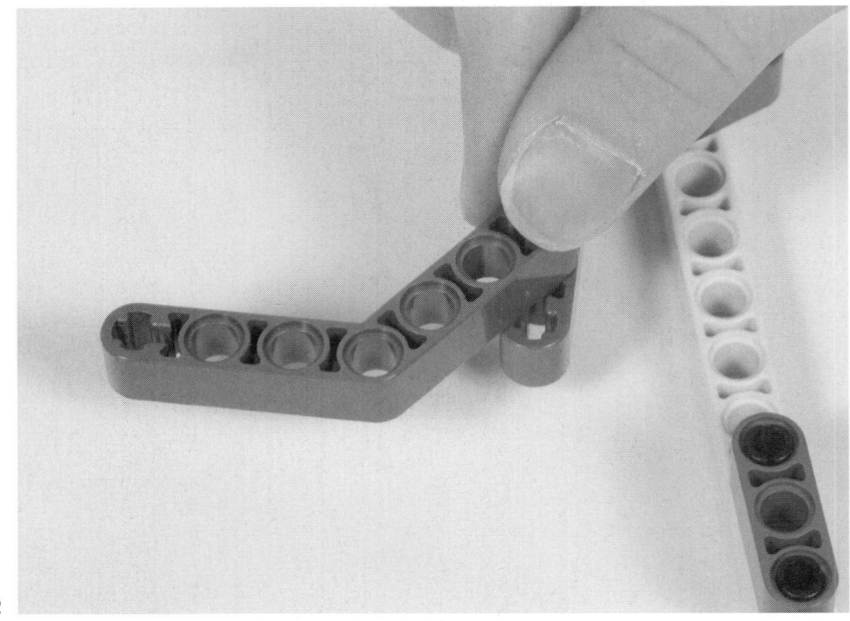

22

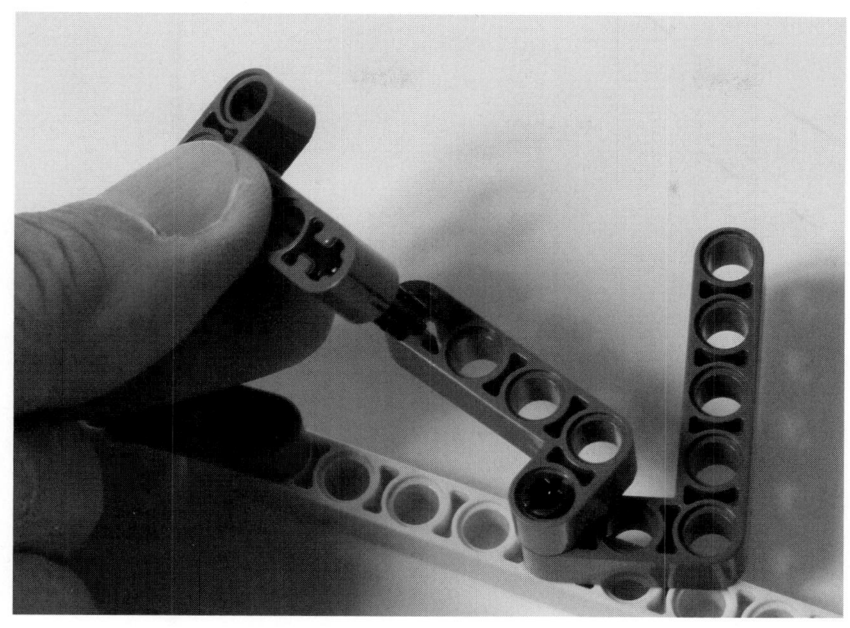

23

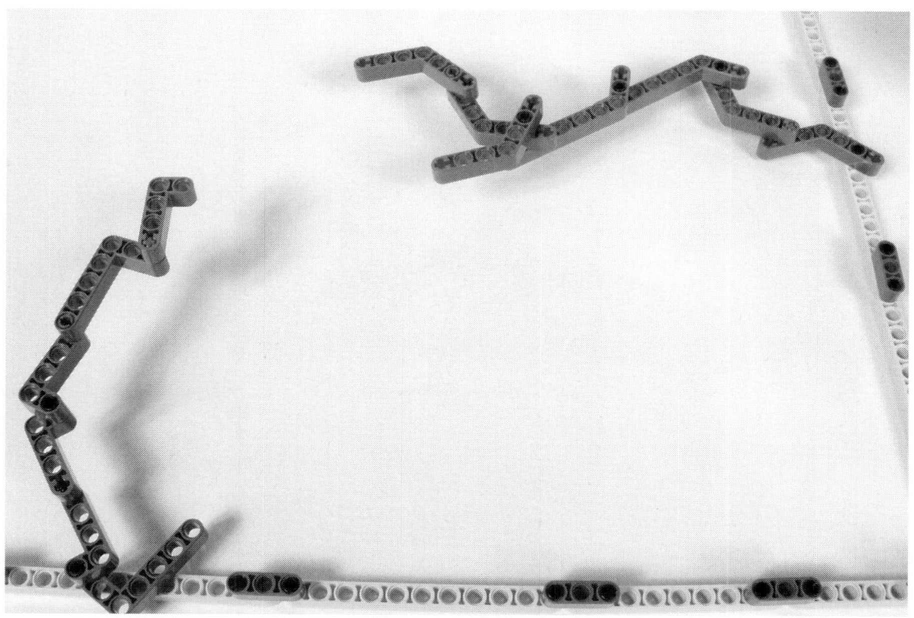

24

25 You can go under or over an opponent.

26 Be careful how you connect to your game pieces. These connections aren't using the full length of each game piece.

27 A good assembly of game pieces begins to look like a winner.

28 You thought rush hour traffic was bad. In this game, there can be only one winner.

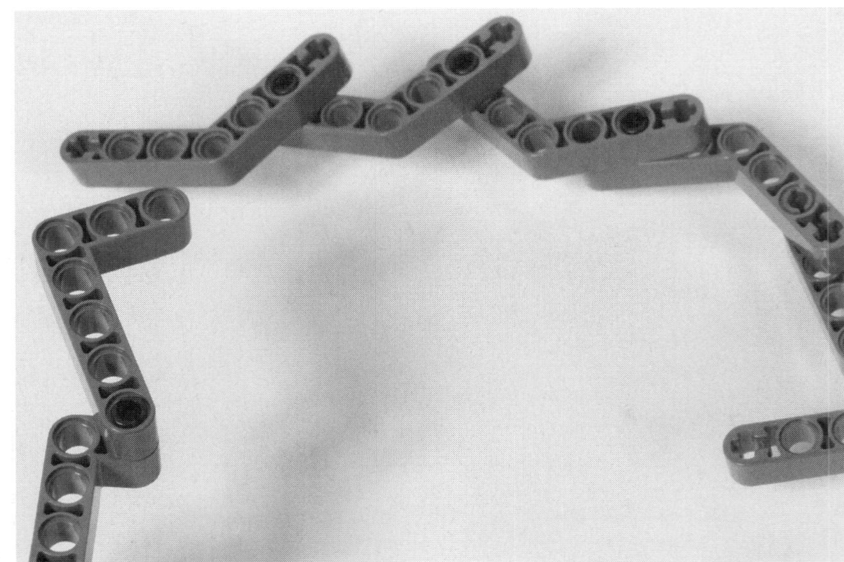

25

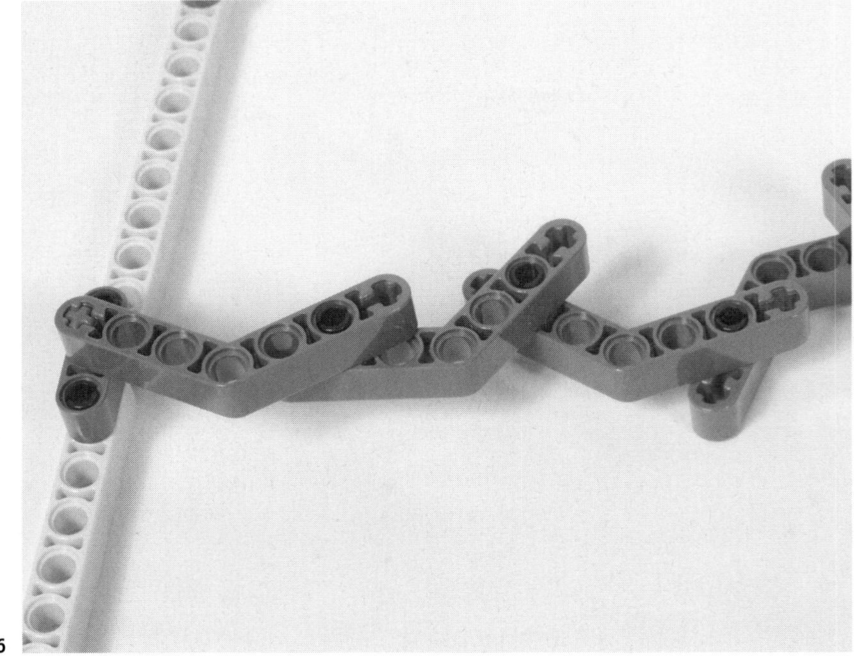

26

27

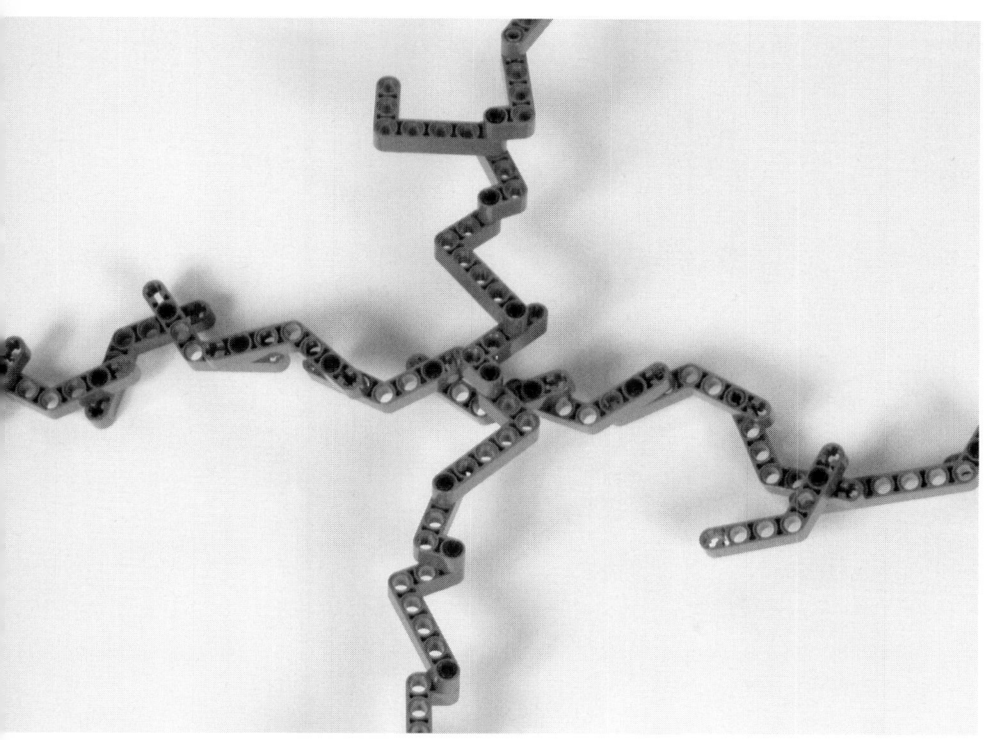

28

29 **This game is finished.**

30 **And the winner is... happy.**

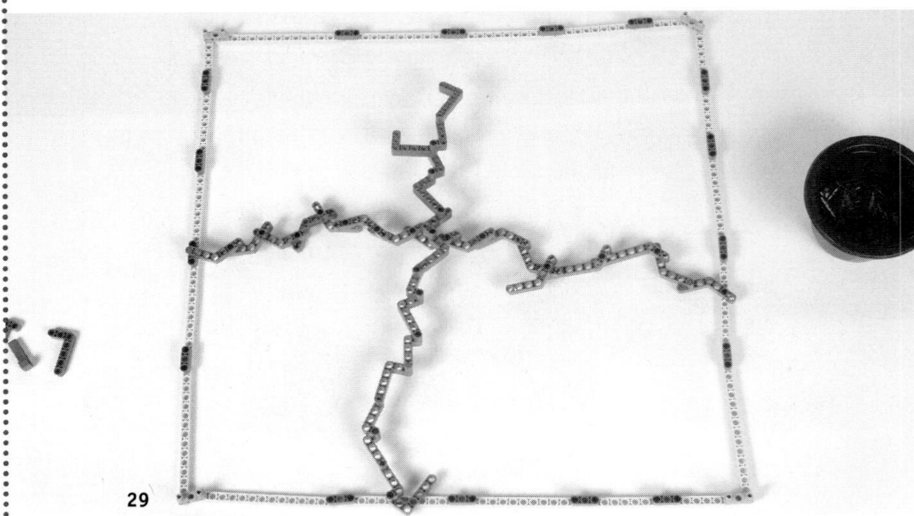

29

30

AND FINALLY, A FIFTH OF APPLE'S JUICE

On May 19, 2006 at 6:00 p.m. a retail juggernaut opened its 147th worldwide store in New York. Sporting a trendy 767 Fifth Avenue address and sprouting in the shadow of the General Motors Building, a gorgeous glass cube has grown into the gateway to a 25,000-square-foot underground shopping space.

Operating 24 hours a day, seven days a week, and 365 days a year, this new retail store will always be open for the residents of the city that never sleeps. Unbelievably the store will be populated with a staff of over 400 personnel. So how does a company hire people to work any day of the year? Yeah, they'll be working Thanksgiving Day, Christmas Day, and New Year's Eve. Remarkably, this company's "cachet" was able to attract several thousand applicants for these 400+ staff slots.

So who is the corporate behemoth? Is it IBM, Porsche, or Donald Trump? No, sorry, are you kidding? It *is* one of the fastest-growing retailers in the world, it *is* Apple Computer.

Officially known as Apple Store® Fifth Avenue, this underground retail space is uniquely marked by a 32-foot glass cube that serves as its entrance. Once you pass through the cube's doors you are whisked into the lower retail shop on a glass circular staircase. Inside the shop you can try and buy any Apple product that fits your pocketbook.

Still, how many people need to buy an iPod at 3:00 a.m. on Christmas Day? Heck, Santa Claus will have already finished his appointed rounds by that time, so if you deserve an iPod you'll have one, otherwise its coal and switches for you. I don't know about you, but it's really tough to believe that the Apple Store Fifth Avenue staff are called "Mac Geniuses" when they agreed to work at 3:00 a.m. Christmas Day.

Regardless of the staff's motives for working such inhumane hours, the Apple Store Fifth Avenue is a magnificent piece of architecture (see Fig. IV-20). If you would like to build a reproduction of Apple Store Fifth Avenue that is able to accommodate the NXT Brick, there is a LEGO Digital Designer model available for readers of this book at the following Web site: www.pco2go.com/lego/ fifth.html.

IV-20 A sketch of Apple Store Fifth Avenue.

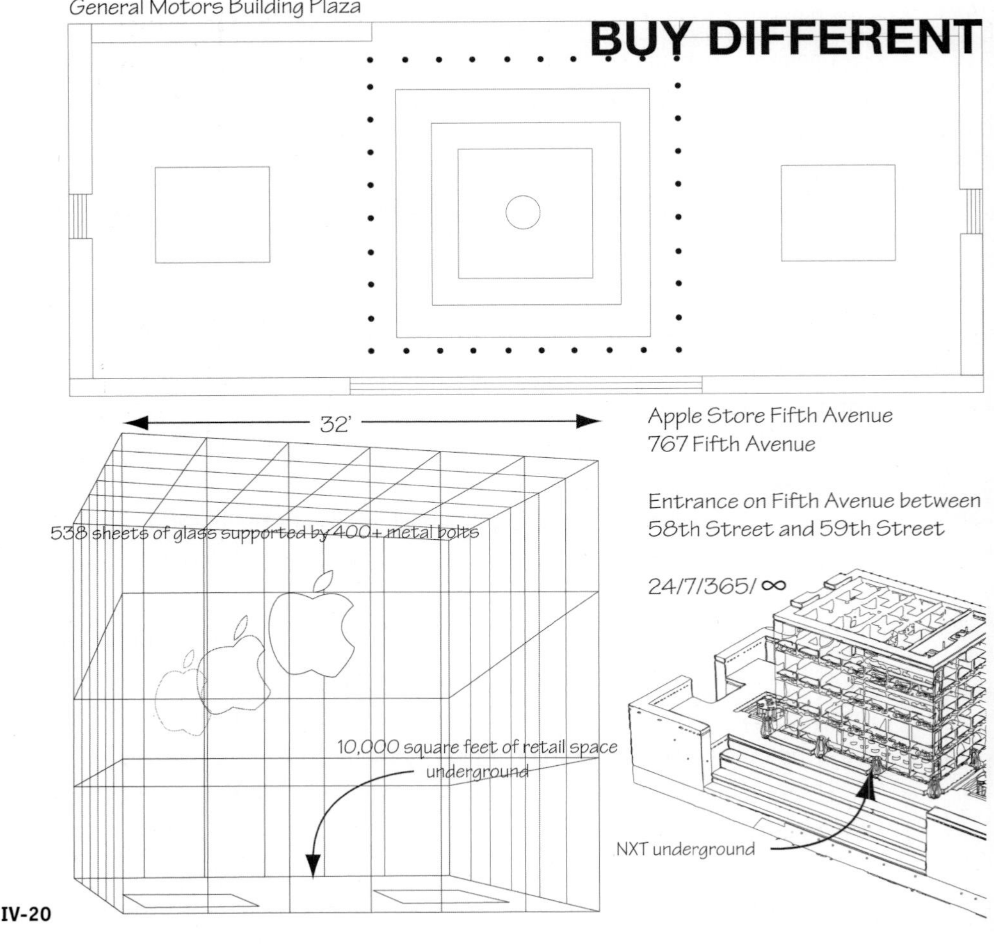

IV-20

Apple Store Fifth Avenue

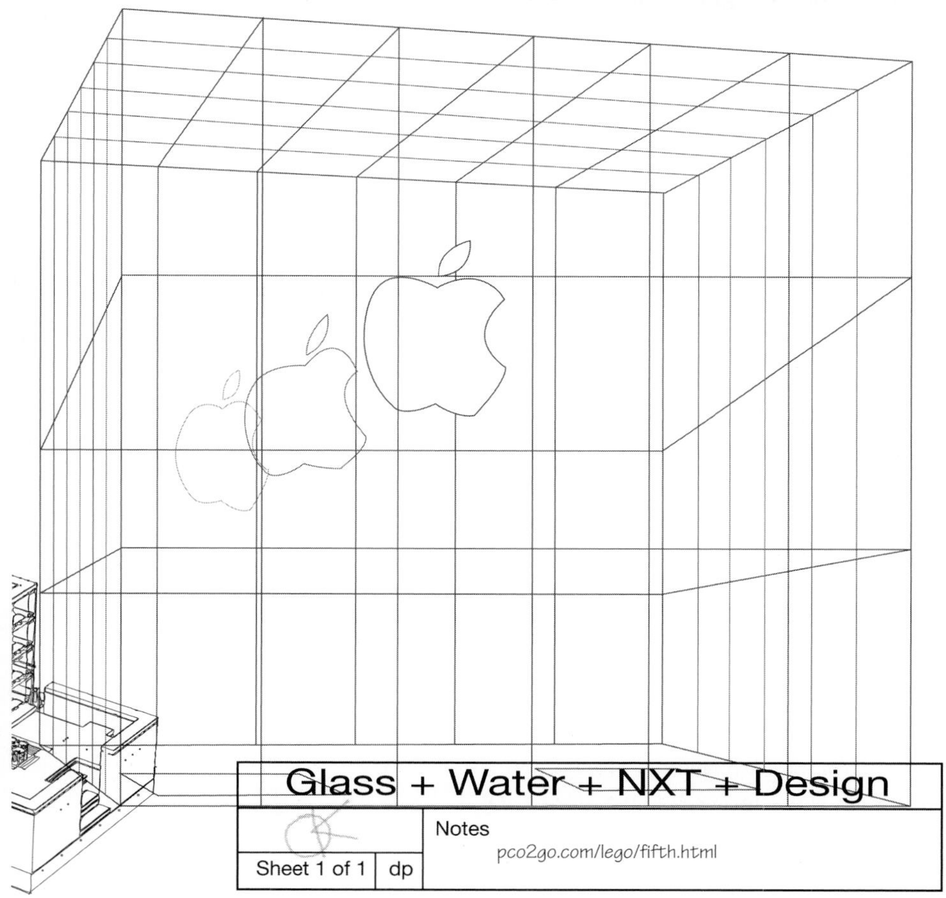

Glass + Water + NXT + Design		
		Notes pco2go.com/lego/fifth.html
Sheet 1 of 1	dp	

FIFTH AVENUE BRICK LIST

NUMBER OF BRICKS: 427

2 - 2436 Angle Plate 1x2/1x4 White
8 - 3004 Brick 1x2 Brick Yellow
2 - 3008 Brick 1x8 Brick Yellow
14 - 3009 Brick 1x6 Brick Yellow
6 - 3010 Brick 1x4 Brick Yellow
26 - 3069 Flat Tile 1x2 Brick Yellow
20 - 3622 Brick 1x3 Brick Yellow
8 - 2412 Radiator Grille 1x2 Bright Blue
2 - 3032 Plate 4x6 Bright Blue
1 - 300 Brick 2x4 Black
11 - 3007 Brick 2x8 Black
2 - 3020 Plate 2x4 Black
5 - 3027 Plate 6x16 Black
1 - 3034 Plate 2x8 Black
2 - 303 Plate 4x8 Black
2 - 379 Plate 2x6 Black
104 - 4864 Wall Element - Tr 1x2x2 Transparent
2 - 6143 Brick Ø16 W. Cross Transparent Yellow
12 - 4589 Nose Cone Small 1x Transparent Bright Orange
20 - 243 Flat Tile 1x4 Medium Stone Grey
4 - 3002 Brick 2x3 Medium Stone Grey
26 - 3070 Flat Tile 1x Medium Stone Grey
6 - 3009 Brick 1x6 Medium Stone Grey
20 - 3010 Brick 1x4 Medium Stone Grey
11 - 3020 Plate 2x4 Medium Stone Grey
2 - 3023 Plate 1x2 Medium Stone Grey
10 - 3034 Plate 2x8 Medium Stone Grey
30 - 3068 Flat Tile 2x2 Medium Stone Grey
4 - 3623 Plate 1x3 Medium Stone Grey
4 - 3666 Plate 1x6 Medium Stone Grey
10 - 3710 Plate 1x4 Medium Stone Grey
6 - 379 Plate 2x6 Medium Stone Grey
8 - 4162 Flat Tile 1x8 Medium Stone Grey
6 - 4282 Plate 2x16 Medium Stone Grey
4 - 44237 Brick 2x6 Medium Stone Grey
13 - 6112 Brick 1x12 Medium Stone Grey
2 - 6576 Plate 4x8 W. 12 Knobs Medium Stone Grey
11 - 6636 Flat Tile 1x6 Medium Stone Grey

I DID IT; NO, I DID IT

Who shot Liberty Valance? Oops, no, that's the wrong question. Rather, who designed the Apple Store Fifth Avenue?

Trying to find out who designed the glass Apple cube on Fifth Avenue in New York is as confusing as identifying the true trigger man in the legendary western film, *The Man Who Shot Liberty Valance*.

Some sources have fingered architect Dan Shannon of the firm Moed de Armas & Shannon while others have suggested that the mega-architectural firm Gensler was responsible for the design. Even more comical, however, was Deborah Schoeneman's attribution from a *New York Magazine* article ("Steve Jobs Loves his Big iCube: It's his; he's not leaving it here;" Dec. 5, 2005) that the "man" himself, Steve Jobs (also known as the iCon), "personally designed" the glass cube store.

The latter possibility is as ludicrous as saying that Brad Pitt could sponsor an architectural design competition. Well, OK that *is* true, the man who played Achilles (*Troy*; 2004) did sponsor a design competition for New Orleans in mid-2006. But, Steve Jobs, an architect?

Seems like a simple enough question doesn't it? Who designed the Apple Store Fifth Avenue? Well, I asked and asked.

After several fruitless days of trying to track down the answer to this question, the only worthwhile (and confirmed) crumb of information that I could obtain was that according to Gensler's spinoff firm "Studio 585," "rollout services" have been provided to Apple Computer. These services served as a prototype for all Apple Stores nationwide—not specifically the Apple Store Fifth Avenue. Both the Design Standards and follow-on rollout management were supplied by Studio 585.

Geesch, all of this corporate obfuscation makes you want to scream, "Hey, pilgrim. Get a rope."

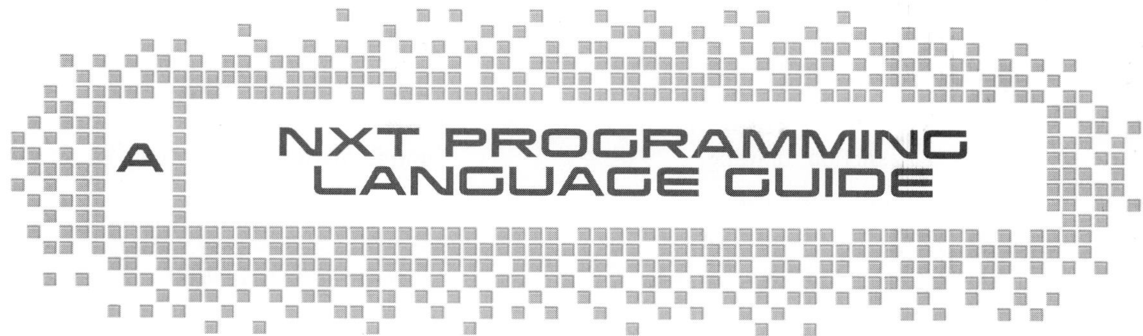

THE FOLLOWING convention was used in the preparation of this guide:

PALETTE NAME

BLOCK NAME

Properties: Parameters

Or,

Additional Properties: Parameters

COMMON PALETTE

MOVE

Port: A, B, C

Direction: Fwd, Back, Stop

Steering: C or B

Power: 0-100

Duration: Unlimited, Degrees, Rotations, Inches

Next: Brake or Coast

RECORD/PLAY

Action: Record or Play

Name:

Recording: A, B, C

Time:

SOUND

Action: Sound File or Tone

Control: Play or Stop

Volume: 0-100

Function: Repeat

File:

Wait: Wait for Completion

DISPLAY

Action: Image

Display: Clear

File:

Position: X & Y

Or,

Action: Text

Display: Clear

Text:
Position: X & Y
Line: 1-8
Or,
Action: Drawing
Display: Clear
Type: Point, Line (start/end), Circle (radius)
Position: X & Y
Or,
Action: Reset

WAIT

Control: Sensor, Time (seconds)
Sensor: Light Sensor
Port: 1-4
Until: Light (<>) 0-100
Function: Generate Light (Tip—used for generating power for hacking projects)
Or,
Control: Sensor, Time (seconds)
Sensor: NXT Buttons
Button: Left, Right
Action: Pressed, Released, Bumped
Or,
Control: Sensor, Time (seconds)
Sensor: Receive
Message: Text, Number, Logic (Compare to)
Mailbox: 1-10
Or,
Control: Sensor, Time (seconds)
Sensor: Rotation Sensor
Port: A, B, C
Action: Read, Reset
Until: Forward, Back (<>) (Degrees, Rotations)
Or,
Control: Sensor, Time (seconds)
Sensor: Sound Sensor
Port: 1, 2, 3, 4
Until: Sound (<>) (0-100)
Or,
Control: Sensor, Time (seconds)
Sensor: Timer
Timer: 1-3
Action: Read, Reset
Until: <>
Or,
Control: Sensor, Time (seconds)
Sensor: Touch Sensor
Port: 1-4
Action: Pressed, Released, Bumped
Or,
Control: Sensor, Time (seconds)
Sensor: Ultrasonic
Port: 1-4
Until: Distance (<>)
Show: Centimeters, Inches

LOOP

Control: Forever, Sensor, Time (Until), Count (Until), Logic (True, False)
Show: Counter

SWITCH

Control: Sensor, Value (Conditions; True, False)
Display: Flat View

ACTION PALETTE

MOTOR

Port: A, B, C
Direction: Fwd, Back, Stop
Action: Constant
Power: 0-100
Control: Motor Power
Duration: Unlimited, Degrees, Rotations, Seconds
Wait: Wait for Completion
Next: Brake, Coast

SOUND

Action: Sound File or Tone
Control: Play or Stop
Volume: 0-100
Function: Repeat
File:
Wait: Wait for Completion

DISPLAY

Action: Image
Display: Clear
File:
Position: X & Y
Or,
Action: Text
Display: Clear
Text:
Position: X & Y
Line: 1-8
Or,
Action: Drawing
Display: Clear
Type: Point, Line (start/end), Circle (radius)
Position: X & Y
Or,
Action: Reset

SEND MESSAGE

Connection: 0-3
Message: Text, Number, Logic (Compare to)
Mailbox: 1-10

SENSOR PALETTE

TOUCH SENSOR

Port: 1-4
Action: Pressed, Released, Bumped

SOUND SENSOR

Port: 1-4
Compare: Sound (<>)

LIGHT SENSOR

Port: 1-4
Compare: Light (<>)
Function: Generate Light (Tip—used for generating power for hacking projects)

ULTRASONIC SENSOR

Port: 1-4
Compare: Distance (<>)
Show: Centimeters, Inches

BLOCK NXT BUTTONS

Button: Enter, Left, Right
Action: Pressed, Released, Bumped

ROTATION SENSOR

Port: A, B, C
Action: Read, Reset

Compare: Forward, Back, Degrees, Rotations (<>)

TIMER

Timer: 1-3

Action: Read, Reset

Compare: <>

RECEIVE MESSAGE

Message: Text, Number, Logic

Compare to:

Mailbox: 1-10

FLOW PALETTE

WAIT

see *Wait* from **Common Palette**

LOOP

see *Loop* from **Common Palette**

SWITCH

see *Switch* from **Common Palette**

STOP

DATA PALETTE

LOGIC

Operation: And, Or, XOr, Not (A, B)

MATH

Operation: Addition, Subtraction, Multiplication, Division (A, B)

COMPARE

Operation: Less than, Greater than, Equals (A, B)

RANGE

Operation: Inside Range, Outside Range (A, B)

Test Value:

RANDOM

Range: Minimum-0, Maximum-32,767 (A, B)

VARIABLE

List: Logic, Number, Text

Action: Read, Write

Value: True, False, Value, Text

ADVANCED PALETTE

TEXT

Text: A, B, C

NUMBER TO TEXT

Number:

KEEP ALIVE

FILE ACCESS

Action: Read, Write, Close, Delete

Name:

File:

Type: Text, Number

CALIBRATE

Port: 1-4

Sensor: Light, Touch

Action: Calibrate, Delete

Value: Maximum, Minimum

RESET

Port: A, B, C

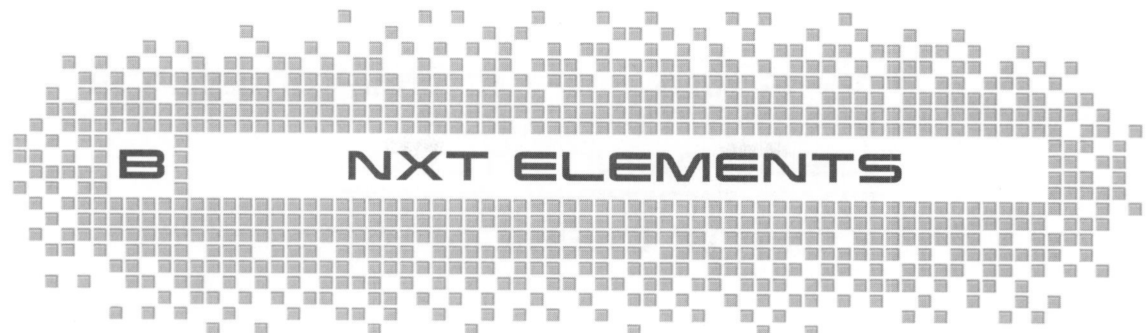

B NXT ELEMENTS

Bundled inside the LEGO MINDSTORMS NXT robot design kit are a number of unique bricks, parts, pieces, and elements. Most of these elements are TECHNIC beams, connectors, and couplers. Rather than illustrate the current commercial version of this kit, here, for the first time published in a major book, is a special look inside the prerelease MDP version of this remarkable robot design kit.

THE MAJOR COMPONENTS FROM THE MDP LEGO MINDSTORMS NXT PRERELEASE ROBOT DESIGN KIT

1 The NXT Brick.

2 Top view of NXT Brick.

3 Side view of NXT Brick.

1

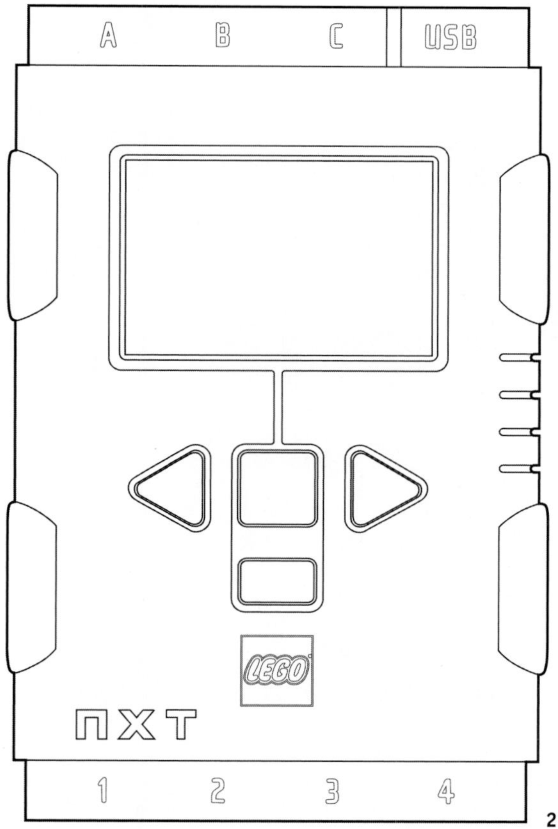

2

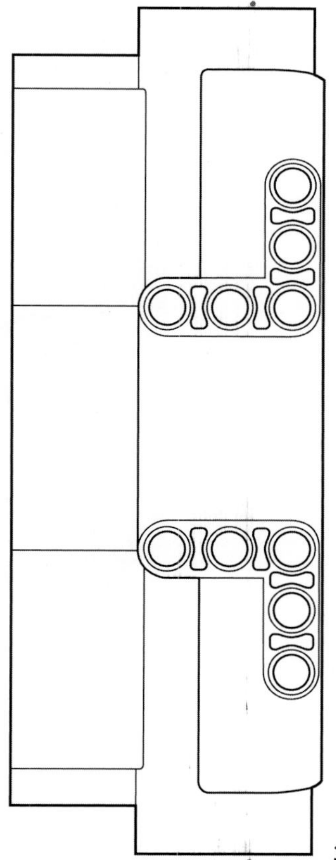

3

4 Motors and sensors.

5 USB cable and NXT Connector cables.

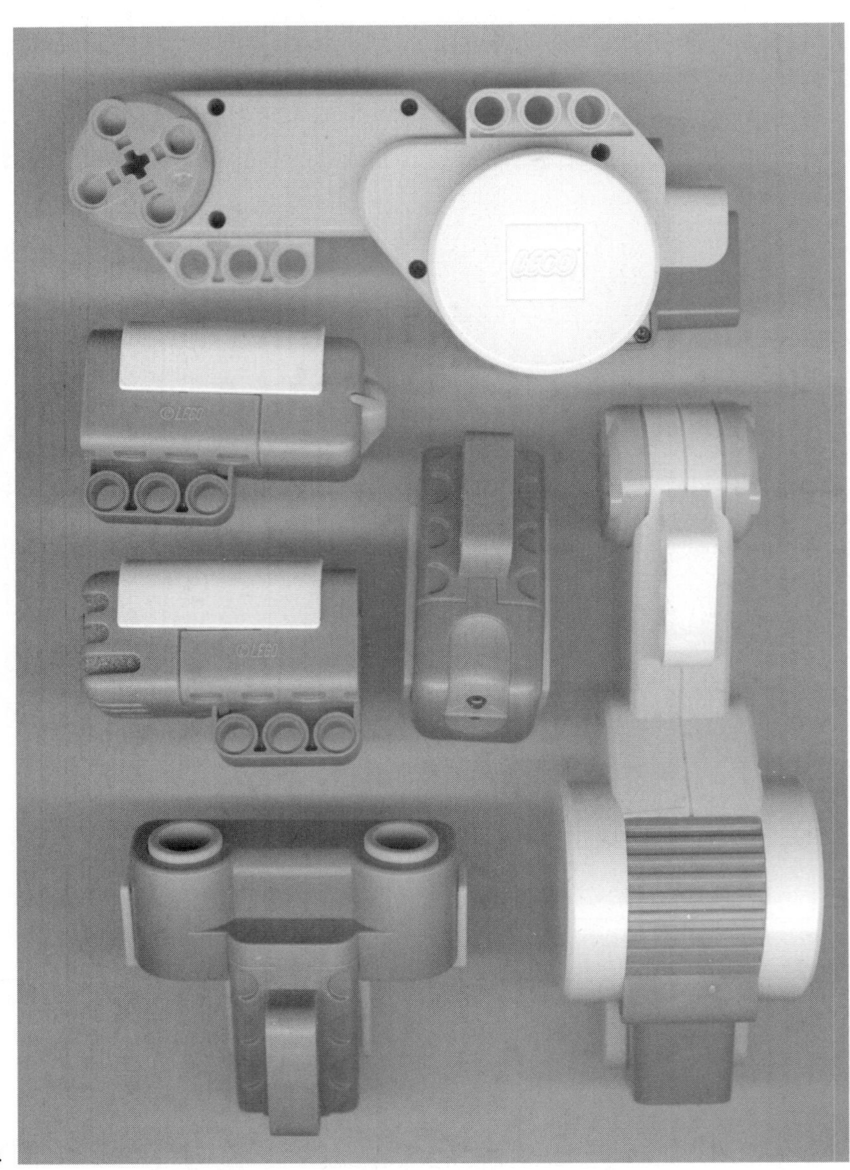

4

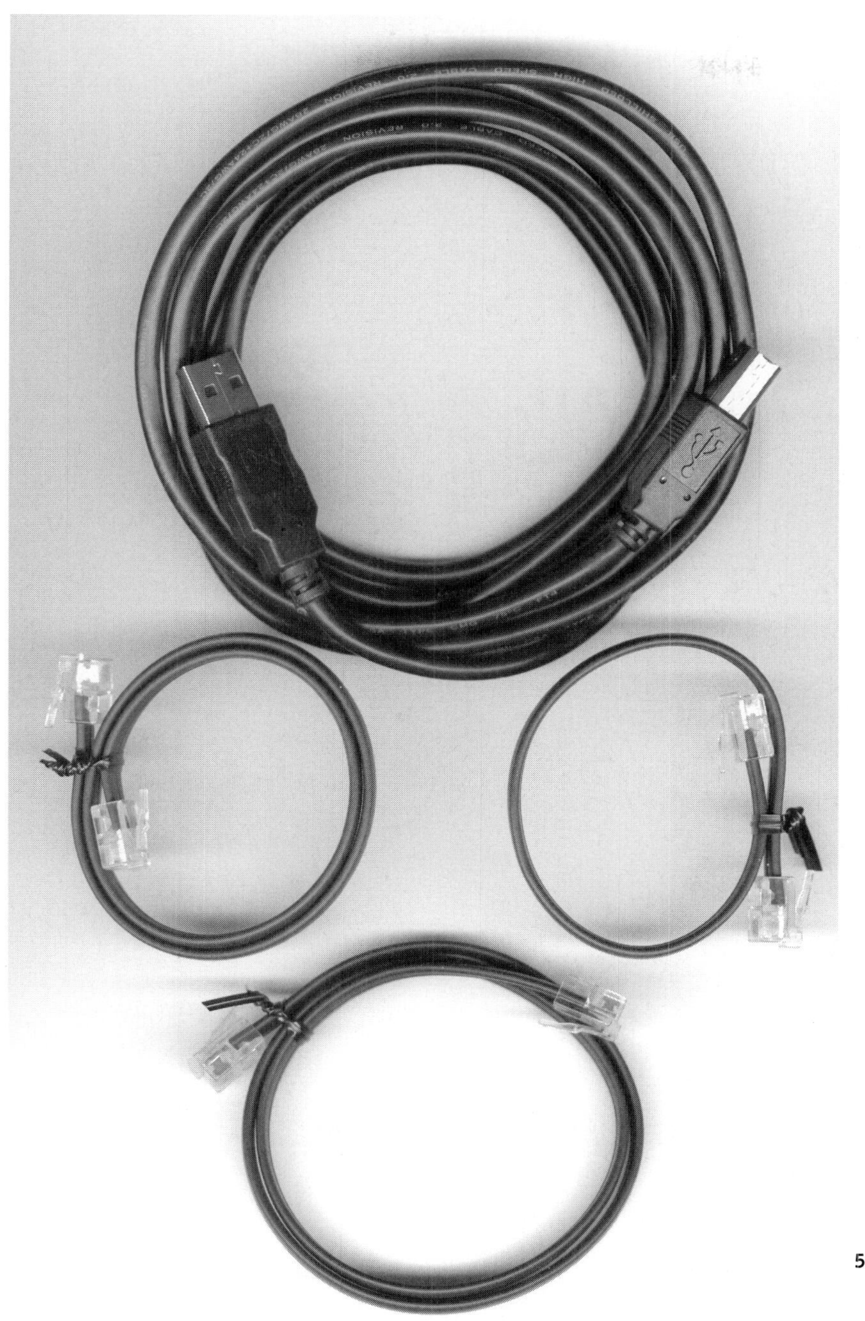

5

6 **Couplers.**

7 **Beams.**

8 **Angled beams.**

9 **Connectors.**

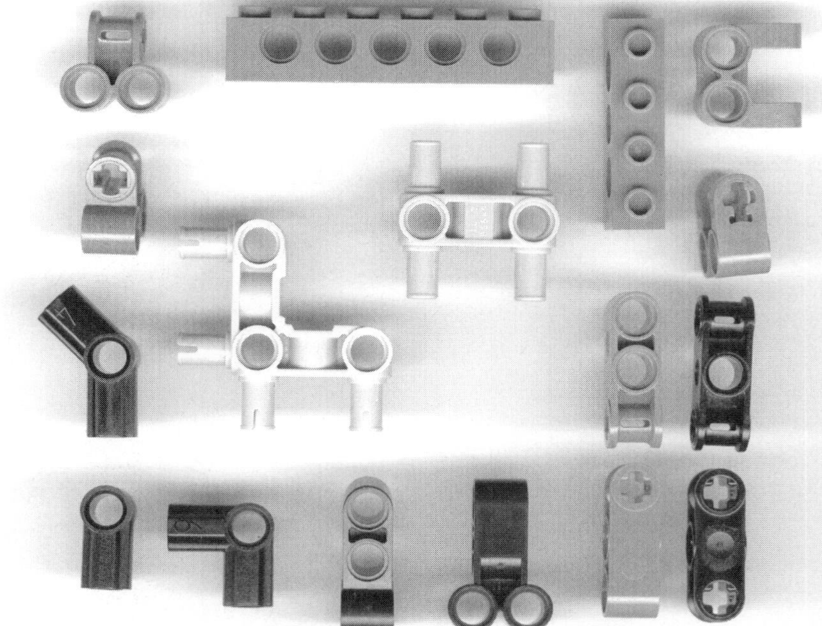

6

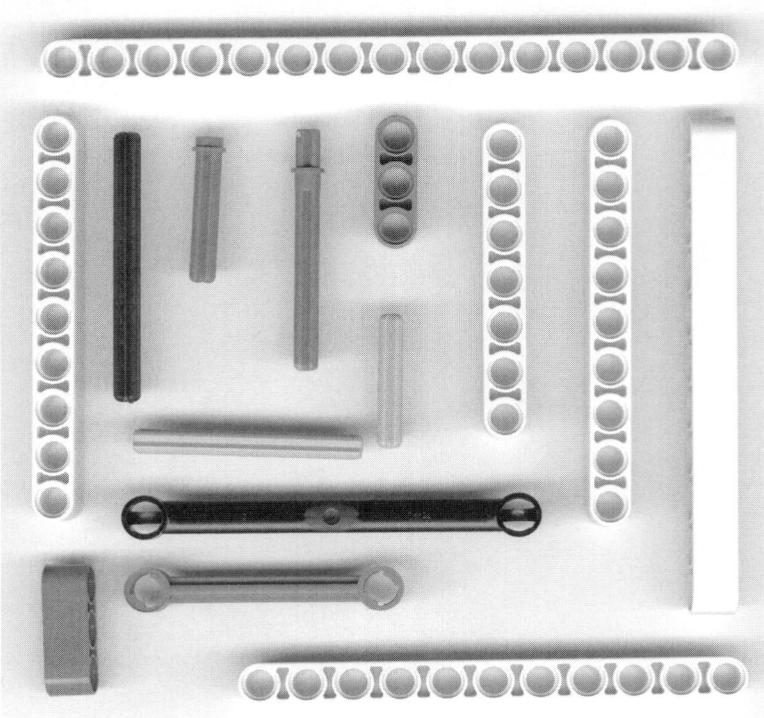

7

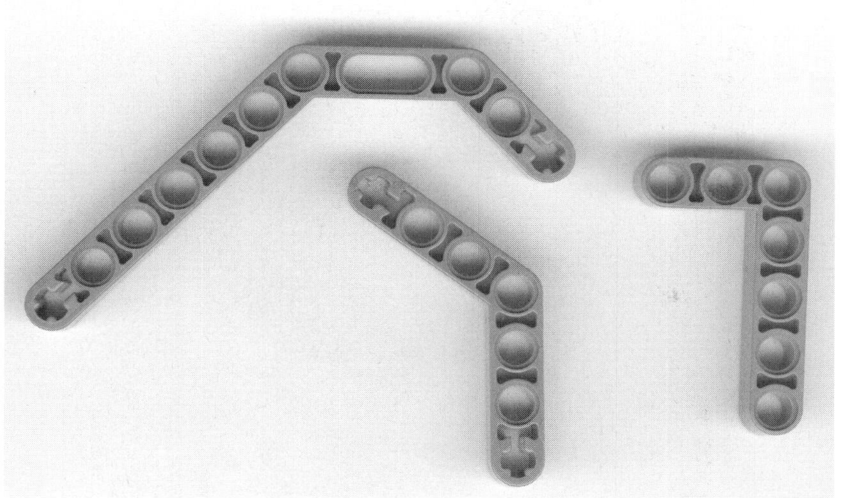

8

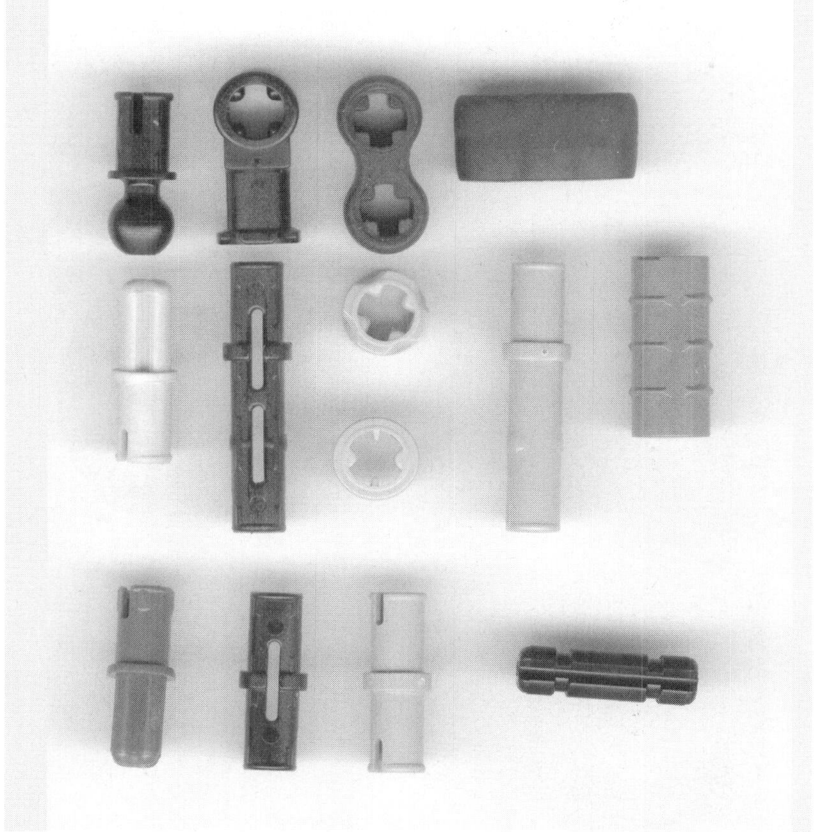

9

10 Gears.

11 Wheel, hubs, and ball.

12 Bionicle.

13 Two CD-Rs.

10

11

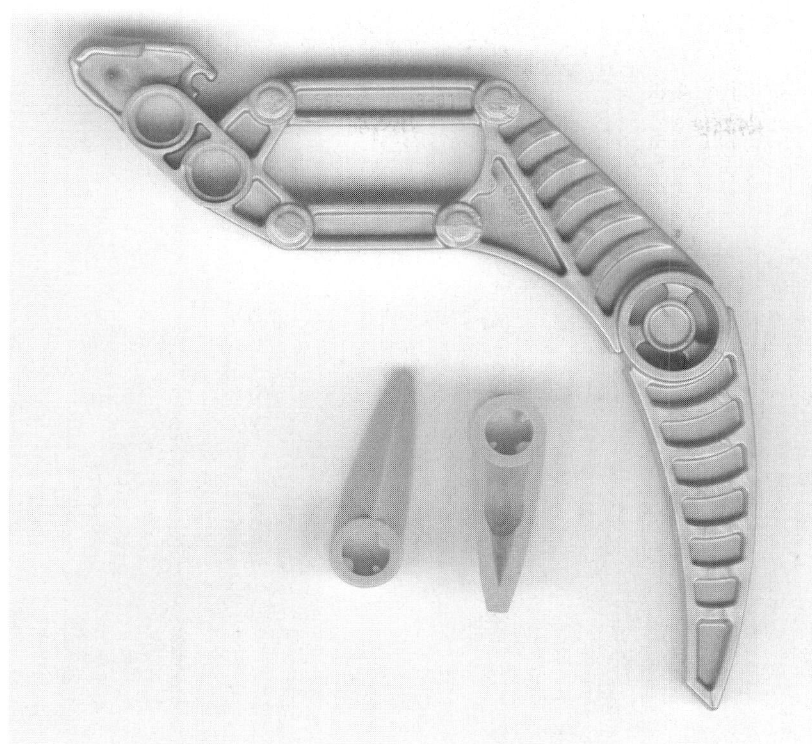

12

in-box
in-store
gis-order
packaging
buildinginstr
miscellaneous
LEGO
date:
no:
in-box
in-store
gis-order
packaging
buildinginstr
miscellaneous
2
LEGO
date:
no:

13

14 The rest of the gang ready to be hacked.

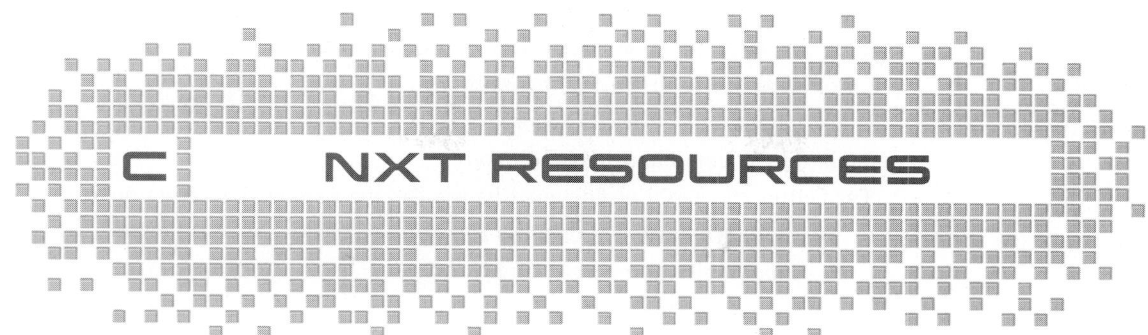

C NXT RESOURCES

HERE IS *THE* collection of every reference, document, product, and Web site that is mentioned in this book. Live and learn.

Academy Hobby—www.academyhobby.com
Acroname—www.acroname.com
American Ingenuity, Inc.—www.aidomes.com
American Wind Energy Association—www.awea.org
Amusements in Mathematics by Henry Ernest Dudeney (unk, March 25, 1917)
The Apple Store Fifth Avenue to Open on Friday, May 19—
 www.apple.com/pr/library/2006/may/18retail.html
"The Art of LEGO Design" by Fred G. Martin The Robotics Practitioner:
 The Journal for Robot Builders, vol. 1, no. 2, Spring 1995—
 www.handyboard.com/techdocs/
AS&C CooLight—www.coolight.com
Atmel Corporation—www.atmel.com
Beal Systems GLOWIRE—www.glowire.com
Bergey Windpower Co.—www.bergey.com
"Bill Gates mocks MIT's $100 laptop" (CNN.com, March 16, 2006; a Reuters report)—
 www.cnn.com/2006/TECH/ptech/03/16/gates.100.laptop.reut/ index.html
Bionicle—www.bionicle.com
Bluetooth SIG—www.bluetooth.org

The Box: How the Shipping Container Made the World Smaller and the World Economy Bigger by Marc Levinson (Princeton University Press, 2006)
Buckminster Fuller Institute—www.bfi.org
Calder Foundation—www.calder.org
Cambridge Silicon Radio—www.csr.com
"Car Culture: Some Cities Convert Historic Parking Garages into Lofts or Lots" by Mary Beth Klatt, *Preservation Magazine*, Oct. 21, 2005—www.nationaltrust.org/magazine/archives/arch_story/102105.htm
Carhenge—www.carhenge.com
Cartier—www.cartier.com
Cirronet, Inc.—www.cirronet.com
Columbia University, Department of Art History and Archaeology, Media Center for Art History and Archaeology—www.mcah.columbia.edu/cgi-bin/dbcourses/item?skip=1540
"A Cube in the Land of the Wheel; Redefining Public Space at the G.M. Building" by David W. Dunlap, The New York Times, March 2, 2005
The Dwell Home—www.thedwellhome.com/winner.html
The Electronic Goldmine—www.goldmine-elec.com
Element Products—www.elementinc.com/
Flambeau, Inc.—www.flambeau.com
Foster and Partners—www.fosterandpartners.com
Freescale Semiconductor—www.freescale.com
Garmin International—www.garmin.com
Gensler—www.gensler.com
Geodesic Math and How to Use It by Hugh Kenner (University of California Press, 2003)
Gleason Research—gleasonresearch.com
Hanbit Electronics Co., Ltd.—www.hbe.co.kr
Hitachi; Renesas Technology Corp.—america.renesas.com
Howard Hughes Medical Institute, "Seeing, Hearing, and Smelling the World"—www.hhmi.org/senses/
iRobot Corporation Hacker—www.irobot.com/sp.cfm?pageid=198
ISOCONTAINERS.com—www.isocontainers.com
Jameco Electronics—www.robotstore.com
"Steve Jobs Loves his Big iCube: It's his; he's not leaving it here" by Deborah Schoeneman, *New York Magazine*, Dec. 5, 2005—newyorkmetro.com/nymetro/news/people/columns/intelligencer/15204/
"K" Line, Ltd.—www.kline.co.jp

Donald E. Knuth personal list of Dudeney puzzles published in *The Weekly Dispatch*—www-cs-staff.stanford.edu/~uno/dudeney-twd.txt

Konvertor—www.konvertor.net

LaCie—www.lacie.com

The LEGO Group—www.lego.com

Logo Foundation—el.media.mit.edu/Logo-foundation/index.html

Richard Long—www.richardlong.org/index.html

LOT-EK—www.lot-ek.com/

Sam Loyd's Cyclopedia of 5000 Puzzles, Tricks, and Conundrums with Answers by Sam Loyd (The Lamb Publishing Company, NY, 1914)

MaMaMedia—www.mamamedia.com

Fred G. Martin—www.cs.uml.edu/~fredm/

Mathematical Puzzles of Sam Loyd by Martin Gardner (Dover Publications, Inc., New York, Volume 1, 1959)

McGuire-Nicholas—www.mcguire-nicholas.com

Micro-X-Tech—www.dragonmodelsusa.com

Mind Control—www.elementdirect.com/product_info.php?products_id=28

Mindstorms: Children, Computers, and Powerful Ideas by Seymour Papert (Basic Books, 2d Edition, 1993)

Modern Puzzles by Henry Ernest Dudeney (unk, 1926)

Moed de Armas & Shannon—www.mdeas.com/Default.htm

Musée national Picasso Paris—www.musee-picasso.fr

Museu Picasso—www.museupicasso.bcn.es

My Best Mathematical and Logic Puzzles by Martin Gardner (Dover Publications, Inc., New York, 1994)

My special Web site for this book—www.pco2go.com/lego

National Automobile Museum—www.automuseum.org

National Instruments Corporation—www.ni.com

NetMedia—www.basicx.com

New Micros, Inc.—www.newmicros.com

Office for Metropolitan Architecture—www.oma.nl

The Official Robosapien Hacker's Guide by Dave Prochnow (McGraw-Hill, 2006)

One Laptop Per Child—laptop.org

Ora-Ïto—www.ora-ito.com

Oregon Dome, Inc.—www.domes.com

OSRAM Opto Semiconductors GmbH—www.osram-os.com

Pacific Domes, Inc.—www.pacificdomes.com

Parallax—www.parallax.com

Patek Philippe—www.patekphilippe.com

Patek Philippe Museum—www.patekmuseum.com

Pepper Computer, Inc.—www.pepper.com

PicoCricket Kit—www.picocricket.com

Plano Molding Company—www.planomolding.com

Playful Invention Company—www.picocricket.com

The Pritzker Architecture Prize—www.pritzkerprize.com

Professor Seymour Papert—www.papert.org

PSP Hacks, Mods, and Expansions by Dave Prochnow (McGraw-Hill, 2006)

The Programmable LEGO Brick by MIT, ca. 1994—lcs.www.media.mit.edu/projects/programmable-brick/

Mitchel Resnick—web.media.mit.edu/~mres/

Resolution: 4 Architecture—www.re4a.com

"Reversing Trains: A Turn of the Century Sorting Problem," by Nancy Amato, Manuel Blum, Sandra Irani, and Ronitt Rubinfeld, *Journal of Algorithms*, 10(3): 413-428, September 1989

ROBOLAB at Tufts University Center for Engineering Education Outreach (CEEO)—http://130.64.87.22/robolabatceeo/

Roomba Dev Tools—www.roombadevtools.com

Roomba SCI Specification—www.irobot.com/sp.cfm?pageid=198

The Sculptures of Picasso by Daniel Henry Kahnweiler (Rodney Phillips, 1949)

Sea Box, Inc.—www.seabox.com

Shenzhen Lianyida Science Co.,Ltd.—www.lyd.cc

Shigeru Ban Architects—www.shigerubanarchitects.com/

The Shock of the New by Robert Hughes (Alfred A. Knopf, 1980)

Eric Sophie—www.biomechanicalbricks.com

Take This Stuff and Hack It! by Dave Prochnow (McGraw-Hill, 2007)

Tamiya—www.tamiyausa.com

Timberline Geodesic Domes—www.domehome.com

Tower Hobbies—www.towerhobbies.com

Turtles, Termites, and Traffic Jams: Explorations in Massively Parallel Microworlds by Mitchel Resnick (The MIT Press, 1997)

U.S. Green Building Council—www.usgbc.org

United States Patent and Trademark Office—www.uspto.gov

Walker Art Center—design.walkerart.org

Handbook of Watch & Clock Repairs by H.G. Harris (Emerson Books, 1961)

ZigBee Alliance—www.zigbee.org

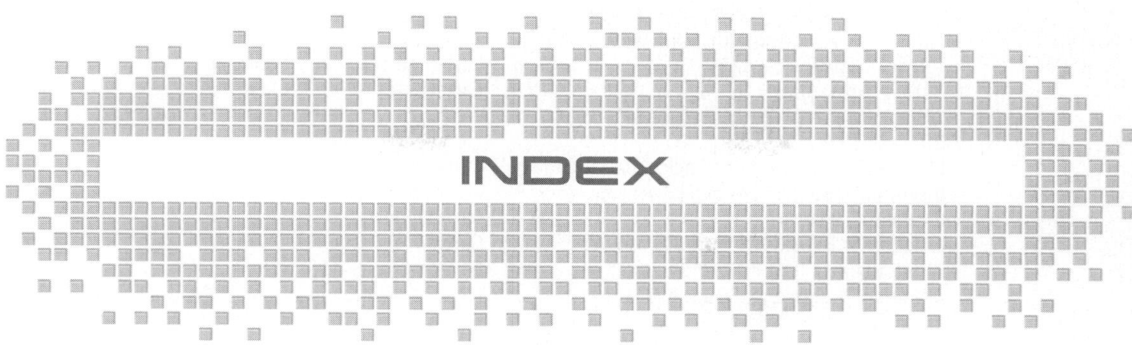

INDEX

Everything a *just slightly wicked* electronics hobbyist could want!

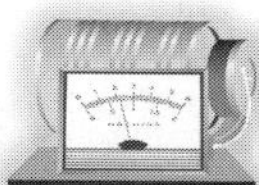

101 Spy Gadgets for the Evil Genius
Brad Graham & Kathy McGowan
0-07-146894-3

50 Model Rocket Projects for the Evil Genius
Gavin Harper
0-07-146984-2

Electronic Sensors for the Evil Genius
Thomas Petruzzellis
0-07-147036-0

MORE Electronic Gadgets for the Evil Genius
Robert Iannini
0-07-145905-7

50 Awesome Auto Projects for the Evil Genius
Gavin Harper
0-07-145823-9

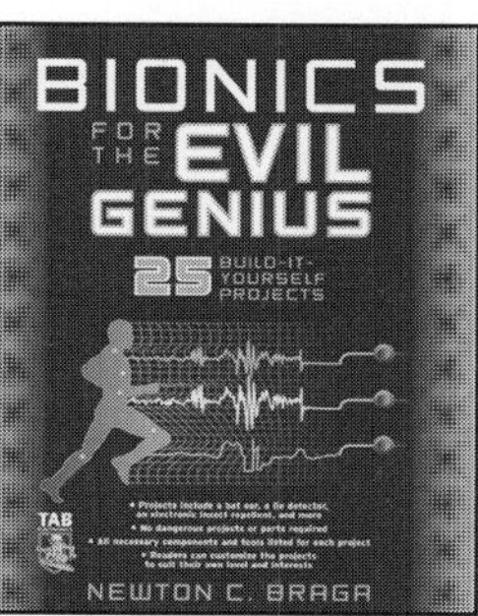

Bionics for the Evil Genius
Newton Braga
0-07-145925-1

Mechatronics for the Evil Genius
Newton Braga
0-07-145759-3

123 PIC Microcontroller Experiments for the Evil Genius
Myke Predko
0-07-145142-0

Other Books Available:

Electronics Circuits for the Evil Genius
Dave Cutcher
0-07-144881-0

Electronic Gadgets for the Evil Genius
Robert Iannini
0-07-142609-4

123 Robotics Experiments for the Evil Genius
Myke Predko
0-07-141358-8

Learn more.  Do more.